HANDBOOK OF

Probability

Wiley Handbooks in

APPLIED STATISTICS

The *Wiley Handbooks in Applied Statistics* is a series of books that present both established techniques and cutting-edge developments in the field of applied statistics. The goal of each handbook is to supply a practical, one-stop reference that treats the statistical theory, formulae, and applications that, together, make up the cornerstones of a particular topic in the field. A self-contained presentation allows each volume to serve as a quick reference on ideas and methods for practitioners, while providing an accessible introduction to key concepts for students. The result is a high-quality, comprehensive collection that is sure to serve as a mainstay for novices and professionals alike.

Chatterjee and Simonoff · *Handbook of Regression Analysis*

Florescu and Tudor · *Handbook of Probability*

Forthcoming *Wiley Handbooks in Applied Statistics*

Montgomery, Johnson, Jones, and Borror · *Handbook in Design of Experiments*

Florescu and Tudor · *Handbook of Stochastic Processes*

Rausand · *Handbook of Reliability*

Vining · *Handbook of Quality Control*

HANDBOOK OF Probability

IONUŢ FLORESCU
Stevens Institute of Technology, Hoboken, NJ

CIPRIAN TUDOR
Université Lille 1, Paris, France

Published by John Wiley & Sons, Inc., Hoboken, New Jersey.
Published simultaneously in Canada.

Library of Congress Cataloging-in-Publication Data

Florescu, Ionut, 1973-
Handbook of probability / Ionut Florescu, Ciprian Tudor.
pages cm
"Published simultaneously in Canada"–Title page verso.
Includes bibliographical references and index.
ISBN 978-0-470-64727-1 (cloth) – 1. Probabilities. I. Tudor, Ciprian, 1973- II. Title.
QA273.F65 2013
519.2–dc23

2013013969

Printed in the United States of America

10 9 8 7 6 5 4 3 2 1

To the memory of
Professor Constantin
Tudor

Contents in Brief

Contents

List of Figures

Preface

Our motivation when we started to write the Handbook was to provide a reference book easily accessible without needing too much background knowledge, but at the same time containing the fundamental notions of the probability theory.

We believe the primary audience of the book is split into two main categories of readers.

The first category consists of students who already completed their graduate work and are ready to start their thesis research in an area that needs applications of probability concepts. They will find this book very useful for a quick reminder of correct derivations using probability. We believe these student should reference the present handbook frequently as opposed to taking a probability course offered by an application department. Typically, such preparation courses in application areas do not have the time or the knowledge to go into the depth provided by the present handbook.

The second intended audience consists of professionals working in the industry, particularly in one of the many fields of application of stochastic processes. Both authors' research is concentrated in the area of applications to Finance, and thus some of the chapters contain specific examples from this field. In this rapidly growing area the most recent trend is to obtain some kind of certification that will attest the knowledge of essential topics in the field. Chartered Financial Analyst (CFA) and Financial Risk Manager (FRM) are better known such certifications, but there are many others (e.g., CFP, CPA, CAIA, CLU, ChFC, CASL, CPCU). Each such certificate requires completion of several exams, all requiring basic knowledge of probability and stochastic processes. All individuals attempting these exams should consult the present book. Furthermore, since the topics tested during these exams are in fact the primary subject of their future work, the candidates will find the Handbook very useful long after the exam is passed.

This manuscript was developed from lecture notes for several undergraduate and graduate courses given by the first author at Purdue University and Stevens Institute of Technology and by the second author at the Université de Paris 1 Panthéon-Sorbonne and at the Université de Lille 1. The authors would like to thank their colleagues at all these universities with whom they shared the teaching load during the past years.

IONUŢ FLORESCU AND CIPRIAN TUDOR

Hoboken, U.S. & Paris, France
April 18, 2013

Introduction

The probability theory began in the seventeenth century in France when two great French mathematicians, Blaise Pascal and Pierre de Fermat, started a correspondence over the games of chance. Today, the probability theory is a well-established and recognized branch of mathematics with applications in most areas of Science and Engineering.

The Handbook is designed from an introductory course in probability. However, as mentioned in the Preface, we tried to make each chapter as independent as possible from the other chapters. Someone in need of a quick reminder can easily read the part of the book he/she is concerned about without going through an entire set of background material.

The present Handbook contains fourteen chapters, the final two being appendices with more advanced material. The sequence of the chapters in the book is as follows. Chapters 1 and 2 introduce the probability space, sigma algebras, and the probability measure. Chapters 3, 4, and 5 contain a detailed study of random variables. After a general discussion of random variables in Chapter 3, we have chosen to separate the discrete and continuous random variables, which are analyzed separately in Chapters 4 and 5 respectively. In Chapter 6 we discuss methods used to generate random variables. In today's world, where computers are part of any scientific activity, it is very important to know how to simulate a random experiment to find out the expectations one may have about the results of the random phenomenon. Random vectors are treated in Chapter 7. We introduce the characteristic function and the moment-generating function for random variables and vectors in Chapters 8 and 9. Chapter 10 describes the Gaussian random vectors that are extensively used in practice. Chapters 11 and 12 describe various types of convergence for sequences of random variables and their relationship (Chapter 11) and some of the most famous limit theorems (the law of large numbers and the central limit theorems), their versions, and their applications (Chapter 12). Appendices A and B (Chapters 13 and 14) focus on the more general integration theory and moments of random variables with any distribution. In this book, we choose to treat separately the discrete and continuous random variables since in applications one will typically use one of these cases and a quick consultation of the book will suffice. However, the contents of the Appendices show that the moments of continuous and discrete random variables are in fact particular cases of a more general theory.

Each chapter of the Handbook has the following general format:

- Introduction
- Historical Notes
- Theory and Applications
- Exercises

The *Introduction* describes the intended purpose of the chapter. The *Historical Notes* section provides numerous historical comments, especially dealing with the development of the theory exposed in the corresponding chapter. We chose to include an introductory section in each chapter because we believe that in an introductory text in probability, the main ideas should be closely related to the fundamental ideas developed by the founding fathers of Probability. The section *Theory and Applications* is the core of each chapter. In this section we introduce the main notion and we state and prove the main results. We tried to include within this section as many basic examples as was possible. Many of these examples contain immediate applications of the theory developed within the chapter. Sometimes, examples are grouped in a special section. Each chapter of the Handbook ends with the *Exercises* section containing problems. We divided this section into two parts. The first part contains solved problems, and the second part has unsolved problems. We believe that the solved problems will be very useful for the reader of the respective chapter. The unsolved problems will challenge and test the knowledge gained through the reading of the chapter.

CHAPTER ONE

Probability Space

1.1 Introduction/Purpose of the Chapter

The most important object when working with probability is the proper definition of the space studied. Typically, one wants to obtain answers about real-life phenomena which do not have a predetermined outcome. For example, when playing a complex game a person may be wondering: What are my chances to win this game? Or, am I paying too much to play this game, and is there perhaps a different game I should rather play? A certain civil engineer wants to know what is the probability that a particular construction material will fail under a lot of stress. To be able to answer these and other questions, we need to make the transition from reality to a space describing what may happen and to create consistent laws on that space. This framework allows the creation of a mathematical model of the random phenomena. This model, should it be created in the proper (consistent) way, will allow the modeler to provide approximate answers to the relevant questions asked. Thus, *the first and the most important step in creating consistent models is to define a probability space which is capable of answering the interesting questions that may be asked.*

We denote with Ω the set that contains all the possible outcomes of a random experiment. The set Ω is often called *sample space* or *universal sample space.* For example, if one rolls a die, $\Omega = \{1, 2, 3, 4, 5, 6\}$. The space Ω does not necessarily contain numbers but rather some representation of the outcomes of the real phenomena. For example, if one looks at the types of bricks which may be used to build a house, a picture of each possible brick is a possible representation of each element of Ω.

Handbook of Probability, First Edition. Ionuţ Florescu and Ciprian Tudor.

A generic element of Ω will be denoted by ω. Any collection of outcomes (elements in Ω) is called an *event*. That is, an event is any subset of the sample space Ω. We will use capital letters from the beginning of the alphabet (A, B, C, etc.) to denote events.

In probability, one needs to measure the *size* of these events. Since an event is just a subset, we need to define those subsets of Ω that *can* be measured. The concept of *sigma algebra* allows us to define a collection of subsets of the sample space on which a measure can be defined. In this chapter we introduce the notion of algebra and sigma algebra and we discuss their basic properties.

1.2 Vignette/Historical Notes

The first recorded notions of Probability Theory appear in 1654 in an exchange of letters between the famous mathematicians Blaise Pascal and Pierre de Fermat. The correspondence was prompted by a simple observation by Antoine Gombaud Chevalier de Méré, a French nobleman with an interest in gambling and who was puzzled by an apparent contradiction concerning a popular dice game. The game consisted of throwing a pair of dice 24 times; the problem was to decide whether or not to bet even money on the occurrence of at least one "double six" during the 24 throws. A seemingly well-established gambling rule led de Méré to believe that betting on a double six in 24 throws would be profitable (based on the payoff of the game). However, his own calculations based on many repetitions of the 24 throws indicated just the opposite. If we translate de Méré problem in todays language, he was trying to establish if the event has probability greater than 0.5. Today the confusion is easy to pinpoint to the proper definition of the probability space. For example, the convention at the time was that rolling a three (two and one showing on the dice) would be the same as rolling a two (a double one). Puzzled by this and other similar gambling problems, de Méré wrote to Pascal. The further correspondence between Fermat and Pascal is the first known documentation of the fundamental principles of the theory of probability.

The first formal treatment of probability theory was provided by Pierre-Simon, marquis de Laplace (1749–1827) in his Théorie Analytique des Probabilités published in 1812. (Laplace, 1886, republished). In 1933 the monograph *Grundbegriffe der Wahrscheinlichkeitsrechnung* by the Russian preeminent mathematician Andrey Nikolaevich Kolmogorov (1903–1987) outlined the axiomatic approach that forms the basis of the modern probability theory as we know it today (Kolmogoroff, 1973, republished).

1.3 Notations and Definitions

The following notations will be used throughout the book for set (event) operations. In the following, ω is any element and A, B are any sets in the sample space Ω.

- We describe a set or a collection of elements using a notation of the form

$$\{x \mid x \text{ satisfy property } \mathcal{P}\}.$$

  The first part before the | describes the basic element in this collection, and the part after | describes the property that all the elements in the collection must satisfy.
- $\emptyset$ is a notation for the set that does not contain any element. This set is called the *empty set*.
- $\omega \in A$ denotes that the element ω is in the set A; we say "ω *belongs* to A." Obviously, $\omega \notin A$ means that the element is not in the set.
- $A \subseteq B$ denotes that A is a *subset* of B; that is, every element in A is also in B. The set A may actually be equal to B. In contrast, the notation $A \subset B$ means that A is a *proper subset* of B; that is, A is strictly included in B. Mathematically, $A \subseteq B$ is equivalent to the following statement: Any $x \in A$ implies $x \in B$.
- *Union* of sets:

$$\begin{aligned} A \cup B &= \{\text{set of elements that are } \textbf{either} \text{ in } A \textbf{ or} \text{ in } B\} \\ &= \{\omega \in \Omega \mid \omega \in A \textbf{ or } \omega \in B\}. \end{aligned}$$

- *Intersection* of sets:

$$\begin{aligned} A \cap B = AB &= \text{set of elements that are } \textbf{both} \text{ in } A \textbf{ and} \text{ in } B \\ &= \{\omega \in \Omega \mid \omega \in A \textbf{ and } \omega \in B\}. \end{aligned}$$

- *Complement* of a set:

$$A^c = \bar{A} = \text{set of elements that are in } \Omega \text{ but } \textbf{not} \text{ in } A = \{\omega \in \Omega \mid \omega \notin A\}.$$

- *Difference* of two sets:

$$\begin{aligned} A \setminus B &= \text{set of elements that are in } A \text{ but } \textbf{not} \text{ in } B \\ &= \{\omega \in \Omega \mid \omega \in A, \omega \notin B\}. \end{aligned}$$

- *Symmetric difference*:

$$A \triangle B = (A \setminus B) \cup (B \setminus A).$$

- Two sets A, B such that $A \cap B = \emptyset$ are called *disjoint or mutually exclusive* sets.
- A collection of sets $A_1, A_2, \ldots, A_n$ such that $A_1 \cup A_2 \cup \cdots \cup A_n = \Omega$ and $A_i \cap A_j = \emptyset$ for any $i \neq j$ is called a *partition of the space* Ω.

Every set operation may be expressed in terms of basic operations. For example,

$$A \setminus B = A \cap B^c \quad \text{and} \quad (A \setminus B)^c = A^c \cup B.$$

There is a distributive law for intersection over union. If A, B, C are included in Ω, then

$$A \cap (B \cup C) = (A \cap B) \cup (A \cap C)$$

and

$$A \cup (B \cap C) = (A \cup B) \cap (A \cup C).$$

Furthermore, $\Omega^c = \emptyset$ and $\emptyset^c = \Omega$, and for all $A \subset \Omega$ we have

$$A \cup A^c = \Omega, \quad A \cap A^c = \emptyset, \quad \text{and} \quad (A^c)^c = A.$$

The De Morgan laws are also very important:

$$\begin{aligned}(A \cup B)^c &= A^c \cap B^c, \\ (A \cap B)^c &= A^c \cup B^c. \end{aligned} \tag{1.1}$$

All of these rules may be extended to any finite number of sets in an obvious way. More details and further references about set operations may be found in Billingsley (1995) or Chung (2000).

1.4 Theory and Applications

1.4.1 ALGEBRAS

We introduce the notion of *σ-algebra* (or *σ-field*) to introduce a collection of sets which we may measure. In other words, we introduce a proper domain of definition for the (soon to be introduced) probability function. First let us denote

$$\mathcal{P}(\Omega) = \{A \mid A \subseteq \Omega\};$$

that is, $\mathcal{P}(\Omega)$ is the collection of all possible subsets of Ω, a set containing all possible sets in Ω. This collection is called the parts of Ω.

An *algebra* on Ω is a collection of such sets in $\mathcal{P}(\Omega)$ (including Ω) which is closed under complementarity and finite union.

Definition 1.1 (Algebra on Ω) *Let Ω be a nonempty space. A collection $\mathscr{A}$ of events in Ω is called an algebra (or field) on Ω if and only if:*

i. $\Omega \in \mathscr{A}$.
ii. *If $A \in \mathscr{A}$, then $A^c \in \mathscr{A}$.*
iii. *If $A, B \in \mathscr{A}$, then $A \cup B \in \mathscr{A}$.*

Let us list some immediate property of an algebra.

Proposition 1.2 *Let $\mathscr{A}$ be an algebra on Ω. Then:*

1. $\emptyset \in \mathscr{A}$.
2. *If $A, B \in \mathscr{A}$, then $A \cap B \in \mathscr{A}$.*
3. *For any n natural number in $\mathbb{N}$, if $A_1, \ldots, A_n \in \mathscr{A}$, then $\bigcup_{i=1}^{n} A_i \in \mathscr{A}$.*

Proof: Note that these are properties of the collection of sets. The collection $\mathscr{A}$ *must* contain these. Specifically, $\emptyset = \Omega^c$ and since $\Omega \in \mathscr{A}$ the second point in Definition 1.1 says that $\emptyset \in \mathscr{A}$. Since

$$(A \cap B)^c = A^c \cup B^c,$$

by the DeMorgan laws (1.1), points ii and iii in Definition 1.1 imply that $(A \cap B)^c$ is in $\mathscr{A}$ and so $A \cap B$ is in $\mathcal{A}$. The proof for the third part is a simple induction on $n \geq 2$, with the verification step already provided in iii of Definition 1.1. ■

1.4.2 SIGMA ALGEBRAS

A σ-algebra on Ω is a generalization of an algebra on Ω.

Definition 1.3 (σ-Algebra on Ω) *A σ-algebra $\mathcal{F}$ on Ω is an algebra on Ω and in addition it is closed under countable unions. That is, the first two points in Definition 1.1 remain the same, and the third property is replaced with:*

iii. *If $n \in \mathbb{N}$ is a natural number and $A_n \in \mathcal{F}$ for all n, then*

$$\bigcup_{n \in \mathbb{N}} A_n \in \mathcal{F}.$$

Definition 1.3 implies other properties of a σ-algebra. For example, if a sigma algebra contains two elements A and B, it also contains $A \cap B$ and $A \cup B$. The fact that the intersection must be in the σ-algebra is an easy consequence of point ii in the above definitions. For the σ-algebra case the intersection can be extended to a countable number of events as the next result shows.

Proposition 1.4 *Let $\mathcal{F}$ be a σ-algebra on Ω. Then:*

1. $\emptyset \in \mathcal{F}$.
2. *If $\{A_i\}_{i \in \mathbb{N}}$ are elements of the σ-algebra $\mathcal{F}$, then*

$$\bigcap_{i \in \mathbb{N}} A_i \in \mathcal{F}.$$

Proof: Since $\emptyset = \Omega^c$ and $\Omega \in \mathcal{F}$, it follows that

$$\emptyset \in \mathcal{F}.$$

For the second part, we note that

$$\left(\bigcap_{i\in\mathbb{N}} A_i\right)^c = \bigcup_{i\in\mathbb{N}} A_i^c$$

and since for every $i \in \mathbb{N}$, $A_i^c \in \mathcal{F}$ from the second part of the definition, their countable union is $\bigcup_{i\in\mathbb{N}} A_i^c \in \mathcal{F}$, and thus its complement is $\bigcap_{i\in\mathbb{N}} A_i \in \mathcal{F}$. ■

The σ-algebra is a very nice collection of sets. Using simple the operations union, intersection, complementarity, and difference applied to any sets in the σ-algebra, we end up with a set still in the σ-algebra. In fact, we may apply these simple operations in any combination a countably infinite number of times and we still end up with a set in the σ-algebra.

However, a noncountable intersection or union of the elements of a σ-algebra does not necessarily belongs to it (although it may). This can only happen when the index n is in some continuous set—for example, $n \in (0, 1)$. Such situations are advanced: however, they become relevant when talking about stochastic processes. Extra care has to be observed in these cases. In the context of stochastic processes, we will introduce filtrations, which are simply increasing σ-algebras.

Because of these nice properties, a σ-algebra provides a suitable domain of definition for the probability function (and thus defines probabilities of random events). However, a σ-algebra is a very abstract concept which, in general, is hard to work with. To simplify notions, we introduce the next definition. It will be much easier to work with the generators of a σ-algebra.

This will be a recurring theme in probability; in order to show a property for a big class, we show the property for a small generating subset of the class and then use standard arguments to extend the property to the entire class.

Definition 1.5 (σ-algebra generated by a class $\mathscr{C}$ of sets in Ω) *Let $\mathscr{C}$ be any collection of sets in Ω. Then we define $\sigma(\mathscr{C})$ as the smallest σ-algebra on Ω that contains $\mathscr{C}$. That is, $\sigma(\mathscr{C})$ satisfies the following properties:*

1. *It contains all the sets in $\mathscr{C}$: $\mathscr{C} \subseteq \sigma(\mathscr{C})$.*
2. *$\sigma(\mathscr{C})$ is a σ-algebra.*
3. *It is the smallest σ-algebra with this property: If $\mathscr{C} \subseteq \mathscr{G}$ and $\mathscr{G}$ is another σ-field, then $\sigma(\mathscr{C}) \subseteq \mathscr{G}$.*

The idea, as mentioned earlier, is to verify some statement on the set $\mathscr{C}$. Then, due to the properties that would be presented later, the said statement will be extended and be valid for all the sets in $\sigma(\mathscr{C})$.

Proposition 1.6 *Let $\mathscr{C} \subset \mathcal{P}(\Omega)$. Then*

$$\sigma(\mathscr{C}) = \bigcap_{\mathcal{F}\sigma\text{-algebra};\mathcal{F}\supset\mathscr{C}} \mathcal{F}.$$

Proof: The proof is left as an exercise (see Exercise 1.12). ■

Proposition 1.7 (Properties of σ-algebras) *The following are true:*

- *$\mathcal{P}(\Omega)$ is a σ-algebra, the largest possible σ-algebra on Ω.*
- *If the collection $\mathscr{C}$ is already a σ-algebra, then $\sigma(\mathscr{C}) = \mathscr{C}$.*
- *If $\mathscr{C} = \{\emptyset\}$ or $\mathscr{C} = \{\Omega\}$, then $\sigma(\mathscr{C}) = \{\emptyset, \Omega\}$, the smallest possible σ-algebra on Ω.*
- *If $\mathscr{C} \subseteq \mathscr{C}'$, then $\sigma(\mathscr{C}) \subseteq \sigma(\mathscr{C}')$.*
- *If $\mathscr{C} \subseteq \mathscr{C}' \subseteq \sigma(\mathscr{C})$, then $\sigma(\mathscr{C}') = \sigma(\mathscr{C})$.*

Again, these properties are easy to derive directly from the definition and are left as exercise.

In general, listing all the elements of a σ-algebra explicitly is hard. Only in simple cases is this is even possible. This explains why we prefer to work with the generating collection $\mathscr{C}$ instead of directly with the σ-algebra $\sigma(\mathscr{C})$.

1.4.3 MEASURABLE SPACES

After introducing σ-algebras, we are now able to give the notion of a space on which we can introduce probability measure.

Definition 1.8 (Measurable space) *A pair $(\Omega, \mathcal{F})$, where Ω is a nonempty set and $\mathscr{F}$ is a σ-algebra on Ω is called a* measurable space.

On this type of space we shall introduce the probability measure.

1.4.4 EXAMPLES

EXAMPLE 1.1 The measurable space generated by a set

Suppose a set $A \subset \Omega$. Let us calculate $\sigma(A)$. Clearly, by definition, Ω is in $\sigma(A)$. Using the complementarity property, we clearly see that A^c and $\emptyset$ are also in $\sigma(A)$. We only need to take unions of these sets and see that there are no more new sets. Thus,

$$\sigma(A) = \{\Omega, \emptyset, A, A^c\}.$$

EXAMPLE 1.2 The space modeling the rolls of a six sided die

Roll a six-sided die. Then use $\Omega = \{1, 2, 3, 4, 5, 6\}$ to denote the space of outcomes of rolling the die. Let $A = \{1, 2\}$. Then

$$\sigma(A) = \{\{1, 2\}, \{3, 4, 5, 6\}, \emptyset, \Omega\}.$$

EXAMPLE 1.3 The measurable set generated by two sets

Suppose that $\mathscr{C} = \{A, B\}$, where A and B are two sets in Ω such that $A \subset B$. List all the sets in $\sigma(\mathscr{C})$.

Solution: A common mistake made by students who learn these notions is the following argument:
$A \subset B$; therefore, using the fourth property in the proposition above, we obtain $\sigma(A) \subseteq \sigma(B)$ and therefore the σ-algebra asked is $\sigma(A, B) = \sigma(B) = \{\Omega, \emptyset, B, B^c\}$.

This argument is wrong on several levels. Firstly, the quoted property refers to collections of sets and not to the sets themselves. While it is true that $A \subset B$, it is not true that $\{A\} \subset \{B\}$ as collections of sets. Instead, $\{A\} \subset \{A, B\}$ and indeed this implies $\sigma(A) \subseteq \sigma(A, B)$ (and similarly for $\sigma(B)$). But this just means that the result should contain all the sets in $\sigma(A)$ (the sets in the previous example).

Second, as this example shows and as the following proposition says, it is not true that $\sigma(A) \cup \sigma(B) = \sigma(A, B)$, so we can't just simply list all the elements in $\sigma(A)$ and $\sigma(B)$ together. The only way to solve this problem is the hard way—that is, actually calculating the sets.

Clearly, $\sigma(A, B)$ should contain the basic sets and their complements; thus

$$\sigma(A, B) \supset \{\Omega, \emptyset, A, B, A^c, B^c\}.$$

The σ-algebra should also contain all their unions according to the definition. Therefore, it must contain

$$\begin{aligned} &A \cup B = B, \\ &A \cup B^c, \\ &A^c \cup B = \Omega, \\ &A^c \cup B^c = A^c, \end{aligned}$$

where the equalities are obtained using that $A \subset B$. So the only new set to be added is $A \cup B^c$ and thus its complement as well: $(A \cup B^c)^c = A^c \cap B$. Now we need to verify that by taking unions we do not obtain any new sets, a task left for the reader. In conclusion, when $A \subset B$ we have

$$\sigma(A, B) = \{\Omega, \emptyset, A, B, A^c, B^c, A \cup B^c, A^c \cap B\}.$$

■

Remark 1.9 *Experimenting with generated σ-algebras is a very useful tool to understand this idea in probability: We typically work with the generating sets of the σ-algebras, (A and B in Example 1.3), and then some standard arguments will take over and generalize the work to the entire σ-algebra automatically.*

Proposition 1.10 (Intersection and union of σ-algebras) *Suppose that $\mathcal{F}_1$ and $\mathcal{F}_2$ are two σ-algebras on Ω. Then:*

1. *$\mathcal{F}_1 \cap \mathcal{F}_2$ is a σ-algebra.*
2. *$\mathcal{F}_1 \cup \mathcal{F}_2$ is **not** a σ-algebra. The smallest σ-algebra that contains both of them is $\sigma(\mathcal{F}_1 \cup \mathcal{F}_2)$ and is denoted $\mathcal{F}_1 \vee \mathcal{F}_2$*

Proof: For part 2 there is nothing to show. A counterexample is immediate from the previous exercise. Take $\mathcal{F}_1 = \sigma(A)$ and $\mathcal{F}_2 = \sigma(B)$. Example 1.3 calculates $\sigma(A, B)$, and it is simple to see that $\mathcal{F}_1 \cup \mathcal{F}_2$ needs more sets to become a σ-algebra. It is worth mentioning that in some trivial cases the union may be a σ-algebra, such as for example when $\mathcal{F}_1 \subseteq \mathcal{F}_2$ and they both contain a countable number of elements.

For part 1 we need to verify the three properties in the definition of the sigma algebra. For example, take A in $\mathcal{F}_1 \cap \mathcal{F}_2$. So A belongs to both collections of sets. Since $\mathcal{F}_1$ is a σ-algebra by definition, we have $A^c \in \mathcal{F}_1$. Similarly, $A^c \in \mathcal{F}_2$. Therefore, $A^c \in \mathcal{F}_1 \cap \mathcal{F}_2$. The rest of the definition is verified in a similar manner. ∎

Remark 1.11 *Part 1 of the proposition above may be generalized to any countable intersection of σ-algebras. That is, if $\{\mathcal{F}_i\}_{i \in I}$ is any sequence of σ-fields and I is countable, then*

$$\bigcap_{i \in I} \mathcal{F}_i$$

is a σ-field.

1.4.5 THE BOREL σ-ALGEBRA

Let Ω be a topological space (think geometry is defined in this space, and this assures us that the open subsets exist in this space).

Definition 1.12 *We define:*

$$\begin{aligned}\mathscr{B}(\Omega) &= \text{The Borel } \sigma\text{-algebra} \\ &= \sigma\text{-algebra generated by the class of open subsets of } \Omega. \end{aligned} \tag{1.2}$$

In the special case when $\Omega = \mathbb{R}$, we denote $\mathscr{B} = \mathscr{B}(\mathbb{R})$, the Borel sets of $\mathbb{R}$. This $\mathscr{B}$ is the most important σ-algebra. The reason is that most experiments on abstract spaces Ω may be made equivalent with experiments on $\mathbb{R}$ (as we shall see later when we talk about random variables). Thus, if we define a probability measure on $\mathscr{B}$, we have a way to calculate probabilities for most experiments.

There is nothing special about using the open sets as generators, except for the fact that the open sets can be defined in any topological space. The Borel sets over $\mathbb{R}$ may be generated by various other classes of intervals; in the end the result is the same σ-algebra. Please see problem 2.10 for these classes of generators.

Pretty much, any subset of $\mathbb{R}$ you can think about is in $\mathscr{B}$. However, it is possible (though very difficult) to explicitly construct a subset of $\mathbb{R}$ which is not in $\mathscr{B}$. See (Billingsley, 1995, page 45) for such a construction in the case $\Omega = (0, 1]$. These sets are generally constructed using a Cantor set argument.

EXAMPLE 1.4

Let $\Omega \neq \emptyset$ and let $A_1, \ldots, A_n$ be a partition of Ω with a finite number of sets. We denote

$$\mathscr{T} = \left\{ \bigcup_{i \in J} A_i \,\middle|\, J \subset \{1, 2, \ldots, n\} \right\}.$$

We will show that $\mathscr{T}$ is the σ-algebra generated by the sets in the partition—that is in fact the same as $\sigma(A_1, \ldots, A_n)$.

Solution: We start by proving that $\mathscr{T}$ is a σ-algebra. First,

$$\Omega = \bigcup_{i=1}^{n} A_i,$$

which implies that $\Omega \in \mathcal{T}$.

Suppose that $A = \bigcup_{i \in J} A_i$ is an element of $\mathscr{T}$. Then its complement is

$$A^c = \bigcap_{i \in J} A_i^c = \bigcup_{i \in J^c} A_i.$$

This is easy to see since A_i form a partition of Ω. Thus $A_i^c = \bigcup_{j \neq i} A_j$ and the intersection of these gives exactly the sets not indexed by J. Therefore, $A^c \in \mathscr{T}$.

Moreover, for any integer $p \in \mathbb{N}$.

$$\bigcup_{p} \bigcup_{i \in J_p} A_i = \bigcup_{i \in \cup_p J_p} A_i$$

which implies that $\mathscr{T}$ is closed under finite number of unions, and $\mathscr{T}$ is an algebra. Furthermore, since the total index set $\{1, 2, \ldots, n\}$ is finite, any countable union $\cup_{p=1}^{\infty} J_p$ of subsets J_p is also included in $\{1, 2, \ldots, n\}$ and $\mathscr{T}$ is closed under countable unions, thus it is in fact a σ-algebra.

Since $\mathscr{T}$ is a σ-algebra which contains the events A_i, it must also contain the sigma algebra generated by $A_1, \ldots, A_n$. This means that $\mathscr{T} \supseteq \sigma(A_1, \ldots, A_n)$.

Conversely, $\sigma(A_1, \ldots, A_n)$ is a σ-algebra, thus by definition it must contain any countable (in particular finite) union of A_i's. But the set of finite unions is precisely $\mathscr{T}$, which shows that $\mathscr{T} \subseteq \sigma(A_1, \ldots, A_n)$.

The double inclusion above then shows that $\mathscr{T}$ is the σ-algebra generated by $A_1, \ldots, A_n$. ■

EXAMPLE 1.5 Examples of algebras and σ-algebras on $\mathbb{R}$

Let $\mathscr{T}$ be the σ-algebra generated by the finite subsets of $\mathbb{R}$ (i.e., all subsets in $\mathbb{R}$ with a finite number of elements). Denote by $\mathscr{C}$ the set of subsets $A \subseteq \mathbb{R}$ such that A is countable or A^c is countable.

1. Show that $\mathscr{C}$ is a σ-algebra.
2. Show that $\mathscr{T}$ and $\mathscr{C}$ coincide.
3. Compare $\mathscr{T}$ and the Borel σ-algebra $\mathcal{B}$.

Solution: 1. We need to verify the three properties of a σ-algebra. Clearly $\emptyset \in \mathscr{C}$, and by the definition of $\mathscr{C}$ we have $A^c \in \mathscr{C}$ if $A \in \mathscr{C}$.

Let A_n be a sequence of events included in $\mathscr{C}$. We have two possibilities: Either all A_n's in the sequence are countable, or there exist an event in the sequence $(A_n)_n$ which is not countable.

If every A_n is countable, then clearly

$$\bigcup_n A_n$$

is countable (a countable union of countable sets is countable) and consequently $\bigcup_n A_n$ is in $\mathscr{C}$.

Suppose that there exists at least some set, say A_p, that is not countable. Since $A_p \in \mathscr{C}$, then A_p^c is countable. By DeMorgan laws (1.1) we have

$$\left(\bigcup_n A_n\right)^c = \bigcap_n A_n^c \subset A_p^c,$$

where the last inclusion holds because A_p^c is countable. Thus, the complement of the set $\left(\bigcup_n A_n\right)^c$ is at most countable and therefore it belongs to $\mathscr{C}$.

2. Clearly $\mathscr{T} \subseteq \mathscr{C}$. Recall that any finite set is also countable and so $\mathscr{C}$ is a σ-algebra that contains all the finite parts of $\mathbb{R}$. Since $\mathscr{T}$ is the smallest σ-algebra that contains these finite parts, the inclusion follows.

Conversely, we will prove that $\mathscr{T}$ contains every set $A \subseteq \mathbb{R}$ such that A or A^c is finite or countable. Indeed, if A is countable, then A can be written as

$$A = \bigcup_{x \in A} \{x\}.$$

Since every singleton $\{x\}$ is in $\mathscr{T}$ and $\mathscr{T}$ is stable under countable union, we deduce that $A \in \mathscr{T}$. If A^c is countable, the same argument applied to it will imply that $A^c \in \mathscr{T}$. But $\mathscr{T}$ is a σ-algebra and so $A \in \mathscr{T}$. In either case we deduce that any set A in $\mathscr{C}$ belongs to $\mathscr{T}$. Therefore $\mathscr{C} \subseteq \mathscr{T}$, and by double inclusion the two are the same.

3. The Borel σ-algebra contains all the finite subsets of $\mathbb{R}$. Therefore, we have $\mathscr{T} \subset \mathscr{B}$. This inclusion is strict. Indeed, the interval $(0, 1)$ is in $\mathscr{B}$, but it is not in $\mathscr{C}$ because neither $(0, 1)$ nor its complement is countable. Since $\mathscr{T}$ is the same as $\mathscr{C}$, it is a proper subset (included strictly) of $\mathscr{B}$. ■

1.5 Summary

In this chapter we introduce the space on which probability is defined. This beginning chapter introduces the notion of σ-algebra, which is going to be the domain of definition for the probability measure. We need to know what we can measure and what we cannot. The sets that can be measured and given a probability will always be in the σ-algebra. Furthermore, this chapter introduces the basic notions of the set operations such as inclusion, intersection, and so on.

EXERCISES

1.1 Show that if A, B are two event in Ω, then

$$A \cup B = (A \setminus B) \cup (B \setminus A) \cup (A \cap B).$$

1.2 If $(A_i)_{i \in I}$ is any collection of events in Ω, then show that

$$\left(\bigcup_{i \in I} A_i\right)^c = \bigcap_{i \in I} A_i^c$$

and

$$\left(\bigcap_{i \in I} A_i\right)^c = \bigcup_{i \in I} A_i^c.$$

1.3 Roll a six-sided die. Use $\Omega = \{1, 2, 3, 4, 5, 6\}$ to denote the possible outcomes. An example of an event in this space Ω is $A = \{\text{Roll an even number}\} = \{2, 4, 6\}$. Find the cardinality (total number of elements) of $\mathcal{P}(\Omega)$.

1.4 Suppose two events A and B are in some space Ω. List the elements of the generated σ algebra $\sigma(A, B)$ in the following cases:

(a) $A \cap B = \emptyset$.

(b) $A \subset B$.

(c) $A \cap B \neq \emptyset$, $A \setminus B \neq \emptyset$, and $B \setminus A \neq \emptyset$.

1.5 **An algebra which is not a σ-algebra**
Let $\mathscr{B}_0$ be the collection of sets of the form $(a_1, a_1'] \cup (a_2, a_2'] \cup \cdots \cup (a_m, a_m']$, for any $m \in \mathbb{N}^* = \{1, 2 \ldots\}$ and all $a_1 < a_1' < a_2 < a_2' < \cdots < a_m < a_m'$ in $\Omega = (0, 1]$. Verify that $\mathscr{B}_0$ is an algebra. Show that $\mathscr{B}_0$ is not a σ-algebra.

1.6 Let $\mathcal{F} = \{A \subseteq \Omega \mid A \text{ finite } \textbf{or } A^c \text{ is finite}\}$.

(a) Show that $\mathcal{F}$ is an algebra.

(b) Show that if Ω is finite, then $\mathcal{F}$ is a σ-algebra.

(c) Show that if Ω is infinite, then $\mathcal{F}$ is **not** a σ-algebra.

1.7 **A σ-algebra does not necessarily contain all the events in Ω**
Let $\mathcal{F} = \{A \subseteq \Omega \mid A \text{ countable } \textbf{or } A^c \text{ is countable}\}$. Show that $\mathcal{F}$ is a σ-algebra.
Note that if Ω is uncountable, this implies that it contains a set A such that both A and A^c are uncountable, thus $A \notin \mathcal{F}$.

1.8 Show that the Borel sets of $\mathbb{R}$ $\mathscr{B} = \sigma(\{(-\infty, x] \mid x \in \mathbb{R}\})$.

Hint: Show that the generating set is the same; that is, show that any set of the form $(-\infty, x]$ can be written as countable union (or intersection) of open intervals and vice versa that any open interval in $\mathbb{R}$ can be written as countable union (or intersection) of sets of the form $(-\infty, x]$.

1.9 Show that the following classes all generate the Borel σ-algebra; or, expressed differently, show the equality of the following collections of sets:

$$\sigma((a, b) : a < b \in \mathbb{R}) = \sigma([a, b] : a < b \in \mathbb{R}) = \sigma((-\infty, b) : b \in \mathbb{R}) \\ = \sigma((-\infty, b) : b \in \mathbb{Q}),$$

where $\mathbb{Q}$ is the set of all rational numbers.

1.10 Let A, B, C be three events in a probability space. Express the following events in terms of elements in A, B, C using union, intersection, complementarity, etc.

(a) Only the event A happens.

(b) At least one of the three events happens.

(c) At least two event happen.

(d) At most one event happen.

(e) Exactly two events happen.

Hint:

(a) $A \cap B^c \cap C^c$

(b) $A \cup B \cup C$

(c) $(A \cap B) \cup (A \cap C) \cup (B \cap C)$

(d) $((A \cap B) \cup (A \cap C) \cup (B \cap C))^c$

(e) $((A \cap B) \cup (A \cap C) \cup (B \cap C)) \cap (A \cap B \cap C)^c$

(Note that these expressions are not the only way to write the sets.)

1.11 Let $\mathcal{B}$, $\mathcal{C}$ be collections of subsets of Ω.

(a) If $\mathcal{B} \subset \mathcal{C}$, prove that $\sigma(\mathcal{B}) \subset \sigma(\mathcal{C})$.

(b) If $\mathcal{B} \subset \sigma(\mathcal{C}) \subset \sigma(\mathcal{B})$, show that $\sigma(\mathcal{B}) = \sigma(\mathcal{C})$.

1.12 Give a proof for Proposition 1.6.

Hint. Use exercise 1.11 for one of the two inclusions (the other one is obvious).

1.13 This problem gives the structure of the Borel σ-algebra on $\mathbb{R}^2$.

(a) Let $\mathcal{R}$ be the set of open rectangles in $\mathbb{R}^2$ with extremities in $\mathbb{Q}$—that is, the set of the form $(p, q) \times (r, s)$, where $p, q, r, s \in \mathbb{Q}$. Show that $\mathcal{R}$ is countable.

(b) Let U be an open set in $\mathbb{R}^2$. Prove that

$$U = \bigcup_{R \in \mathcal{R}, R \subset U} R;$$

that is, the open sets may be written as countable unions of elements in $\mathcal{R}$.

(c) Finally, conclude that the σ-algebra generated by the open rectangles in $\mathbb{R}^2$ coincides with the Borel σ-algebra on $\mathbb{R}^2$.

1.14 Prove the claim stated in Remark 1.11.

CHAPTER TWO

Probability Measure

2.1 Introduction/Purpose of the Chapter

The previous chapter presents the concept of measurable space. A measurable space is a couple $(\Omega, \mathcal{F})$ with Ω a non-empty set and $\mathcal{F}$ a σ-algebra. Such a measurable space is ready for a probability measure. This is a measure defined on the events in the σ-algebra $\mathcal{F}$, taking values in the unit interval $[0, 1]$. This setup then can be applied to any random phenomena, from simple ones—for example, rolling a dice or tossing a coin—to complex ones such as running a survey to ascertain whether or not each person in a sample supports the death penalty, or modeling a laboratory investigation to study the best amount of a certain chemical and its effects on the yield of a product.

The probability measures the *size* of the events. Any probability measure is defined by two properties: the probability of Ω must equal to 1 and the measure must be countably additive. What is the reason for these two properties? The first one implies that all possible outcomes of the experiment are accounted for in Ω—the *universe* of the experiment. The second property is very logical. It is natural to have $\mathbf{P}(A \cup B) = \mathbf{P}(A) + \mathbf{P}(B)$ for two disjoint events A and B in $\mathcal{F}$; that is, probability that A or B is happening is equal to the sum of the probabilities of A and B. The countable additivity is needed to extend this property to the entire σ-algebra as well as provide a way to go from calculating probability on the generators to the entire σ-algebra.

Handbook of Probability, First Edition. Ionuţ Florescu and Ciprian Tudor.

2.2 Vignette/Historical Notes

Any probability model has two essential components: (a) the sample space and its associated σ-algebra which defines *what can be measured* and (b) a probability law which defines *how to measure*. A natural question to ask is, Where do the numerical values for such a probability law come from? There are primarily three points of view on this aspect of how to connect the theory and reality. The probabilities could be assigned directly from data, following some models, or be assigned in a subjective manner. The frequentist approach considers any given experiment as one of a possible infinite replications of the experiment. The objectivist approach sees probabilities as adapting to the real aspects of the universe; thus the experiment in fact keeps changing and adapting itself to these laws of the universe. Finally, the subjectivist view describes probabilities as a way of characterizing beliefs about the world, rather than as having any external physical significance. The later two view of probability adhere to the Bayesian principles.

Throughout this book we will present views and examples primarily from the frequentist and objectivist perspective. In these historical notes we shall give an example related to subjectively assigning probabilities. The example is reproduced from Wild and Seber (1999).

The Tenerife airport disaster which occurred on March 27, 1977, when a Pan Am jet collided with a KLM Boeing 747 jet on the airport's runway in the Canary Islands is the deadliest recorded accident in the aviation history. Five hundred and eighty-three lives were lost in the accident. Naturally, the aviation officials at the time were trying to calm potential passengers and educate them about the dangers of flying. The famous Australian statistician, Terence Paul Speed, noticed the following wire service report in "The West Australian" published shortly after the accident.

> "NEW YORK, Mon: Mr. Webster Todd, Chairman of the American National Transportation Safety Board said today statistics showed that the chances of two jumbo jets colliding on the ground were about 6 million to one .....-AAP."

It seems clear from the report that the National Transportation Safety Board has performed a scientifically based assessment based upon hard data ("... statistics showed..."). The conclusion likewise is very definite. Terry Speed, who has strong research interests in probability, was intrigued by this and wondered how the Board had calculated their figure. So Terry wrote to the Chairman. He received the following reply from a high government official which Wild and Seber (1999) reproduce in their book with Terry Speed's permission:

> Dear Professor Speed, In response to your aerogram of April 5, 1977, the Chairman's statement concerning the chances of two jumbo jets colliding (6 million to one) has no statistical validity nor was it intended to be a rigorous or precise probability statement. The statement was made to

emphasize the intuitive feeling that such an occurrence indeed has a very remote but not impossible chance of happening.

2.3 Theory and Applications

2.3.1 DEFINITION AND BASIC PROPERTIES

Definition 2.1 (Probability measure, probability space) *Given a measurable space $(\Omega, \mathcal{F})$, a probability measure is any function $\mathbf{P} : \mathcal{F} \to [0, 1]$ with the following properties:*

(i) $\mathbf{P}(\Omega) = 1$.

(ii) (Countable additivity) For any sequence $\{A_n\}_{n\in\mathbb{N}}$ of disjoint events in $\mathcal{F}$ (i.e. $A_i \cap A_j = \varnothing$, for all $i \neq j$):

$$\mathbf{P}\left(\bigcup_{n=1}^{\infty} A_n\right) = \sum_{n=1}^{\infty} \mathbf{P}(A_n).$$

The triple $(\Omega, \mathcal{F}, \mathbf{P})$ is called a probability space.

Note that the probability measure is a set function (i.e., a function defined on sets). In many ways, probability measures behave in a way similar to that of length, area, and volume measures. These analogies provide clues about modeling experiments and choosing the appropriate numbers between 0 and 1. Many other reasonable properties follow from the definition of probability measure. For example:

Proposition 2.2 (Elementary properties of probability measure) *Let $(\Omega, \mathcal{F}, \mathbf{P})$ be a probability space. Then:*

1. *For all sets $A, B \in \mathcal{F}$ with $A \subseteq B$, then*

$$\mathbf{P}(B \setminus A) = \mathbf{P}(B) - \mathbf{P}(A).$$

2. *We have $\forall A, B \in \mathcal{F}$ with $A \subseteq B$,*

$$\mathbf{P}(A) \leq \mathbf{P}(B), \qquad \mathbf{P}(A^c) = 1 - \mathbf{P}(A), \qquad \mathbf{P}(\emptyset) = 0.$$

3. $\mathbf{P}(A \cup B) = \mathbf{P}(A) + \mathbf{P}(B) - \mathbf{P}(A \cap B)$, $\forall A, B \in \mathcal{F}$.

4. *(General inclusion–exclusion formula, also named Poincaré's formula):*

$$\mathbf{P}(A_1 \cup A_2 \cup \cdots \cup A_n) = \sum_{I \subseteq \{1,2,..,n\}} (-1)^{|I|+1} \mathbf{P}\left(\bigcap_{i \in I} A_i\right) \tag{2.1}$$

$$= \sum_{i=1}^{n} \mathbf{P}(A_i) - \sum_{i<j \leq n} \mathbf{P}(A_i \cap A_j)$$

$$+ \sum_{i<j<k \leq n} \mathbf{P}(A_i \cap A_j \cap A_k) - \cdots + (-1)^n \mathbf{P}(A_1 \cap A_2 \cdots \cap A_n) \tag{2.2}$$

Note that successive partial sums are alternating between over-and-under estimating.

5. *(Finite subadditivity, sometimes called Boole's inequality):*

$$\mathbf{P}\left(\bigcup_{i=1}^{n} A_i\right) \leq \sum_{i=1}^{n} \mathbf{P}(A_i), \qquad \forall A_1, A_2, \ldots, A_n \in \mathcal{F}.$$

Proof: 1. Since

$$B = A \cup (B \setminus A)$$

and A, $B \setminus A$ are disjoint, we immediately obtain

$$\mathbf{P}(B) = \mathbf{P}(A) + \mathbf{P}(B \setminus A).$$

Property 2. follows from the first part directly.

For part 3., let $A, B \in \mathcal{F}$ and note that

$$A \cup B = (A \setminus B) \cup (B \setminus A) \cup (A \cap B)$$

and since the sets $(A \setminus B)$, $(B \setminus A)$, $(A \cap B)$ are disjoint,

$$\mathbf{P}(A \cup B) = \mathbf{P}(B \setminus A) + \mathbf{P}(A \setminus B) + \mathbf{P}(A \cap B). \tag{2.3}$$

Moreover,

$$\mathbf{P}(A \setminus B) = \mathbf{P}(A) - \mathbf{P}(A \cap B)$$

and

$$\mathbf{P}(B \setminus A) = \mathbf{P}(B) - \mathbf{P}(A \cap B).$$

Putting them together these last two relations into (2.3), we obtain the inclusion exclusion with two sets.

Part 4. is an extension of part 3 to n sets. It is proven using induction over n. We skip this part as it does not bring much in terms of ideas.

To show finite subadditivity, set

$$B_1 = A_1, \quad B_2 = A_2 \cap A_1^c, \quad \ldots, \quad B_n = A_n \cap A_{n-1}^c \cap \ldots \cap A_1^c.$$

Then,

$$B_i \cap B_j = \emptyset, \qquad \text{if } i \neq j$$

and

$$\bigcup_{i=1}^{n} B_i = \bigcup_{i=1}^{n} A_i.$$

Now, using point 2. we have $\mathbf{P}(B_i) \leq \mathbf{P}(A_i)$ for every i and

$$\mathbf{P}\left(\bigcup_{i=1}^{n} A_i\right) = \mathbf{P}\left(\bigcup_{i=1}^{n} B_i\right) = \sum_{i=1}^{n} \mathbf{P}(B_i) \leq \sum_{i=1}^{n} \mathbf{P}(A_i).$$

■

EXAMPLE 2.1 Simple operations with probabilities

A random experiment with a finite set of outcomes is modeled by the set Ω and the probability $\mathbf{P}$. Let A, B be two events in $\mathbf{P}(\Omega)$ such that

$$\mathbf{P}(A) = 0.6, \quad \mathbf{P}(B) = 0.4, \quad \text{and} \quad \mathbf{P}(A \cap B) = 0.2.$$

Calculate the probabilities of the following events:

$$A \cup B, \quad A^c, \quad B^c, \quad B \cap A^c, \quad A \cup B^c, \quad \text{and} \quad A^c \cup B^c.$$

Solution: We have

$$\mathbf{P}(A \cup B) = \mathbf{P}(A) + \mathbf{P}(B) - \mathbf{P}(A \cap B) = 0.8,$$

$$\mathbf{P}(A^c) = 1 - \mathbf{P}(A) = 0.4, \mathbf{P}(B^c) = 1 - \mathbf{P}(B) = 0.6.$$

Since

$$B = B \cap \Omega = B \cap (A \cup A^c) = (B \cap A) \cup (B \cap A^c),$$

we have

$$\mathbf{P}(B) = \mathbf{P}(B \cap A) + \mathbf{P}(B \cap A^c)$$

so

$$\mathbf{P}(B \cap A^c) = \mathbf{P}(B) - \mathbf{P}(A \cap B) = 0.2.$$

Also

$$\mathbf{P}(A \cup B^c) = \mathbf{P}((B \cap A^c)^c) = 1 - \mathbf{P}(B \cap A^c) = 0.8$$

and

$$\mathbf{P}(A^c \cup B^c) = \mathbf{P}(A \cap B)^c = 1 - \mathbf{P}(A \cap B) = 0.8.$$

■

EXAMPLE 2.2 A coin example

Two fair coins are tossed; find the probability that two heads are obtained.

Solution: Each coin has two possible outcomes let us denote them H (heads) and T (tails). The sample space Ω is given by

$$\Omega = \{(H, T), (H, H), (T, H), (T, T)\}.$$

Since we know that the coin is fair, then it is equally likely to land on either outcome H or T. Let E be the event "two heads are obtained."

Then $E = \{(H, H)\}$. Since all are equally likely, we have

$$\mathbf{P}(E) = \frac{|E|}{|\Omega|} = \frac{1}{4}.$$

■

EXAMPLE 2.3 Rolling two dies

In an experiment, two fair six-sided dice are rolled. Find the probability that the sum of the numbers shown on the dice is

(a) equal to 1
(b) equal to 4
(c) less than 13

Solution: (a) The sample space Ω of the outcomes when rolling the two dice is shown below.

$$\begin{aligned}\Omega = \{&(1, 1), (1, 2), (1, 3), (1, 4), (1, 5), (1, 6), (2, 1), (2, 2), (2, 3), (2, 4), (2, 5),\\ &(2, 6), (3, 1), (3, 2), (3, 3), (3, 4), (3, 5), (3, 6), (4, 1), (4, 2), (4, 3), (4, 4),\\ &(4, 5), (4, 6), (5, 1), (5, 2), (5, 3), (5, 4), (5, 5), (5, 6), (6, 1), (6, 2), (6, 3),\\ &(6, 4), (6, 5), (6, 6)\}.\end{aligned}$$

As in the previous example, since the dice are fair, all outcomes are equally likely (have the same probability of occurrence).

Let E be the event "sum equal to 1." There are no outcomes which correspond to a sum equal to 1, hence

$$\mathbf{P}(E) = \frac{|E|}{|\Omega|} = \frac{0}{36} = 0.$$

(b), three possible outcomes give a sum equal to 4:

$$E = \{(1,3),(2,2),(3,1)\}$$

hence

$$\mathbf{P}(E) = \frac{|E|}{|\Omega|} = 3/36 = 1/12.$$

(c) All possible outcomes, $E = S$, give a sum less than 13, hence.

$$\mathbf{P}(E) = \frac{|E|}{|\Omega|} = 36/36 = 1.$$

Next, we introduce the notion of partition of a space. ■

Definition 2.3 (A partition of a set) *A partition of any set Ω is any collection of sets $\{\Omega_i\}$, which are disjoint (i.e., $\Omega_i \cap \Omega_j = \emptyset$, if $i \neq j$) such that their union recreates the original set:*

$$\bigcup_i \Omega_i = \Omega.$$

The probability of an event can be expressed in terms of the intersections of the event with a partition of the entire sample space Ω. This rule is also known as the "law of total probability."

Proposition 2.4 (The law of total probability) *Let $(A_i)_{i\in I}$ be a partition of Ω. Then for every event $B \in \mathcal{F}$, one has*

$$\mathbf{P}(B) = \sum_{i\in I} \mathbf{P}(B \cap A_i).$$

Proof: We can write

$$B = B \cap \Omega = B \cap \left(\bigcup_{i\in I} A_i\right) = \bigcup_{i\in I} (B \cap A_i)$$

by the distributive property of the intersection and union. Since the sets $(B \cap A_i)_i$ are disjoint, we obtain

$$\mathbf{P}(B) = \sum_{i \in I} \mathbf{P}(B \cap A_i).$$

■

2.3.2 UNIQUENESS OF PROBABILITY MEASURES

The results in this section are concerned with equality of two probability measures. Let us start with the following definition.

Definition 2.5 *A family of events $\mathcal{U} \subseteq \mathcal{P}(\Omega)$ is called a λ-system if:*

1. *$\Omega \in \mathcal{U}$.*
2. *For any $A, B \in \mathcal{U}$, with $A \subset B$, we also have $B \setminus A \in \mathcal{U}$.*
3. *If $(A_n)_n \in \mathcal{U}$ is any sequence of disjoint events ($A_i \cap A_j = \emptyset$ if $i \neq j$), then*

$$\bigcup_n A_n \in \mathcal{U}.$$

Sometimes, a λ-system is called a Dynkin system (named after Eugene Borisovich Dynkin, who introduced the concept).

Proposition 2.6 *The following are true:*

(a) A σ-field is a λ-system.
(b) Every λ-system closed to finite intersections is a sigma-field.

Proof: Both parts a and b are immediate consequences of the definition of a σ-field. ■

Definition 2.7 *If a collection of sets $\mathcal{M} \subseteq P(\Omega)$, then we denote by $\mathcal{U}(\mathcal{M})$ the smallest λ-system which contains $\mathcal{M}$.*

Theorem 2.8 *If $\mathcal{M} \subseteq P(\Omega)$ and $\mathcal{M}$ is closed under finite intersections, then*

$$\mathcal{U}(\mathcal{M}) = \sigma(\mathcal{M}).$$

Proof: The inclusion

$$\sigma(\mathcal{M}) \supseteq \mathcal{U}(\mathcal{M})$$

is immediate because $\sigma(\mathcal{M})$ is a λ-system, being a σ-algebra (Proposition 2.6) and it contains $\mathcal{M}$. For the converse inclusion, it suffices to check that $\mathcal{U}(\mathcal{M})$ is a σ-algebra (and so it will contain the smallest σ-algebra $\sigma(\mathcal{M})$). Due do

Proposition 2.6, it is enough to show that $\mathcal{U}(\mathcal{M})$ is closed under finite intersections. Let $A \in \mathcal{U}(\mathcal{M})$ and denote

$$\mathcal{U}_A = \{B \in \mathcal{U}(\mathcal{M}) \mid A \cap B \in \mathcal{U}(\mathcal{M}\},$$

the collection of sets such that their intersection with A is in the λ-system. To show that $\mathcal{U}(\mathcal{M})$ is closed under finite intersections, it is enough to show that $A \cap B \in \mathcal{U}(\mathcal{M})$ for any A, $B \in \mathcal{U}(\mathcal{M})$. We note that by definition we have

$$B \in \mathcal{U}_A \qquad \text{if and only if} \quad A \in \mathcal{U}_B.$$

It can be easily shown that the collection of sets $\mathcal{U}_A$ is a λ-system for any set A.

Now, take a generic set $A \in \mathcal{M}$. Since we know from the hypothesis that $\mathcal{M}$ is closed under finite intersections, by definition we then have $\mathcal{U}_A \supset \mathcal{M}$. Since $\mathcal{U}_A$ is a λ-system, we must then have

$$\mathcal{U}_A \supset \mathcal{U}(\mathcal{M}),$$

since $\mathcal{U}(\mathcal{M})$ is the smallest possible with this property. Hence $\mathcal{U}_A \supset \mathcal{U}(\mathcal{M})$ for any $A \in \mathcal{M}$. One can further show that $\mathcal{U}_A \supset \mathcal{U}(\mathcal{M})$ for any $A \in \mathcal{U}(\mathcal{M})$. We conclude that $\mathcal{U}(\mathcal{M})$ is stable under finite intersections. ■

Theorem 2.9 *Consider* $\mathbf{P}_1$, $\mathbf{P}_2$ *two probability measures on* $(\Omega, \mathcal{F})$. *Assume that a collection of sets* $\mathcal{M}$ *closed to finite intersections is given by*

$$\mathcal{F} = \sigma(\mathcal{M}).$$

Assume that $\mathbf{P}_1$ *and* $\mathbf{P}_2$ *coincide on* $\mathcal{M}$. *Then* $\mathbf{P}_1$ *and* $\mathbf{P}_2$ *coincide on* $\mathcal{F} = \sigma(\mathcal{M})$.

Proof: Let us define

$$\mathcal{U} = \{A \in \mathscr{F} \mid \mathbf{P}_1(A) = \mathbf{P}_2(A)\},$$

that is, the sets on which the measures coincide. This set $\mathcal{U}$ satisfies the following properties:

1. $\Omega \in \mathcal{U}$.
2. For every A, $B \in \mathcal{U}$, with $A \subseteq B$, we have $B \setminus A \in \mathcal{U}$.
3. If $(A_n)_n \in \mathcal{U}$, with $A_i \cap A_j = \emptyset$ if $i \neq j$ then $\bigcup_n A_n \in \mathcal{U}$.

Indeed, part 1 is clear since

$$\mathbf{P}_1(\Omega) = \mathbf{P}_2(\Omega) = 1.$$

Take two sets A, $B \in \mathcal{U}$ with $A \subseteq B$. Then

$$\mathbf{P}_1(B \setminus A) = \mathbf{P}_1(B) - \mathbf{P}_1(A) = \mathbf{P}_2(B) - \mathbf{P}_2(A) = \mathbf{P}_2(B \setminus A),$$

which implies $B \setminus A \in \mathcal{U}$.

For the last part consider $(A_n)_n \in \mathcal{U}$, with $A_i \cap A_j = \emptyset$ if $i \neq j$. Then

$$\mathbf{P}_1\left(\bigcup_n A_n\right) = \sum_n \mathbf{P}_1(A_n) = \sum_n \mathbf{P}_2(A_n) = \mathbf{P}_2\left(\bigcup_n A_n\right)$$

and so $\bigcup_n A_n \in \mathcal{U}$.

Properties 1–3 show that $\mathcal{U}$ is a λ-system (by definition). Recall that the probabilities agree on $\mathcal{M}$ by hypothesis, thus $\mathcal{U}$ includes $\mathcal{U}(\mathcal{M})$ (the smallest λ-system containing $\mathcal{M}$). Recall the result in Theorem 2.8 and that $\mathcal{M}$ is closed for finite intersections. Therefore,

$$\mathcal{U} \supseteq \mathcal{U}(\mathcal{M}) = \sigma(M).$$

Since $\mathcal{M}$ generates the entire σ-algebra, we thus have

$$\mathcal{U} = \sigma(\mathcal{M}) = \mathcal{F}.$$

■

Corollary 2.10 *Let $\mathbf{P}_1, \mathbf{P}_2$ be two probability measures on $\mathcal{F}$. Then if*

$$\mathcal{F} = \sigma(\Delta)$$

where $\Delta = (A_j)_j$ is a partition with a countable number of sets in Ω then $\mathbf{P}_1 = \mathbf{P}_2$ on the entire $\mathcal{F}$ if and only if

$$\mathbf{P}_1(A_j) = \mathbf{P}_2(A_j)$$

for every j.

Proof: We apply the previous theorem for $\mathcal{M}$ containing the sets in the partition Δ augmented with the $\emptyset$. ■

2.3.3 MONOTONE CLASS

This concept is a variant of the notion of λ-system presented before in our exposition. The concept of monotone class plays an important role in the probability theory. For example, it is used in the proof of the Carathèodory theorem, and it is also related to the uniqueness of probability measures.

Definition 2.11 (Monotone class) *A class (collection) $\mathscr{M}$ of subsets in Ω* is monotone *if it is closed under the formation of monotone unions and intersections, i.e.:*

(i) *$A_1, A_2, \ldots \in \mathscr{M}$ and $A_n \subseteq A_{n+1}$, $\bigcup_n A_n = A \Rightarrow A \in \mathscr{M}$.*
(ii) *$A_1, A_2, \ldots \in \mathscr{M}$ and $A_n \supseteq A_{n+1} \Rightarrow \bigcap_n A_n \in \mathscr{M}$.*

The next theorem is only needed for the proof of the Carathèodory theorem. However, the proof is interesting, and that is why it is presented here.

Theorem 2.12 *If $\mathcal{F}_0$ is an algebra and $\mathscr{M}$ is a monotone class, then if $\mathcal{F}_0 \subseteq \mathscr{M}$, then $\sigma(\mathcal{F}_0) \subseteq \mathscr{M}$.*

Proof: Let $m(\mathcal{F}_0)$ = minimal monotone class over $\mathcal{F}_0$ = the intersection of all monotone classes containing $\mathcal{F}_0$.

We will prove that

$$\sigma(\mathcal{F}_0) \subseteq m(\mathcal{F}_0).$$

To show this, it is enough to prove that $m(\mathcal{F}_0)$ is an algebra. Then exercise 2.20 will show that $m(\mathcal{F}_0)$ is a σ-algebra. Since $\sigma(\mathcal{F}_0)$ is the smallest σ-algebra which contains $\mathscr{M}$, the conclusion follows.

To this end, let

$$\mathscr{G} = \{A : A^c \in m(\mathcal{F}_0)\}.$$

(i) Since $m(\mathcal{F}_0)$ is a monotone class, so is $\mathscr{G}$.

(ii) Since $\mathcal{F}_0$ is an algebra, its elements are in $\mathscr{G} \Rightarrow \mathcal{F}_0 \subseteq \mathscr{G}$.

(i) and (ii) $\Rightarrow m(\mathcal{F}_0) \subseteq \mathscr{G}$. Thus $m(\mathcal{F}_0)$ is closed under complementarity.

Now define

$$\mathscr{G}_1 = \{A : A \cup B \in m(\mathcal{F}_0), \forall B \in \mathcal{F}_0\}.$$

We show that $\mathscr{G}_1$ is a monotone class:

Let $A_n \nearrow$ an increasing sequence of sets, $A_n \in \mathscr{G}_1$. By definition of $\mathscr{G}_1$, for all n, $A_n \cup B \in m(\mathcal{F}_0), \forall B \in \mathcal{F}_0$.

But

$$(A_n \cup B) \supseteq (A_{n-1} \cup B)$$

and thus the definition of $m(\mathcal{F}_0)$ implies

$$\bigcup_n (A_n \cup B) \in m(\mathcal{F}_0), \quad \forall B \in \mathcal{F}_0 \Rightarrow \left(\bigcup_n A_n\right) \cup B \in m(\mathcal{F}_0), \quad \forall B,$$

and thus

$$\bigcup_n A_n \in \mathscr{G}_1.$$

This shows that $\mathscr{G}_1$ is a monotone class. But since $\mathcal{F}_0$ is an algebra, its elements (the contained sets) are in $\mathscr{G}_1$,[1] thus $\mathcal{F}_0 \subseteq \mathscr{G}_1$. Since $m(\mathcal{F}_0)$ is the smallest monotone class containing $\mathcal{F}_0$, we immediately have $m(\mathcal{F}_0) \subseteq \mathscr{G}_1$.

Let $\mathscr{G}_2 = \{B : A \cup B \in m(\mathcal{F}_0), \forall A \in m(\mathcal{F}_0)\}$

$\mathscr{G}_2$ is a monotone class. (identical proof—see exercise 2.18).

[1] One can just verify the definition of $\mathscr{G}_1$ for this.

Let $B \in \mathcal{F}_0$. Since $m(\mathcal{F}_0) \subseteq \mathscr{G}_1$ for any set, we have

$$A \in m(\mathcal{F}_0) \Rightarrow A \cup B \in m(\mathcal{F}_0).$$

Thus, by the definition of $\mathscr{G}_2 \Rightarrow B \in \mathscr{G}_2 \Rightarrow \mathcal{F}_0 \subseteq \mathscr{G}_2$.

The previous implication and the fact that $\mathscr{G}_2$ is a monotone class implies that $m(\mathcal{F}_0) \subseteq \mathscr{G}_2$.

Therefore,

$$\forall A, \qquad B \in m(\mathcal{F}_0) \Rightarrow A \cup B \in m(\mathcal{F}_0) \Rightarrow m(\mathcal{F}_0)$$

is an algebra. ■

2.3.4 EXAMPLES

EXAMPLE 2.4 An "equally likely outcomes" probability measure

Let Ω be a finite set, $\mathcal{F} = \mathscr{P}(\Omega)$, and define $\mathbf{P} : \mathcal{F} \to [0, 1]$ by

$$\mathbf{P}(A) = \frac{|A|}{|\Omega|}.$$

Then $\mathbf{P}$ is a probability measure.

Please note that in the previous example, since the space is finite, the number of elements in Ω is finite, $|\Omega| < \infty$. Thus, the probability of any outcome in the space is the same; that is, for any $\omega \in \Omega$ we have

$$\mathbf{P}(\{\omega\}) = \frac{|\{\omega\}|}{|\Omega|} = \frac{1}{|\Omega|}.$$

EXAMPLE 2.5 Discrete probability space

Let Ω be a countable space. Let $\mathcal{F} = \mathscr{P}(\Omega)$. Let $p : \Omega \to [0, N)$ be a function on Ω such that

$$\sum_{\omega \in \Omega} p(\omega) = N < \infty,$$

where N is a finite constant. Define:

$$\mathbf{P}(A) = \frac{1}{N} \sum_{\omega \in A} p(\omega)$$

We can show that $(\Omega, \mathcal{F}, \mathbf{P})$ is a Probability Space for any such $p(\omega)$. In fact this way we may define any discrete probability space.

Solution: Indeed, from the definition we obtain

$$\mathbf{P}(\Omega) = \frac{1}{N} \sum_{\omega \in \Omega} p(\omega) = \frac{1}{N} N = 1.$$

To show the countable additivity property, let A be a set in Ω such that $A = \bigcup_{i=1}^{\infty} A_i$, with A_i disjoint sets in Ω. Since the space is countable, we may write

$$A_i = \{\omega_1^i, \omega_2^i, \ldots\},$$

where any of the sets may be finite, but $\omega_j^i \neq \omega_l^k$ for all i, j, k, l where either $i \neq k$ or $j \neq l$. Then using the definition, we have

$$\begin{aligned}
\mathbf{P}(A) &= \frac{1}{N} \sum_{\omega \in \bigcup_{i=1}^{\infty} A_i} p(\omega) = \frac{1}{N} \sum_{i \geq 1, j \geq 1} p(\omega_j^i) \\
&= \frac{1}{N} \sum_{i \geq 1} \left(p(\omega_1^i) + p(\omega_2^i) + \cdots \right) \\
&= \sum_{i \geq 1} \mathbf{P}(A_i).
\end{aligned}$$

■

These simple examples show how to use the probability properties learned thus far.

Remark 2.13 *The previous exercise gives a way to construct discrete probability measures (distributions).*

For example, let us take $\Omega = \mathbb{N}$ the natural numbers and take $N = 1$ in the definition of probability of an event. Then

$$p(\omega) = \begin{cases} 1 - p & \text{if } \omega = 0 \\ p & \text{if } \omega = 1 \\ 0 & \text{otherwise} \end{cases} \qquad \text{gives the Bernoulli(p) distribution.}$$

$$p(\omega) = \begin{cases} \binom{n}{\omega} p^{\omega}(1-p)^{n-\omega} & \textit{if } \omega \leq n \\ 0 & \textit{otherwise} \end{cases}$$ *gives the Binomial(n, p) distribution.*

$$p(\omega) = \begin{cases} \binom{\omega-1}{r-1} p^{r}(1-p)^{\omega-r} & \textit{if } \omega \geq r \\ 0 & \textit{otherwise} \end{cases}$$ *gives the Negative Binomial(r, p) distribution.*

$$p(\omega) = \frac{\lambda^{\omega}}{\omega!}e^{-\lambda}, \textit{ gives the Poisson } (\lambda) \textit{ distribution.}$$

See Chapter 5 for more details on these probability distributions.

EXAMPLE 2.6 Uniform distribution on (0,1)

As another example let $\Omega = (0, 1)$ and $\mathcal{F} = \mathscr{B}((0, 1))$ the Borel σ-algebra. Define a probability measure λ as follows. For any open interval $(a, b) \subseteq (0, 1)$ let

$$\lambda((a, b)) = b - a,$$

which is in fact the length of the interval. We can expand this definition to any other open interval O as

$$\lambda(O) = \lambda(O \cap (0, 1)).$$

Note that we did not specify $\lambda(A)$ for all Borel sets A. Rather, we specified the measure only for the generators of the Borel σ-field. This illustrates the probabilistic concept mentioned before. In our specific situation, under very mild conditions on the generators of the σ-algebra any probability measure defined only on the generators can be uniquely extended to a probability measure on the whole σ-algebra (Caratheòdory extension theorem). In particular when the generators are open sets, these conditions are true and we can restrict the definition to the open sets alone. This example is going to be expanded further in Section 2.4.

2.3.5 MONOTONE CONVERGENCE PROPERTIES OF PROBABILITY

The σ-algebra differs from the regular algebra in that it allows us to deal with countable (not finite) number of sets. In fact, this is a recurrent theme in probability: learning to deal with infinity. On finite spaces, things are more or less simple. One has to define the probability of each individual outcome and everything proceeds from there. However, even in these simple cases imagine that one repeats an experiment (such as a coin toss) over and over. Again, we are forced to cope with infinity. This section introduces a way to deal with this infinity problem.

Let $(\Omega, \mathcal{F}, \mathbf{P})$ be a probability space.

Lemma 2.14 *The following are true:*

1. *If $A_n, A \in \mathcal{F}$ and $A_n \uparrow A$ (that notation means, $A_1 \subseteq A_2 \subseteq \dots A_n \subseteq \dots$, and $A = \bigcup_{n\geq 1} A_n$), then $\mathbf{P}(A_n) \uparrow \mathbf{P}(A)$ as a sequence of numbers.*
2. *If $A_n, A \in \mathcal{F}$ and $A_n \downarrow A$ (i.e., $A_1 \supseteq A_2 \supseteq \dots A_n \supseteq \dots$ and $A = \bigcap_{n\geq 1} A_n$), then $\mathbf{P}(A_n) \downarrow \mathbf{P}(A)$ as a sequence of numbers.*
3. *(Countable subadditivity) If $A_1, A_2, \dots,$ and $\bigcup_{i=1}^{\infty} A_n \in \mathscr{F}$ with A_i's not necessarily disjoint, then*

$$\mathbf{P}\left(\bigcup_{n=1}^{\infty} A_n\right) \leq \sum_{n=1}^{\infty} \mathbf{P}(A_n).$$

Proof: 1. Let $B_1 = A_1, B_2 = A_2 \setminus A_1, \dots, B_n = A_n \setminus A_{n-1}$. Because the sequence is increasing, we have that the B_i's are disjoint:

$$\mathbf{P}(A_n) = \mathbf{P}(B_1 \cup B_2 \cup \dots \cup B_n) = \sum_{i=1}^{n} \mathbf{P}(B_i).$$

Thus using countable additivity we obtain

$$\begin{aligned}\mathbf{P}\left(\bigcup_{n\geq 1} A_n\right) &= \mathbf{P}\left(\bigcup_{n\geq 1} B_n\right)\\ &= \sum_{i=1}^{\infty} \mathbf{P}(B_i) = \lim_{n\to\infty} \sum_{i=1}^{n} \mathbf{P}(B_i)\\ &= \lim_{n\to\infty} \mathbf{P}(A_n)\end{aligned}$$

2. Note that $A_n \downarrow A \;\Leftrightarrow\; A_n^{\,c} \uparrow A^c$, and from part 1 this means

$$1 - \mathbf{P}(A_n) \uparrow 1 - \mathbf{P}(A).$$

3. Let

$$B_1 = A_1, \qquad B_2 = A_1 \cup A_2, \dots, B_n = A_1 \cup \dots \cup A_n, \dots .$$

Recall that we proved the **finite** (not countable) subadditivity property in Proposition 2.2. Applying that result, we obtain

$$\mathbf{P}(B_n) = \mathbf{P}(A_1 \cup \dots \cup A_n) \leq \mathbf{P}(A_1) + \dots + \mathbf{P}(A_n).$$

But $\{B_n\}_{n\geq 1}$ is an increasing sequence of events, thus from part 1 we get that

$$\mathbf{P}(\bigcup_{n=1}^{\infty} B_n) = \lim_{n\to\infty} \mathbf{P}(B_n).$$

Combining the two relations above, we obtain

$$\mathbf{P}(\bigcup_{n=1}^{\infty} A_n) = \mathbf{P}(\bigcup_{n=1}^{\infty} B_n) \leq \lim_{n\to\infty} (\mathbf{P}(A_1) + \cdots + \mathbf{P}(A_n))$$
$$= \sum_{n=1}^{\infty} \mathbf{P}(A_n).$$

■

Definition 2.15 (Null set) *Any set N in the probability space $(\Omega, \mathcal{F}, \mathbf{P})$ which has the property*

$$\mathbf{P}(N) = 0$$

is called a ***P-null set****. If it is clear which probability measure is referred, the set may just be named a* ***null set****.*

Clearly, $\emptyset$ is a null set. However, it may not be the only one in the space Ω. To see an example of this situation, we need to look at probability spaces which are not finite. For instance, in Example 2.7 we introduced the Uniform measure. For any set $A = \{p\}$ where p is some number between zero and one, the measure of A is equal to zero (because the length of the interval $[p, p]$ is zero).

Lemma 2.16 *The union of a countable number of* **P***-null sets is a* **P***-null set.*

Proof: This lemma is a direct consequence of the countable subadditivity (Proposition 2.14). ■

The lemma is important in that it tells us that no matter how many zeros, they never add to anything else but zero. In $\mathbb{R}$ and its associated Borel σ-algebra $\mathcal{B}$, it also allows us to deal with those troublesome sets that could not be included in the Borel sets. All those sets have probability zero, and by extending the probability space definition a little we just include them all into the σ-algebra, which means we do not have to worry about them at all. The following definition is formalizing this concept.

Definition 2.17 (Complete probability space) *A probability space $(\Omega, \mathcal{F}, \mathbf{P})$ is said to be a complete probability space if all subsets of sets of probability 0 are in $\mathcal{F}$. Mathematically, if $M \subseteq N$ for some $N \in \mathcal{F}$ with $\mathbf{P}(N) = 0$, then $M \in \mathcal{F}$.*

2.3.6 CONDITIONAL PROBABILITY

In some cases the probability of an event happening depends not just on the experiment itself but on other information as well. Conditional probability forms a framework in which this additional information can be incorporated.

Let $(\Omega, \mathcal{F}, \mathbf{P})$ be a probability space. Then for $A, B \in \mathcal{F}$ with $\mathbf{P}(B) \neq 0$ we define the conditional probability of A given B by

$$\mathbf{P}(A|\, B) = \frac{\mathbf{P}(A \cap B)}{\mathbf{P}(B)}.$$

We note that the condition $\mathbf{P}(B) \neq 0$ is important not only mathematically but also conceptually. The conditioning part means that B is given to have happened. But when we condition by a set of probability 0 (which cannot happen), then any result is possible (even crazy and contradicting conclusions).

We can immediately rewrite the formula above to obtain the *multiplicative rule*:

$$\begin{aligned}
&\mathbf{P}\ (A \cap B) = \mathbf{P}(A|\, B)\mathbf{P}(B),\\
&\mathbf{P}\ (A \cap B \cap C) = \mathbf{P}(A|\, B \cap C)\mathbf{P}(B|\, C)\mathbf{P}(C)\\
&\mathbf{P}\ (A \cap B \cap C \cap D) = \mathbf{P}(A|\, B \cap C \cap D)\mathbf{P}(B|\, C \cap D)\mathbf{P}(C|D)\mathbf{P}(D), \quad \text{etc.}
\end{aligned}$$

This multiplicative rule is very useful for stochastic processes and estimation of parameters of a distribution.

EXAMPLE 2.7 An urn problem

We have two urns, I and II. Urn I contains 2 black balls and 3 white balls. Urn II contains 1 black ball and 1 white ball. One of the two urns is chosen at random and a ball is drawn at random from it. If a back ball is chosen, what is the probability it came from urn I?

Solution: To start formalizing this problem, let B be the event "a black ball is drawn," and I the event "urn I is chosen." To calculate the conditional, we need the numerator in the formula.

The joint probability of both these events happening is

$$\mathbf{P}(B \cap I) = \mathbf{P}(B \mid I)\mathbf{P}(I) = \frac{2}{5}\frac{1}{2} = \frac{1}{5}.$$

Here we used the multiplicative rule above since this simplifies the problem considerably. Note that if we know the ball came from urn I, then it is very simple to calculate the probability that the ball is black—that probability is exactly the conditional probability $\mathbf{P}(B \mid I)$.

However, we need to calculate the other conditional ($\mathbf{P}(I \mid B)$) using the definition

$$\mathbf{P}(I \mid B) = \frac{\mathbf{P}(I \cap B)}{\mathbf{P}(B)} = \frac{\mathbf{P}(I \cap B)}{\mathbf{P}(B \cap I) + \mathbf{P}(B \cap II)}$$
$$= \frac{\frac{1}{5}}{\frac{1}{5} + \frac{1}{4}}$$
$$= \frac{4}{9}.$$

■

The expression above is in fact Bayes' formula, which we will see in a minute. In the denominator we also used a particular case of the next proposition.

Proposition 2.18 (Total probability formula) *Given $A_1, A_2, \ldots, A_n$ a partition of Ω (i.e., the sets A_i are disjoint and $\Omega = \bigcup_{i=1}^{n} A_i$), and suppose $\mathbf{P}(A_i) > 0$ for every $i = 1, \ldots, n$. Then*

$$\mathbf{P}(B) = \sum_{i=1}^{n} \mathbf{P}(B | A_i)\mathbf{P}(A_i), \qquad \forall B \in \mathcal{F} \tag{2.4}$$

Proof: The proof is immediate using Proposition 2.4 and the multiplicative rule. ■

EXAMPLE 2.8 Another conditional example

Stanley takes an oral exam in statistics by answering 3 questions written on an examination card. There are 20 such examination cards and Stanley will receive one of them drawn at random. Of the 20 there are 8 favorable cards (Stanley knows the answers for all 3 questions written on the card), all the others contain at least a question that Stanley has no clue how to answer. Stanley will get an A if he answers all 3 questions on the card correctly.

What is the probability that Stanley gets an A if he draws the card

(i) first?
(ii) second?
(iii) third?

Solution: Let A denote the event that Stanley draws a favorable card (and consequently gets an A).

(i) If he draws the card first, then clearly $\mathbf{P}(A) = 8/20 = 2/5$.

(ii) If Stanley draws second, then one card was already taken by the student in front of him. That first card taken might have been favorable (call that hypothesis H_1) or unfavorable (hypothesis H_2). Obviously, the hypotheses H_1 and H_2 partition the sample space since no other type of cards is possible. Note that the probabilities of H_1 and H_2 for that first card are 8/20 and 12/20, respectively.

Now, Stanley goes ahead and draws a second card after one card has already been taken. If H_1 had happened, the probability of A is 7/19, and if H_2 had happened, the probability of A is 8/19. But these are just conditional probabilities $\mathbf{P}(A|H_1) = 7/19$ and $\mathbf{P}(A|H_2) = 8/19$, and using the total probability formula, we obtain

$$\mathbf{P}(A) = 7/19 \times 8/20 + 8/19 \times 12/20 = 8/20 = 2/5$$

(iii) This is very similar with the previous case but there are 4 possible hypotheses.

- $H_1 =$ both cards taken in front of Stanley were favorable,
- $H_2 =$ exactly one card was favorable
- $H_3 =$ none of the cards taken before him were favorable.

Using the same rule as before we readily see:

$$\mathbf{P}(H_1) = \frac{8}{20}\frac{7}{19}, \quad \mathbf{P}(H_3) = \frac{12}{20}\frac{11}{19}, \quad \text{and} \quad \mathbf{P}(H_2) = 1 - \mathbf{P}(H_1) - \mathbf{P}(H_3).$$

Furthermore, $\mathbf{P}(A|H_1) = 6/18$, $\mathbf{P}(A|H_2) = 7/18$, and $\mathbf{P}(A|H_3) = 8/18$. Finally,

$$\mathbf{P}(A) = 6/18 \times 7/19 \times 8/20 + 7/18 \times \cdots + 8/18 \times 11/19 \times 12/20 = 8/20.$$

Moral of this example: Stanley should concentrate more on the exam rather than worrying about the order in which the examination cards are drawn.

Note that this analysis is done before any drawing takes place. If Stanley waits and then checks with the person in front of him and finds out if the card was favorable or not that changes the samples space to worse (7 good out of 19) if the guy in front of him was lucky or better (8 out of 19) in the other situation. The whole sample space is changed from the moment the situation is clarified. ■

Proposition 2.19 (Bayes' formula) *Let $A_1, A_2, \ldots, A_n$ form a partition of Ω with $\mathbf{P}(A_i) > 0$ for every $i = 1, \ldots, n$. Then for any set $B \in \mathcal{F}$ we can write*

$$\mathbf{P}(A_j|\, B) = \frac{\mathbf{P}(B|\, A_j)\mathbf{P}(A_j)}{\sum_{i=1}^{n} \mathbf{P}(B|\, A_i)\mathbf{P}(A_i)}. \tag{2.5}$$

Proof: Using the definition of conditional probability and

$$\mathbf{P}(B \cap A_j) = \mathbf{P}(B)\mathbf{P}(A_j|B) = \mathbf{P}(A_j)\mathbf{P}(B|A_j),$$

we obtain

$$\mathbf{P}(A_j|\, B) = \frac{\mathbf{P}(A_j)\mathbf{P}(B|A_j)}{\mathbf{P}(B)} = \frac{\mathbf{P}(B|A_j)\mathbf{P}(A_j)}{\sum_{i=1}^{n} \mathbf{P}(B|\, A_i)\mathbf{P}(A_i)}.$$

■

This theorem can be interpreted as: All of a sudden we find out that B happened. In the light of this new information there should be a simple way to update all the probabilities of the sets in the partition ($\mathbf{P}(A_j|B)$). Indeed, the Bayes rule provides just that.

EXAMPLE 2.9

There are two urns each containing two types of colored balls. The first urn contains 50 red balls and 50 blue balls. The second urn contains 30 red balls and 70 blue balls. One of the two urns is randomly chosen (both urns have probability 50 percent of being chosen), and then a ball is drawn at random from one of the two urns. If a red ball is drawn, what is the probability that it comes from the first urn?

Solution: In probabilistic terms, what we know about this problem can be formalized as follows:

$$\mathbf{P}(red \mid urn1) = \frac{1}{2}, \qquad \mathbf{P}(red \mid urn2) = \frac{3}{10},$$
$$\mathbf{P}(urn1) = \mathbf{P}(urn2) = \frac{1}{2}.$$

The unconditional probability of drawing a red ball can be derived using the law of total probability:

$$\begin{aligned}\mathbf{P}(red) &= \mathbf{P}(red \mid urn1)\mathbf{P}(red) + \mathbf{P}(red|urn2)\mathbf{P}(urn2)\\ &= \frac{1}{2}\frac{1}{2} + \frac{3}{10}\frac{1}{2} = \frac{2}{5}.\end{aligned}$$

Using Bayes' rule, we obtain

$$\mathbf{P}(urn1 \mid red) = \frac{\mathbf{P}(red \mid urn1)\mathbf{P}(urn1)}{\mathbf{P}(red)} = \frac{\frac{1}{2}\frac{1}{2}}{\frac{2}{5}} = \frac{5}{8}.$$

■

EXAMPLE 2.10 De Mére's paradox

As a result of extensive observation of dice games the French gambler Chevaliér De Mére noticed that the total number of spots showing on 3 dice thrown simultaneously turn out to be 11 more often than 12. However, from his point of view this is not possible since 11 occurs in six ways:

$$(6:4:1); (6:3:2); (5:5:1); (5:4:2); (5:3:3); (4:4:3),$$

while 12 also in six ways:

$$(6:5:1); (6:4:2); (6:3:3); (5:5:2); (5:4:3); (4:4:4).$$

What is the fallacy in the argument?

Solution due to Pascal: The argument would be correct if these "ways" would have the same probability. However, this is not true. For example, (6:4:1) occurs in 3! ways, (5:5:1) occurs in 3, ways and (4:4:4) occurs in 1 way.

If we keep this in mind, we can easily calculate

$$\mathbf{P}(11) = 27/216,$$
$$\mathbf{P}(12) = 25/216.$$

Indeed De Mére's observation is correct and he should bet on 11 rather than on 12 if they have the same game payoff. ∎

EXAMPLE 2.11 Another De Mére's paradox

What is more probable?

1. Throw 4 dice and obtain at least one 6, or
2. Throw 2 dice 24 times and obtain at least once a double 6?

Solution: For option 1: $1 - \mathbf{P}(\text{No 6's}) = 1 - (5/6)^4 = 0.517747$.

For option 2: $1 - \mathbf{P}(\text{None of the 24 trials has a double 6}) = 1 - (35/36)^{24} = 0.491404$ ∎

EXAMPLE 2.12 Bertrand's box paradox

This problem was first formulated by Joseph Louis François Bertrand in his Calcul de Probabilités (Bertrand, 1889). Solving this problem is an exercise on understanding Bayes' formula.

> Suppose that we know that three boxes contain the following: One box contains two gold coins, a second box contains two silver coins, and a third box has one of each. We chose a box at random, and from that box we chose a coin also at random. Then we look at the coin chosen. Given that the coin chosen was gold, what is the probability that the other coin in the box chosen is also gold? At a first glance it may seem that this probability is 1/2, but after calculation this probability turns out to be 2/3.

Solution: We plot the sample space in Figure 2.1. Using this tree, we can calculate the probability:

$$\mathbf{P}(\text{Second coin is gold}|\,\text{First coin is gold}) = \frac{\mathbf{P}(\text{Second coin is gold and First coin is gold})}{\mathbf{P}(\text{First coin is gold})}.$$

Now, using the probabilities from the tree, we continue:

$$= \frac{\frac{1}{3}\frac{1}{2}1 + \frac{1}{3}\frac{1}{2}1}{\frac{1}{3}\frac{1}{2}1 + \frac{1}{3}\frac{1}{2}1 + \frac{1}{3}\frac{1}{2}1} = \frac{2}{3}.$$

Now that we have seen the solution, we can recognize a logical solution to the problem as well. Given that the coin seen is gold, we can throw away the middle box. If this would be box 1 then we have two possibilities that the other coin is

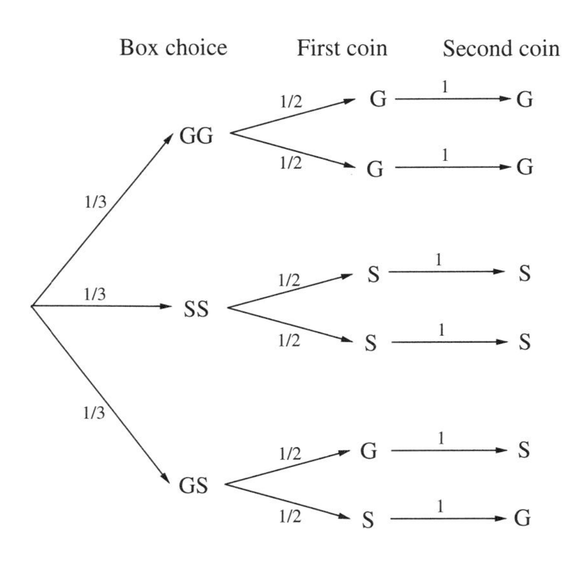

FIGURE 2.1 **The tree diagram of conditional probabilities.**

gold (depending on which we have chosen in the first place). If this is box 2, then there is one possibility (the remaining coin is silver). Thus the probability should be 2/3 since we have two out of three chances. Of course this "logical" argument does not work if we do not choose the boxes with the same probability. ■

EXAMPLE 2.13 Disease testing instrument

A blood test is 95% effective in detecting a certain disease when it is in fact present. However, the test yields also a false-positive result for 1% of the people tested. If 0.5% of the population actually has the disease, what is the probability that the person is diseased given that the test is positive?

Solution: This problem illustrates once again the application of Bayes' rule. One does not need to use the rule literally. Instead if we work from first principles we will obtain Bayes' rule every time without memorizing anything. We start by describing the sample space. Refer to Figure 2.2 for this purpose.

So, "given that the test is positive" means that we have to calculate a conditional probability. We may write

$$\mathbf{P}(D|+) = \frac{\mathbf{P}(D \cap +)}{\mathbf{P}(+)} = \frac{\mathbf{P}(+|D)\mathbf{P}(D)}{\mathbf{P}(+)} = \frac{0.95(0.005)}{0.95(0.005) + 0.01(0.995)} = 0.323.$$

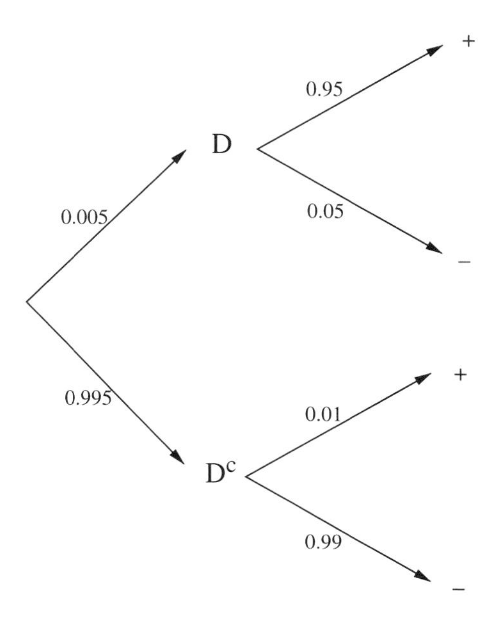

FIGURE 2.2 Blood test probability diagram.

How about if only 0.05% (i.e., 0.0005) of the population has the disease?

$$\mathbf{P}(D|+) = \frac{0.95(0.0005)}{0.95(0.0005) + 0.01(0.9995)} = 0.0454$$

This problem is an exercise in thinking. It is the same test device. In the first case the disease is relatively common, and thus the test device is more or less reliable (though 32% right is very low). In the second case, however, the disease is very rare and thus the precision of the device goes way down. ■

We shall see the next example many times throughout this book. We do not know who to credit with the invention of the problem since it is so mentioned so often in every probability treaties.[2]

EXAMPLE 2.14 Gambler's ruin problem

The formulation of this problem is simple. A gambler plays a game of heads or tails with a fair coin. The player wins 1 dollar if he successfully calls the side of the coin which lands upwards and loses \$1 otherwise. Suppose the initial capital is N dollars and he intends to play until he wins M dollars but no longer. What is the probability that the gambler will be ruined?

Solution: We will perform what is sometimes called a first step analysis. The idea is to see to define a proper quantity and analyze what happens with this quantity if one random step is taking place.

Let $p(x)$ denote the probability that the player is going to be eventually ruined if he starts with x dollars.

If he wins the next game, then he will have $x + 1$ dollars and he is ruined from this position with prob $p(x + 1)$.

If he loses the next game, then he will have $x - 1$ dollars so he is ruined from this position with prob $p(x - 1)$.

Let R be the event he is eventually ruined. Let W be the event he wins the next toss. Let L be the event he loses the next toss. Using the total probability formula we can write

$$\mathbf{P}(R) = \mathbf{P}(R|W)\mathbf{P}(W) + \mathbf{P}(R|L)\mathbf{P}(L).$$

Now from position x the same equation is

$$p(x) = p(x+1)(1/2) + p(x-1)(1/2).$$

However, this is true for almost all x. More precisely the relation is true for $x \geq 1$ and $x \leq M - 1$. When $x = 0$ or $x = M$ we obviously have $p(0) = 1$ and $p(M) = 0$, which give the boundary conditions for the equation above.

[2] The formalization may be due to Huygens (1629–1695).

This is a linear difference equation with constant coefficients. The theory on how to solve these is similar to the theory on differential equations with constant coefficients. One looks for solutions of the type $p(x) = y^x$. Substituting gives a characteristic equation which is then solved.

Applying this method in our case gives the characteristic equation

$$y = \frac{1}{2}y^2 + \frac{1}{2} \Rightarrow y^2 - 2y + 1 = 0 \Rightarrow (y-1)^2 = 0 \Rightarrow y_1 = y_2 = 1.$$

In our case the two solutions are equal; thus we seek a solution of the form $p(x) = (C + Dx)1^n = C + Dx$. Using the initial conditions, we get $p(0) = 1 \Rightarrow C = 1$ and $p(M) = 0 \Rightarrow C + DM = 0 \Rightarrow D = -C/M = -1/M$, thus the general probability of ruin starting with wealth x is

$$p(x) = 1 - \frac{x}{M}.$$

■

2.3.7 INDEPENDENCE OF EVENTS AND σ-FIELDS

Let us now introduce the concept of probabilistic independence.

Definition 2.20 *Two events A and B are called independent if and only if*

$$\mathbf{P}(A \cap B) = \mathbf{P}(A)\mathbf{P}(B).$$

The events $A_1, A_2, A_3, \ldots$ are called mutually independent *(or sometimes simply independent) if for every subset J of $\{1, 2, 3, \ldots\}$ we have*

$$\mathbf{P}\left(\bigcup_{j\in J} A_j\right) = \prod_{j\in J} \mathbf{P}(A_j).$$

The events $A_1, A_2, A_3, \ldots$ are called pairwise independent *(sometimes jointly independent) if*

$$\mathbf{P}\left(A_i \cup A_j\right) = \mathbf{P}(A_i)\mathbf{P}(A_j), \qquad \forall i, j.$$

Remark 2.21 *Note that jointly independent does not imply independence. See the Example 2.16 for a counterexample.*

We now define the concept of independent for two σ-algebras.

Definition 2.22 *Two* σ-fields $\mathscr{G}, \mathscr{H} \in \mathcal{F}$ are **P**–independent *if*

$$\mathbf{P}(G \cup H) = \mathbf{P}(G)\mathbf{P}(H), \qquad \forall G \in \mathscr{G}, \forall H \in \mathscr{H}.$$

The σ-algebras $\mathcal{F}_1, \ldots, \mathcal{F}_k$ are mutually independent (or simply independent) if

$$\mathbf{P}\left(\bigcup_{i=1}^{k} A_i\right) = \prod_{i=1}^{k} \mathbf{P}(A_i)$$

for every $A_i \in \mathcal{F}_i$, $i = 1, .., k$.

It is also possible to define the independent for an arbitrary (not necessary finite or countable) family of σ-fields.

Definition 2.23 *An arbitrary family of σ-fields $(\mathcal{F}_i)_{i\in I}$ is independent if any finite sub-family is independent in the sense of Definition 2.23.*

We refer to Billingsley (1995) for the precise mathematical formulae for independence of $k \geq 2$ σ-algebras.

EXAMPLE 2.15 Unbalanced Coin

Suppose that we have a coin which comes up heads with probability p, and tails with probability $q = 1 - p$. Now suppose that this coin is tossed twice. Using a frequency interpretation of probability, it is reasonable to assign to the outcome (H, H) the probability p^2, to the outcome (H, T) the probability pq, and so on. Let E be the event that heads turns up on the first toss and F the event that tails turns up on the second toss. We will now check that with the above probability assignments, these two events are independent, as expected. We have

$$\mathbf{P}(E) = p^2 + pq = p, \mathbf{P}(F) = pq + q^2 = q.$$

Finally $\mathbf{P}(E \cap F) = pq$, so

$$\mathbf{P}(E \cap F) = \mathbf{P}(E)\mathbf{P}(F).$$

The following criteria is useful in order to check the independence of a family with an infinite number of elements.

Theorem 2.24 *Let $(\mathcal{F}_i)_{i\in I}$ be a family of σ-fields on the probability space $(\Omega, \mathcal{F}, \mathbf{P})$. Assume that for every $i \in I$ there exists a family $\mathcal{M}_i \subseteq \mathcal{F}_i$ satisfying the following:*

i. *$\mathcal{M}_i$ is closed to finite intersections and*

$$\mathcal{F}_i = \sigma(\mathcal{M}_i)$$

for every $i \in I$.

ii. *For every $n \geq 1$, for every $i_1, \ldots, i_n \in I$, and for every $A_j \in \mathcal{M}_{i_j}, j = 1, \ldots, n$, it holds that*

$$\mathbf{P}\left(\bigcup_{i=1}^{k} A_i\right) = \prod_{i=1}^{k} \mathbf{P}(A_i).$$

Then the family of σ-fields $(\mathcal{F}_i)_{i \in I}$ is independent.

Proof: Note first that we can assume that

$$\Omega \in \mathcal{M}_i, \qquad \forall i \in I.$$

Indeed, if we replace $\mathcal{M}_i$ by $\mathcal{M}_i \cap \Omega$, then properties i and ii are still satisfied.

Consider a finite subset of indices

$$J = \{j_1, \ldots, j_n\} \subseteq I$$

and let us show that the indexed family $(\mathcal{F}_j)_{j \in J}$ is independent. We consider the events

$$A_i \in \mathcal{M}_{j_i}, i = 1, \ldots, n-1 \text{ and } A \in \mathcal{F}_{j_n}.$$

We will prove that

$$\mathbf{P}(A_1 \cap \ldots \cap A_{n-1} \cup A) = \mathbf{P}(A_1) \ldots \mathbf{P}(A_{n-1})\mathbf{P}(A). \tag{2.6}$$

If we have

$$\mathbf{P}(A_1 \cap \ldots \cap A_{n-1}) = \mathbf{P}(A_1) \ldots \mathbf{P}(A_{n-1}) = 0,$$

then relation (2.6) is clearly satisfied. Therefore, suppose that

$$\mathbf{P}(A_1 \cap \ldots \cap A_{n-1}) > 0.$$

We will show that the two probabilities $\mathbf{P}$ and $\mathbf{P}(\cdot|A_1 \cap \ldots \cap A_{n-1})$ coincide on the σ-field $\mathcal{F}_{j_n}$. It suffices to check that they coincide on $\mathcal{M}_{j_n}$. This is due to Theorem 2.9 since $\mathcal{M}_{j_n}$ generates $\mathcal{F}_{j_n}$ and is closed under finite intersections.

We can write using property ii, for every $A \in \mathcal{M}_{j_n}$

$$\mathbf{P}(A|A_1 \cap \ldots \cap A_{n-1}) = \frac{\mathbf{P}(A \cap A_1 \cap \ldots \cap A_{n-1})}{\mathbf{P}(A_1 \cap \ldots \cap A_{n-1})}$$
$$= \frac{\mathbf{P}(A)\mathbf{P}(A_1) \ldots \mathbf{P}(A_{n-1})}{\mathbf{P}(A_1) \ldots \mathbf{P}(A_{n-1})} = \mathbf{P}(A),$$

and this shows that $\mathbf{P}$ and $\mathbf{P}(\cdot|A_1 \cap \ldots \cap A_{n-1})$ coincide on $\mathcal{M}_{j_n}$. Therefore relation (2.6) is satisfied. Following the same procedure, we can replace in (2.6) $A_{n-1} \in \mathcal{M}_{j_{n-1}}$ by $A_{n-1} \in \mathcal{F}_{j_{n-1}}$ and we will obtain, step by step, that the equality (2.6) is valid for $A_i \in \mathcal{F}_{j_i}$, for every $i = 1, \ldots, n$. ■

We will show that the independence of σ-fields is associative.

Theorem 2.25 *Let $(\mathcal{F}_i)_{i\in I}$ be an independent family of σ-fields and let $(I_j)_{j\in J}$ be a partition of the set I (that is, $I_k \cup I_l = \emptyset$ if $j \neq l$ and $\bigcup_{j\in J} I_j = I$). For every $j \in J$ we define*

$$\mathcal{G}_j = \sigma\left(\bigcup_{i\in I_j} \mathcal{F}_i\right).$$

Then the family of σ-fields $(\mathcal{G}_j)_{j\in J}$ is independent.

Thus no matter how we associate (or group) σ-algebras which are independent the groups still are independent as long as there is no overlap.

Proof: To prove the theorem, define for every $j \in J$

$$\mathcal{M}_j := \{A_1 \cap \ldots \cap A_{n_j} \mid A_i \in \mathcal{F}_i, i \in I_j\},$$

where n_j is the number of σ algebras $\mathcal{F}_i$ indexed by I_j. Then we have

$$\sigma(\mathcal{M}_j) = \mathcal{G}_j$$

for every $j \in J$. Since the family $(\mathcal{M}_j)_j$ satisfies assumptions i and ii from Theorem 2.24), we only need to check that

$$\mathbf{P}(B_{j_1} \cap \ldots B_{j_n}) = \mathbf{P}(B_{j_1}) \ldots \mathbf{P}(B_{j_n}) \tag{2.7}$$

for every $B_{j_r} \in \mathcal{M}_{j_r}$ for $r = 1, \ldots, n$. Let us consider

$$B_{j_r} = A_1^r \cap \ldots \cup A_{i(r)}^r$$

with $i_1, \ldots, i_r \in I_{j_r}, A_{j_r} \in \mathcal{F}_j, j \in \mathcal{F}_{j_r}$. Then

$$\begin{aligned}
\mathbf{P}\big(B_{j_1} \cap \ldots \cap B_{j_n}\big) &= \mathbf{P}\left(\bigcap_{r=1}^{n}\bigcap_{u=1}^{i(r)} A_u^r\right) \\
&= \prod_{r=1}^{n}\prod_{u=1}^{i(r)} \mathbf{P}(A_u^r) = \prod_{r=1}^{n} \mathbf{P}\left(\bigcap_{u=1}^{i(r)} A_u^r\right) \\
&= \prod_{r=1}^{n} \mathbf{P}(B_{j_r})
\end{aligned}$$

and thus (2.7) is true. ■

The following property is usually called the disassociativity of the independence.

Theorem 2.26 *Let $(\mathcal{G}_j)_{j\in J}$ be an independent family of σ-fields and for every $j \in J$ let $(\mathcal{F}_i)_{i\in I_j} \subseteq \mathcal{G}_j$ be an independent family of σ-fields where*

$$I_{j_1} \cap I_{j_2} = \emptyset \qquad \text{if } j_1 \neq j_2.$$

Then the family

$$(\mathcal{F}_i)_{i \in \bigcup_{j \in J} I_j}$$

is an independent family of σ-fields.

Proof: Thus independent families included in bigger independent families remain independent. It is in some way the reverse of the associativity property presented earlier.

Let us set

$$I := \bigcup_{j \in J} I_j$$

and let us consider the sets $A_i \in \mathcal{F}_i$ such that the set

$$\{i \in I \mid A_i \neq \Omega\}$$

is finite. Then

$$\bigcap_{i \in I_j} A_i \in \mathcal{G}_j \text{ for every } j \in J$$

and the set

$$\{j \in J \mid \bigcap_{i \in I_j} A_i \neq \Omega\}$$

is also finite. We then have

$$\begin{aligned}\mathbf{P}\left(\bigcap_{i \in I} A_i\right) &= \mathbf{P}\left(\bigcap_{j \in J} \bigcap_{i \in I_j} A_i\right) \\ &= \prod_{j \in J} \mathbf{P}(\bigcap_{i \in I_j} A_i) = \prod_{j \in J} \prod_{i \in I_j} \mathbf{P}(A_i) = \prod_{i \in I} \mathbf{P}(A_i)\end{aligned}$$

which shows the independence of the family $(\mathcal{F}_i)_{i \in \bigcup_{j \in J} I_j}$. ■

Let us return to the notion of independence for events. We next give an extension of Definition 2.20 using what we learned in the meantime about independence of σ-algebras.

Definition 2.27 *We will say that the events $(A_i)_{i \in I}$ are independent if and only if the family of σ-fields $(\sigma(A_i))_{i \in I}$ is independent.*

Remark 2.28 *Recall that for any event $A \in \mathcal{F}$ its generated σ-algebra is*

$$\sigma(A) = \{\emptyset, \Omega, A, A^c\}.$$

It follows that the events $(A_i)_{i \in I}$ are independent if and only if

$$\mathbf{P}(B_{i_1} \cap \ldots \cap B_{i_k}) = \mathbf{P}(B_{i_1}) \ldots \mathbf{P}(B_{i_k})$$

for every $i_1, \ldots, i_k \in I$ *and for every* $B_{i_j} \in \{A_{i_j}, A_{i_j}^c, \Omega\}$. *We avoid the* $\emptyset$ *because the result is trivially true in that case.*

Remark 2.29 *In the case of finite families, Definitions 2.20 and 2.27 are coherent. That is, the events*

$$A_1, \ldots, A_n, \ldots.$$

are independent if and only if

$$\mathbf{P}(A_{i_1} \cap \ldots \cap A_{i_k}) = \mathbf{P}(A_{i_1}) \ldots \mathbf{P}(A_{i_k}) \tag{2.8}$$

for every $i_1, \ldots, i_k$. *To see this, apply Theorem 2.24 to the sets* $\mathcal{M}_i = \{\emptyset, A_i\}$.

Recall the definition of a partition of Ω, Definition 2.3.

Proposition 2.30 *Let* $\Delta_1, \ldots, \Delta_n$ *be partitions of* Ω *with a countable number of elements. Then the* σ*-fields* $(\sigma(\Delta_i))_{i \geq 1}$ *are independent if and only if*

$$\mathbf{P}(A_1 \cap \ldots \cap A_n) = \mathbf{P}(A_1) \ldots \mathbf{P}(A_n) \tag{2.9}$$

for every $A_i \in \Delta_i$, $i = 1, 2, \ldots$.

Proof: The "if" part is an immediate consequence of the definition, while the "only if" part is a consequence of Theorem 2.24 with $\mathcal{M}_i = \{\Delta_i, \emptyset\}$. ■

Remark 2.31 *As a particular case, we obtain that three events* A, B, C *are independent if and only if*

$$\mathbf{P}(A \cap B) = \mathbf{P}(A)\mathbf{P}(B), \mathbf{P}(A \cap C) = \mathbf{P}(A)\mathbf{P}(C), \mathbf{P}(B \cap C) = \mathbf{P}(B)\mathbf{P}(C)$$

and

$$\mathbf{P}(A \cap B \cap C) = \mathbf{P}(A)\mathbf{P}(B)\mathbf{P}(C),$$

because these are all the possibilities.

An explicit example of three events is considered below. It shows that even if events are pairwise independent they are not necessary mutually independent.

EXAMPLE 2.16 Pairwise independent is not independent

In relation (2.8) it is crucial the fact that the equality is satisfied for every possible subset of indices $i_1, \ldots, i_k$. Indeed, consider

$$\Omega = \{a, b, c, d\}, A = \{a, b\}, B = \{a, c\}, C = \{a, d\}$$

and let $\mathbf{P}$ be the uniform probability, that is $\mathbf{P}(X) = \frac{|X|}{|\Omega|}$. Clearly,

$$\mathbf{P}(A \cap B) = \mathbf{P}(A \cap C) = \mathbf{P}(C \cap B) = \frac{1}{4},$$

which is equal to

$$\mathbf{P}(A)\mathbf{P}(B) = \mathbf{P}(B)\mathbf{P}(C) = \mathbf{P}(A)\mathbf{P}(C).$$

However,

$$\mathbf{P}(A \cap B \cap C) = \frac{1}{4} \neq \mathbf{P}(A)\mathbf{P}(B)\mathbf{P}(C) = \frac{1}{8}.$$

In this case, the events A, B, C are pairwise independent but not mutually independent.

We finish this section on independence by discussing the relationship between independence and a finite product of probability measures.

If $(\Omega_1, \mathcal{F}_1, \mathbf{P}_1), \ldots, (\Omega_n, \mathcal{F}_n, \mathbf{P}_n)$ are n probability spaces, then the product probability measure

$$\mathbf{P} = \mathbf{P}_1 \otimes \mathbf{P}_2 \otimes \ldots \otimes \mathbf{P}_n$$

is defined as the unique probability on $(\Omega_1 \times \cdots \times \Omega_n)$ such that

$$\mathbf{P}(A_1 \times \cdots \times A_n) = \mathbf{P}_1(A_1) \ldots \mathbf{P}_n(A_n)$$

for every $A_i \in \mathcal{F}_i$, $i = 1, \ldots, n$.

Proposition 2.32 *Let $(\Omega, \mathcal{F}, \mathbf{P})$ be a probability space and let $\mathcal{F}_1, \ldots, \mathcal{F}_n$ be sub-σ-fields of $\mathcal{F}$. Then the σ-algebras $\mathcal{F}_1, \ldots, \mathcal{F}_n$ are independent if and only if*

$$\mathbf{P} \circ d^{-1} = \mathbf{P}_1 \otimes \cdots \otimes \mathbf{P}_n, \tag{2.10}$$

where $\mathbf{P}_i$ denotes the restriction of $\mathbf{P}$ to $\mathcal{F}_i$ for every $i = 1, \ldots, n$ and

$$d : (\Omega, \mathcal{F}, \mathbf{P}) \to (\Omega \times \cdots \times \Omega, \mathcal{F}_1 \otimes \cdots \otimes \mathcal{F}_n, \mathbf{P}_1 \otimes \cdots \otimes \mathbf{P}_n)$$

is the diagonal mapping defined by

$$d(\omega) = (\omega, \ldots, \omega).$$

Proof: We know that the set

$$\mathcal{M} = \{A_1 \times \cdots \times A_n \mid A_i \in \mathcal{F}_i\}$$

is a generator for the σ-field $\mathcal{F}_1 \otimes \dots \mathcal{F}_n$, which is closed under finite intersections. Therefore, it suffices to check the equality of the probabilities (2.10) for all the sets in $\mathcal{M}$. We have

$$\begin{aligned}(\mathbf{P} \circ d^{-1})(A_1 \times \cdots \times A_n) &= \mathbf{P}(d^{-1}(A_1 \times \cdots \times A_n)) \\ &= \mathbf{P}(A_1 \cap \ldots \cap A_n)\end{aligned}$$

since

$$d^{-1}(A_1 \times \cdots \times A_n) = A_1 \cap \ldots \cap A_n.$$

∎

2.3.8 BOREL–CANTELLI LEMMAS

Recall from analysis: For a sequence of numbers, $\{x_n\}_n$ lim sup and lim inf are defined:

$$\begin{aligned}\limsup x_n &= \inf_m \{\sup_{n \geq m} x_n\} = \lim_{m \to \infty} (\sup_{n \geq m} x_n), \\ \liminf x_n &= \sup_m \{\inf_{n \geq m} x_n\} = \lim_{m \to \infty} (\inf_{n \geq m} x_n),\end{aligned}$$

and they represent the highest (respectively lowest) limiting point of a subsequence included in $\{x_n\}_n$.

Note that if z is a number such that $z > \limsup x_n$, then $x_n < z$ eventually.[3]

Likewise, if $z < \limsup x_n$, then $x_n > z$ infinitely often.[4]

These notions are translated to probability in the following way.

Definition 2.33 *Let $A_1, A_2, \ldots$ be an infinite sequence of events, in some probability space $(\Omega, \mathcal{F}, \mathbf{P})$. We define the events:*

$$\begin{aligned}\limsup_{n \to \infty} A_n &= \bigcap_{n \geq 1} \bigcup_{m=n}^{\infty} A_m = \{\omega : \omega \in A_n \textit{ for infinitely many } n\} \\ &= \{A_n \textit{ infinitely often}\}, \\ \liminf_{n \to \infty} A_n &= \bigcup_{n \geq 1} \bigcap_{m=n}^{\infty} A_m = \{\omega : \omega \in A_n \textit{ for all } n \textit{ large enough}\} \\ &= \{A_n \textit{ eventually}\}.\end{aligned}$$

Let us clarify the notions of "infinitely often" and "eventually" a bit more. We say that an outcome ω happens infinitely often for the sequence

[3] That is, there is some n_0 very large so that $x_n < z$ for all $n \geq n_0$.

[4] That is, for any n there exists an $m \geq n$ such that $x_m > z$.

$A_1, A_2, \ldots, A_n, \ldots$ if ω is in the set $\bigcap_{n=1}^{\infty} \bigcup_{m \geq n} A_m$. This means that for any n (no matter how big) there exist an $m \geq n$ and $\omega \in A_m$.

We say that an outcome ω happens eventually for the sequence $A_1, A_2, \ldots, A_n, \ldots$ if ω is in the set $\bigcup_{n=1}^{\infty} \bigcap_{m \geq n} A_m$. This means that there exists an n such that for all $m \geq n$, $\omega \in A_m$; so from this particular n and up, ω is in all the sets.

Why do we chose to give such complicated definitions? The basic intuition is the following: Say you roll a die infinitely many times, then it is obvious what it means for the outcome 1 to appear infinitely often. Also, we can say the average of the rolls will eventually be arbitrarily close to 3.5 (this will be shown later). It is not so clear-cut in general. The framework above provides a generalization to these notions.

2.3.8.1 The Borel Cantelli Lemmas. With this definitions we are now capable to give two important lemmas.

Lemma 2.34 (First Borel–Cantelli) *If $A_1, A_2, \ldots$ is any infinite sequence of events with the property $\sum_{n \geq 1} \mathbf{P}(A_n) < \infty$, then*

$$\mathbf{P}\left(\bigcap_{n=1}^{\infty} \bigcup_{m \geq n} A_m\right) = \mathbf{P}\left(A_n \textit{ events are true infinitely often}\right) = 0.$$

This lemma essentially says that if the probabilities of events go to zero and the sum is convergent, then necessarily A_n will stop occurring. However, the reverse of the statement is not true. To make it hold, we need a very strong condition (independence).

Lemma 2.35 (Second Borel–Cantelli) *If $A_1, A_2, \ldots$ is an infinite sequence of* ***independent*** *events, then*

$$\sum_{n \geq 1} \mathbf{P}(A_n) = \infty \quad \Leftrightarrow \quad \mathbf{P}(A_n \ i.o.) = 1.$$

First Borel–Cantelli:

$$\mathbf{P}\left(A_n \ i.o.\right) = \mathbf{P}\left(\bigcap_{n \geq 1} \bigcup_{m=n}^{\infty} A_m\right) \leq \mathbf{P}\left(\bigcup_{n=m}^{\infty} A_m\right) \leq \sum_{m=n}^{\infty} \mathbf{P}(A_m), \qquad \forall n,$$

where we used the definition and countable subadditivity. By the hypothesis the sum on the right is the tail end of a convergent series, and therefore it converges to zero as $n \to \infty$. Thus, we are done. ■

Second Borel–Cantelli: "$\Rightarrow$" Clearly, showing that $\mathbf{P}\left(A_n \ i.o.\right) = \mathbf{P}(\limsup A_n) = 1$ *is the same as showing that* $\mathbf{P}\left((\limsup A_n)^c\right) = 0$.

By the definition of lim sup and the DeMorgan's laws, we have

$$(\limsup A_n)^c = \left(\bigcap_{n\geq 1}\bigcup_{m=n}^{\infty} A_m\right)^c = \bigcup_{n\geq 1}\bigcap_{m=n}^{\infty} A_m^c.$$

Therefore, it is enough to show that $\mathbf{P}(\bigcap_{m=n}^{\infty} A_m^c) = 0$ for all n (recall that a countable union of null sets is a null set). However,

$$\mathbf{P}\left(\bigcap_{m=n}^{\infty} A_m^c\right) = \lim_{r\to\infty} \mathbf{P}\left(\bigcap_{m=n}^{r} A_m^c\right) = \underbrace{\lim_{r\to\infty}\prod_{m=n}^{\infty}\mathbf{P}\left(A_m^c\right)}_{\text{by independence}}$$

$$= \lim_{r\to\infty}\prod_{m=n}^{r}(1-\mathbf{P}(A_m)) \leq \underbrace{\lim_{r\to\infty}\prod_{m=n}^{r} e^{-\mathbf{P}(A_m)}}_{1-x\leq e^{-x} \text{ if } x\geq 0}$$

$$= \lim_{r\to\infty} e^{-\sum_{m=n}^{r}\mathbf{P}(A_m)} = e^{-\sum_{m=n}^{\infty}\mathbf{P}(A_m)} = 0.$$

The last equality follows since $\sum \mathbf{P}(A_n) = \infty$.
Note that we have used the inequality $1 - x \leq e^{-x}$, which is true if $x \in [0, \infty)$. One can prove this inequality with elementary analysis.

"$\Leftarrow$" This implication is the same as in the first lemma. Indeed, assume by absurd that $\sum \mathbf{P}(A_n) < \infty$. By the First Borel–Cantelli Lemma, this implies that $\mathbf{P}(A_n \text{ i.o.}) = 0$, a contradiction with the hypothesis. ∎

2.3.9 FATOU'S LEMMAS

Again assume that $A_1, A_2, \ldots$ is a sequence of events.

Lemma 2.36 (Fatou's lemma for sets) *Given any measure (not necessarily finite) μ, we have*

$$\mu(A_n \textit{ eventually}) = \mu(\liminf_{n\to\infty} A_n) \leq \liminf_{n\to\infty} \mu(A_n).$$

Proof: Recall that

$$\liminf_{n\to\infty} A_n = \bigcup_{n\geq 1}\bigcap_{m=n}^{\infty} A_m,$$

and denote this set with A. Let

$$B_n = \bigcap_{m=n}^{\infty} A_m,$$

which is an increasing sequence (the sets contain less intersections as n increases) and $B_n \uparrow A =$. By the monotone convergence property of measure (Lemma 2.14), we have $\mu(B_n) \to \mu(A)$. However,

$$\mu(B_n) = \mu(\bigcap_{m=n}^{\infty} A_m) \leq \mu(A_m), \qquad \forall m \geq n,$$

thus $\mu(B_n) \leq \inf_{m \geq n} \mu(A_m)$. Therefore,

$$\mu(A) \leq \lim_{n \to \infty} \inf_{m \geq n} \mu(A_m) = \liminf_{n \to \infty} \mu(A_n).$$

■

Lemma 2.37 (The reverse of Fatou's lemma) *If* $\mathbf{P}$ *is a finite measure (e.g., probability measure), then*

$$\mathbf{P}(A_n \text{ i.o.}) = \mathbf{P}(\limsup_{n \to \infty} A_n) \geq \limsup_{n \to \infty} \mathbf{P}(A_n).$$

Proof: This proof is entirely similar. Recall that $\limsup_{n \to \infty} A_n = \bigcap_{n \geq 1} \bigcup_{m=n}^{\infty} A_m$, and denote this set with A. Let $B_n = \bigcup_{m=n}^{\infty} A_m$. Then clearly B_n is a decreasing sequence and $B_n \downarrow A =$. By the monotone convergence property of measure (Lemma 2.14) and since the measure is finite, we obtain $\mathbf{P}(B_1) < \infty$ so $\mathbf{P}(B_n) \to \mathbf{P}(A)$. However,

$$\mathbf{P}(B_n) = \mathbf{P}(\bigcup_{m=n}^{\infty} A_m) \geq \mathbf{P}(A_m), \qquad \forall m \geq n,$$

thus $\mathbf{P}(B_n) \geq \sup_{m \geq n} \mathbf{P}(A_m)$, again since the measure is finite . Therefore,

$$\mathbf{P}(A) \geq \lim_{n \to \infty} \sup_{m \geq n} \mathbf{P}(A_m) = \limsup_{n \to \infty} \mathbf{P}(A_n).$$

■

Remark 2.38 *Please note that the Fatou's lemma is applicable for any measurable set. However, the reverse of the lemma works only in probability spaces (i.e., spaces that have the total probability equal to 1).*

2.3.10 KOLMOGOROV'S ZERO–ONE LAW

We like to present this theorem since it introduces the concept of *a sequence of* σ*-algebras*, a notion which is essential for stochastic processes.

For a sequence $A_1, A_2, \ldots$ of events in the probability space $(\Omega, \mathcal{F}, \mathscr{P})$ consider the generated σ-algebras

$$\mathscr{T}_n = \sigma(A_n, A_{n+1}, \ldots),$$

as well as their intersection

$$\mathscr{T} = \bigcap_{n=1}^{\infty} \mathscr{T}_n = \bigcap_{n=1}^{\infty} \sigma(A_n, A_{n+1}, \ldots),$$

which is called "the tail" σ-field.

Theorem 2.39 (Kolmogorov's 0–1 law) *If $A_1, A_2, \ldots$ are independent, then for every event A in the tail σ field ($A \in \mathscr{T}$) its probability $\mathbf{P}(A)$ is either 0 or 1.*

Proof: Here we only present the sketch of the proof. The idea is to show that A is independent of itself and thus $\mathbf{P}(A \cap A) = \mathbf{P}(A)\mathbf{P}(A) \Rightarrow \mathbf{P}(A) = \mathbf{P}(A)^2 \Rightarrow \mathbf{P}(A)$ is either 0 or 1. The steps of this proof are as follows:

1. First define $\mathscr{A}_n = \sigma(A_1, \ldots, A_n)$ and show that is independent of $\mathscr{T}_{n+1}$ for all n.
2. Since $\mathscr{T} \subseteq \mathscr{T}_{n+1}$ and $\mathscr{A}_n$ is independent of $\mathscr{T}_{n+1}$, then $\mathscr{A}_n$ and $\mathscr{T}$ are independent for all n.
3. Define $\mathscr{A}_\infty = \sigma(A_1, A_2, \ldots)$. Then from the previous step we deduce that $\mathscr{A}_\infty$ and $\mathscr{T}$ are independent.
4. Finally since $\mathscr{T} \subseteq \mathscr{A}_\infty$ by the previous step, $\mathscr{T}$ is independent of itself and the result follows. ■

Note that $\limsup A_n$ and $\liminf A_n$ are tail events. However, it is only in the case when the original events are independent that we can apply Kolmogorov's theorem. Thus in that case $\mathbf{P}\{A_n \text{ i.o.}\}$ is either 0 or 1.

2.4 Lebesgue Measure on the Unit Interval (0,1]

We conclude this chapter with the most important probability measure. This is the unique measure that makes things behave in a normal way (e.g., the interval (0.2, 0.5) has measure 0.3).

Let $\Omega = (0, 1]$. Let $\mathcal{F}_0$ = class of semiopen subintervals $(a, b]$ of Ω. For an interval $I = (a, b] \in \mathcal{F}_0$ define $\lambda(I) = |I| = b - a$. Let $\varnothing \in \mathcal{F}_0$ the element of length 0. Let $\mathscr{B}_0$ = the algebra of finite disjoint unions of intervals in (0,1]. We also note that problem 1.5 shows that this algebra is not a σ-algebra.

If $A = \sum_{i=1}^{n} I_n \in \mathscr{B}_0$ with I_n disjoint $\mathcal{F}_0$ sets, then

$$\lambda(A) = \sum_{i=1}^{n} \lambda(I_i) = \sum_{i=1}^{n} |I_i|.$$

The goal is to show that λ is countably additive on the algebra $\mathscr{B}_0$. This will allow us to construct a measure (actually a probability measure since we are working on (0,1]) using the next result (Carathéodory's theorem). The constructed measure is well-defined and will be called the Lebesgue measure.

Theorem 2.40 (Theorem for the length of intervals) *Let $I = (a, b] \subseteq (0, 1]$ be an interval and let I_k denote intervals of the form $(a_k, b_k]$ which are bounded but not necessarily in $(0, 1]$.*

(i) If $\bigcup_k I_k \subseteq I$ and I_k are disjoint, then $\sum_k |I_k| \leq |I|$.
(ii) If $I \subseteq \bigcup_k I_k$ (with the I_k not necessarily disjoint), then $|I| \leq \sum_k |I_k|$.
(iii) If $I = \bigcup_k I_k$ and I_k disjoint, then $|I| = \sum_k |I_k|$.

Proof: Exercise (*Hint:* use induction) ∎

Note: Part iii shows that the function λ is well-defined.

Theorem 2.41 *The measure λ defined previously is a (countably additive) probability measure on the field $\mathscr{B}_0$. λ is called the Lebesgue measure restricted to the algebra $\mathscr{B}_0$.*

Proof: Let

$$A = \bigcup_{k=1}^{\infty} A_k,$$

where A_k are disjoint $\mathscr{B}_0$ sets. By definition of $\mathscr{B}_0$, we have

$$A_k = \bigcup_{j=1}^{m_k} J_{k_j}, \qquad A = \bigcup_{i=1}^{n} I_i,$$

where the J_{k_j} are disjoint. Then,

$$\lambda(A) = \sum_{i=1}^{n} |I_i| = \sum_{i=1}^{n} (\sum_{k=1}^{\infty} \sum_{j=1}^{m_k} |I_i \cap J_{k_j}|)$$

$$= \sum_{k=1}^{\infty} \sum_{j=1}^{m_k} (\sum_{i=1}^{n} |I_i \cap J_{k_j}|)$$

and since

$$A \cap J_{k_j} = J_{k_j} \Rightarrow |A \cap J_{k_j}| = \sum_{i=1}^{n} |I_i \cap J_{k_j}| = |J_{k_j}|.$$

The expression above is continued:

$$= \sum_{k=1}^{\infty} \underbrace{\sum_{j=1}^{m_k} |J_{k_j}|}_{=|A_k|} = \sum_{k=1}^{\infty} \lambda(A_k).$$

■

The next theorem will extend the Lebesgue measure to the whole (0, 1]. In this way we define the most used probability space: $((0, 1], \mathscr{B}((0, 1]), \lambda)$. The same construction with minor modifications works in the $(\mathbb{R}, \mathscr{B}(\mathbb{R}), \lambda)$ case.

Theorem 2.42 (Carathéodory's extension theorem) *A probability measure on an algebra has a unique extension to the generated σ-algebra.*

Note: Carathéodory's theorem practically constructs all the interesting probability models. However, once we construct our models, we have no further need of the theorem. It also reminds us of the central idea in the theory of probabilities: If one wants to prove something for a big set, one needs to look first at the generators of that set.

Proof: Skipped here because it is in the exercises. ■

EXERCISES

Problems with Solution

2.1 We roll two dies and we consider the events:

A: "The result of the first die is odd."
B: "The result of the second die is even."
C: "The results of the two dies have the same parity."

Study the pairwise independence of the events A, B, C, then the mutual independence.

Solution: Clearly

$$\Omega = \{1, 2, \ldots, 6\}^2$$

and $|\Omega| = 36$. Also

$$|A| = |B| = |C| = 18$$

while

$$|A \cap B| = |A \cap C| = |B \cap C| = 9.$$

Therefore, A, B, C are pairwise independent but they are not mutually independent since

$$A \cap B \cap C = \emptyset.$$

■

2.2 Let $(\Omega, \mathcal{F}, \mathbf{P})$ be a probability space and let $A, B \in \mathcal{F}$. Prove that

$$\mathbf{P}(A \cup B) + \mathbf{P}(A \cup B^c) + \mathbf{P}(A^c \cup B) + \mathbf{P}(A^c \cup B^c) = 3.$$

Solution: One has

$$\mathbf{P}(A \cup B) = \mathbf{P}(A) + \mathbf{P}(B) - \mathbf{P}(A \cap B),$$

$$\mathbf{P}(A \cup B^c) = \mathbf{P}(A) + \mathbf{P}(B^c) - \mathbf{P}(A \cap B^c),$$

$$\mathbf{P}(A^c \cup B) = \mathbf{P}(A^c) + \mathbf{P}(B) - \mathbf{P}(A^c \cap B),$$

and

$$\mathbf{P}(A^c \cup B^c) = \mathbf{P}(A^c) + \mathbf{P}(B^c - \mathbf{P}(A^c \cap B^c).$$

By adding the above probabilities,

$$\begin{aligned}
&\mathbf{P}(A \cup B) + \mathbf{P}(A \cup B^c) + \mathbf{P}(A^c \cup B) + \mathbf{P}(A^c \cup B^c) \\
&= 2\mathbf{P}(A) + 2\mathbf{P}(B) + 2\mathbf{P}(A^c) + 2\mathbf{P}(B^c) \\
&-\mathbf{P}(A \cap B) - \mathbf{P}(A \cap B^c) - \mathbf{P}(A^c \cap B) - \mathbf{P}(A^c \cap B^c) \\
&= 4 - \mathbf{P}(A \cap B) - \mathbf{P}(A \cap B^c) - \mathbf{P}(A^c \cap B) - \mathbf{P}(A^c \cap B^c).
\end{aligned}$$

But

$$A = A \cap \Omega = A \cap (B \cup B^c) = (A \cap B) \cup (A \cap B^c)$$

so

$$\mathbf{P}(A) = \mathbf{P}(A \cap B) + \mathbf{P}(A \cap B^c).$$

Similarly,

$$\mathbf{P}(A^c) = \mathbf{P}(A^c \cap B) + \mathbf{P}(A^c \cap B^c).$$

Finally

$$\begin{aligned}
&\mathbf{P}(A \cup B) + \mathbf{P}(A \cup B^c) + \mathbf{P}(A^c \cup B) + \mathbf{P}(A^c \cup B^c)\\
&= 4 - \mathbf{P}(A \cap B) - \mathbf{P}(A \cap B^c) - \mathbf{P}(A^c \cap B) - \mathbf{P}(A^c \cap B^c)\\
&= 4 - \mathbf{P}(A) - \mathbf{P}(A^c)\\
&= 3.
\end{aligned}$$

■

2.3 Let A, B, C be three events on $(\Omega, \mathcal{F}, \mathbf{P})$. Prove that

$$\mathbf{P}(A \cap B) \geq \mathbf{P}(A) - \mathbf{P}(B^c)$$

and

$$\mathbf{P}(A \cap B \cap C) \geq 1 - \mathbf{P}(A^c) - \mathbf{P}(B^c) - \mathbf{P}(C^c).$$

Solution: As in the previous exercise,

$$\mathbf{P}(A) = \mathbf{P}(A \cap B) + \mathbf{P}(A \cap B^c).$$

Thus

$$\mathbf{P}(A) - \mathbf{P}(A \cap B^c) = \mathbf{P}(A \cap B).$$

Since $A \cap B^c \subseteq B^c$, we have

$$\mathbf{P}(A \cap B^c) \leq \mathbf{P}(B^c)$$

so

$$\mathbf{P}(A \cap B) = \mathbf{P}(A) - \mathbf{P}(A \cap B^c) \geq \mathbf{P}(A) - \mathbf{P}(B^c).$$

To prove the second inequality, we use the above one twice:

$$\begin{aligned}
\mathbf{P}(A \cap B \cap C) &\geq \mathbf{P}(A \cap B) - \mathbf{P}(C^c)\\
&\geq \mathbf{P}(A) - \mathbf{P}(B^c) - \mathbf{P}(C^c)\\
&= 1 - \mathbf{P}(A^c) - \mathbf{P}(B^c) - \mathbf{P}(C^c).
\end{aligned}$$

■

2.4 A box contains five white balls and 3 red balls. We extract at random one ball and we put it back in the box together with two balls of the same color with the ball that has been not extracted. Given that the second ball extracted is red, what is the most likely color of the first ball?

Solution: Denote by R_i the event "the ith extracted ball is red." We have

$$\mathbf{P}(R_1|R_2) = \frac{\mathbf{P}(R_1)\mathbf{P}(R_2|R_1)}{\mathbf{P}(R_2)}$$

and

$$\begin{aligned}\mathbf{P}(R_2) &= \mathbf{P}(R_1)\mathbf{P}(R_2|R_1) + \mathbf{P}(R_1^c)\mathbf{P}(R_2|R_1^c)\\ &= \frac{3}{8}\frac{3}{10} + \frac{5}{8}\frac{5}{10}\\ &= \frac{34}{80}.\end{aligned}$$

Then

$$\mathbf{P}(R_1|R_2) = \frac{9}{80}\frac{80}{34} = \frac{9}{34}.$$

■

2.5 A medical laboratory test ensures with a probability 0.95 the detection of a certain disease M, when the disease effectively exists. Meanwhile, the test also indicates a positive result for one percent of the people without the disease that have tried the test. We assume that 0.5 percent of the population has the disease M. We choose at random a person for which the test was positive. What is the probability that the person is ill?

Solution: Denote "M = the person has the disease M" and "T = the test is positive." We have

$$\mathbf{P}(T|M) = 0.95, \mathbf{P}(M) = 0.005, \mathbf{P}(T|M^c) = 0.01.$$

Then

$$\begin{aligned}\mathbf{P}(T) &= \mathbf{P}(M)\mathbf{P}(T|M) + \mathbf{P}(M^c)\mathbf{P}(T|M^c)\\ &= 0.0147\end{aligned}$$

and

$$\mathbf{P}(M|T) = \frac{\mathbf{P}(M)\mathbf{P}(T|M)}{\mathbf{P}(T)} = \frac{0.00475}{0.0147}.$$

■

2.6 From previous data analysis, it was found that:

1. When the machine operates well, the probability of producing good-quality products is 0.98.
2. When the machine has certain problems, the probability of producing good-quality products is 0.55.
3. Every morning, when the machine is launched, the probability of the machine adjust to normal operation is 0.95.

If on one morning, the first product produced by the machine is of good quality, what is the probability that the machine is adjusted to normal operation in that morning?

Solution: Define the following events:
Event A = "The product is of good-quality."
Event B = "The machine is adjusted to normal operation." Then

$$\mathbf{P}(A/B) = 0.98, \quad \mathbf{P}(B) = 0.95,$$
$$\mathbf{P}(A/B^c) = 0.55 \quad \mathbf{P}(B^c) = 0.05.$$

We need to compute

$$\mathbf{P}(B/A) = \frac{\mathbf{P}(A/B)\mathbf{P}(B)}{\mathbf{P}(A)}$$

$$= \frac{\mathbf{P}(A/B)\mathbf{P}(B)}{\mathbf{P}(A/B)\mathbf{P}(B) + \mathbf{P}(A/B^c)\mathbf{P}(B^c)} = \frac{0.98 \times 0.95}{0.98 \times 0.95 + 0.55 \times 0.05} = 0.97.$$

■

2.7 A secretary sent N letters to N persons, but it turns out she put the letters in the envelopes randomly. To model this situation, we choose a probability space Ω_N which is the set of all permutations of $\{1, 2, \ldots, N\}$ endowed with the uniform probability $\mathbf{P}_N$ (this is every permutation has the same chance $1/N!$).

For $1 \leq j \leq N$ we denote by A_j the event "the jth letter is in the good envelope."

(a) Calculate $\mathbf{P}_N(A_j)$.

(b) We fix k integers $i_1 < i_2 < \ldots. < i_k$ between 1 and N. Compute the number of permutations σ of the set $\{1, \ldots, N\}$ such that $\sigma(i_1) = i_1, \ldots.., \sigma(i_k) = i_k$. Use this number to deduce

$$\mathbf{P}_N(A_{i_1} \cap \ldots.. \cap A_{i_k}).$$

(c) Denote by B the event "at least one letter is in the good envelope." Express B in terms of A_j.

(d) Use Poincaré's formula to compute $\mathbf{P}_N(B)$ and its limit when $N \to \infty$.

Proof: **(a)** Let's compute the cardinality of the set A_j. The probability required is then simply this number divided by $N!$. The problem is equivalent to counting the number of permutation $\sigma \in \Omega_N$ with $\sigma(j) = j$. This is the same as the number of permutations of the other elements, and that number is $(N-1)!$, thus

$$\mathbf{P}_N(A_j) = \frac{|A_j|}{|\Omega_N|} = \frac{(N-1)!}{N!} = \frac{1}{N}.$$

(b) Now we have to count the number of permutations of $\{1, \ldots, N\}$ with fixed k elements. Again this is the same as permuting all the other elements and that number is $(N-k)!$. Therefore

$$\mathbf{P}_N(A_{i_1} \cap \ldots.. \cap A_{i_k}) = \frac{(N-k)!}{N!}.$$

(c) This set is written as

$$B = \bigcup_{i=1}^{N} A_i.$$

(d) From Poincaré's (inclusion exclusion) formula

$$\begin{aligned}
\mathbf{P}_N(B) &= \sum_{k=1}^{N} (-1)^{k+1} \sum_{1 \le i_1 < \ldots. < i_k \le N} \mathbf{P}_N(A_{i_1} \cap \ldots.. \cap A_{i_k}) \\
&= \sum_{k=1}^{N} (-1)^{k+1} \sum_{1 \le i_1 < \ldots. < i_k \le N} \frac{(N-k)!}{N!} \\
&= \sum_{k=1}^{N} (-1)^{k+1} \frac{(N-k)!}{N!} \binom{N}{k} \\
&= \sum_{k=1}^{N} (-1)^{k+1} \frac{1}{k!}
\end{aligned}$$

and this converges as $N \to \infty$ to

$$\sum_{k=1}^{\infty} (-1)^{k+1} \frac{1}{k!} = 1 - e^{-1}.$$

■

2.8 Do the following parts.

(a) Prove that

$$\binom{2n}{n} = \sum_{k=0}^{n} \binom{n}{k}^2.$$

(b) Each one of two people toss a coin n times. What is the probability $\mathbf{P}_n$ that they obtain the same number of tails?

(c) Give the limit of $\mathbf{P}_n$ when n goes to infinity.

Solution: **(a)** We use the identity

$$(1+t)^n (1+t)^n = (1+t)^{2n}.$$

We develop both sides using the formula

$$(a+b)^n = \sum_{k=0}^{n} \binom{n}{k} a^k b^{n-k}$$

and we identify the coefficient of t^n in both sides.

(b) We define the probability space

$$\Omega = \{h, t\}^{2n}$$

endowed with the uniform probability. Let A be the event "both have the same number of tails." Then

$$\begin{aligned}|A| &= |\{\text{ both zero tails }\}| + |\{\text{ both one tail }\}| + ... + |\{\text{ both } n \text{ tails }\}| \\ &= \binom{n}{0}\binom{n}{0} + \binom{n}{1}\binom{n}{1} + \cdots + \binom{n}{n}\binom{n}{n} \\ &= \sum_{k=0}^{n} \binom{n}{k}^2.\end{aligned}$$

So

$$\mathbf{P}_n = \frac{|A|}{|\Omega|} = \frac{\binom{2n}{n}}{2^n}.$$

(c) We recall Stirling's formula

$$n! = \sqrt{2\pi} n^{n+\frac{1}{2}} \exp(-n(1+\varepsilon_n))$$

with

$$\lim_n n\varepsilon_n = 0.$$

It will follow that

$$\mathbf{P}_n \to_{n\to\infty} 0.$$

■

Problems without Solution

2.9 Show that the Borel sets of $\mathbb{R}$ is generated by intervals of the type $(-\infty, x]$, i.e., $\mathscr{B} = \sigma(\{(-\infty, x] | x \in \mathbb{R}\})$.

Hint: Show that the generating set is the same; that is, show that any set of the form $(-\infty, x]$ can be written as a countable union (or intersection) of open intervals and, vice versa, that any open interval in $\mathbb{R}$ can be written as a countable union (or intersection) of sets of the form $(-\infty, x]$.

2.10 Show that the following classes all generate the Borel σ-algebra; or, put differently, show the equality of the following collections of sets:

$$\begin{aligned}\sigma\left((a,b) : a < b \in \mathbb{R}\right) &= \sigma\left([a,b] : a < b \in \mathbb{R}\right)\\ &= \sigma\left((-\infty,b) : b \in \mathbb{R}\right)\\ &= \sigma\left((-\infty,b) : b \in \mathbb{Q}\right),\end{aligned}$$

where $\mathbb{Q}$ is the set of rational numbers.

2.11 **Properties of probability measures**
Prove properties 1–4 in Proposition 2.2.

Hint: You only have to use the definition of probability. The only thing nontrivial in the definition is the countable additivity property.

2.12 Let $(\Omega, \mathcal{F}, P)$ be a probability space and let A, B, C be events in $\mathcal{F}$. Prove the following particular case of Poincaré's formula:

$$\begin{aligned}P(A \cup B \cup C) = {} & P(A) + P(B) + P(C)\\ & -P(A \cap B) - P(A \cap C) - P(B \cap C)\\ & +P(A \cap B \cap C).\end{aligned}$$

2.13 Suppose that $(A_j)_{j\geq 1}$ is a collection of events in $\mathcal{F}$. Prove that

$$\mathbf{P}\left(\bigcap A_j\right) \geq 1 - \sum_{j\geq 1}\left[1 - \mathbf{P}(A_j)\right).$$

Hint: Apply the subadditivity property to $(A_j^c)_{j\geq 1}$.

2.14 Let $(A_n)_{n\geq 1}$ be a sequence included in $\mathcal{F}$. Prove that if $\mathbf{P}(A_n) = 1$ for every $n \geq 1$, then

$$\mathbf{P}\left(\bigcap_{i\geq 1} A_i\right) = 1.$$

Hint: Use problem 2.13.

2.15 Suppose that A and B are events with $\mathbf{P}(A) = 1$. Prove that

$$\mathbf{P}(A \cap B) = \mathbf{P}(B).$$

2.16 Let $(A_i)_{i\in I}$ be an infinite family of mutually disjoint events in $\mathcal{F}$. Prove that the set

$$S = \{i \in I, \mathbf{P}(A_i) > 0\}$$

is countable.

2.17 **No matter how many zeros, do not add to more than zero.**
Prove Lemma 2.16.

Hint: You may use countable subadditivity.

2.18 If $\mathcal{F}_0$ is an algebra, $m(\mathcal{F}_0)$ is the minimal monotone class over $\mathcal{F}_0$ and $\mathscr{G}_2$ is defined as

$$\mathscr{G}_2 = \{B : A \cup B \in m(\mathcal{F}_0), \forall A \in m(\mathcal{F}_0)\}.$$

Then show that $\mathscr{G}_2$ is a monotone class.

Hint: Look at the proof of Theorem 2.12, and repeat the arguments therein.

2.19 A student answers a multiple choice examination question that has 4 possible answers. Suppose that the probability that the student knows the answer to the question is 0.80 and the probability that the student guesses is 0.20. If student guesses, probability of correct answer is 0.25.

(a) What is the probability that the question is answered correctly?

(b) If the question is answered correctly, what is the probability that the student in fact knew the correct answer and was not guessing.

2.20 **A monotone algebra is a σ-algebra.**
Let $\mathcal{F}$ be an algebra that is also a monotone class. Show that $\mathcal{F}$ is a σ-algebra.

2.21 Prove the *total probability formula* [equation (2.4)] and *Bayes' formula* [equation (2.5)].

2.22 If two events are such that $A \cap B = \emptyset$, are A and B independent? Justify.

2.23 Show that $\mathbf{P}(A|B) = \mathbf{P}(A)$ is the same as independence of the events A and B.

2.24 Prove that if two events A and B are independent, then so are their complements.

2.25 Generalize the previous problem to n sets using induction.

2.26 One urn contains w_1 white balls and b_1 black balls. Another urn contains w_2 white balls and b_2 black balls. A ball is drawn at random from each urn, and then one of the two balls is selected at random.

(a) What is the probability that the final ball selected is white?

(b) Given that the final ball selected was white, what is the probability that in fact it came from the first urn (with w_1 and b_1 balls)?

2.27 At the end of a well-known course, the final grade is decided with the help of an oral examination. There are a total of m possible subjects listed on some pieces of paper. Of them, n are generally considered "easy."

Each student enrolled in the class, one after another, draws a subject at random and then presents it. Of the first two students, who has the better chance of drawing a "favorable" subject?

2.28 Using Cantelli's lemma, show that when you roll a die the outcome $\{1\}$ will appear infinitely often. Also show that eventually the average of all rolls up to roll n will be within ε of 3.5 where $\varepsilon > 0$ is any arbitrary real number.

2.29 Two (now retired) famous tennis players Andre Agassi and Pete Sampras decide to play a number of games together. They play non-stop and at the end it turns out that Sampras won n games while Agassi won m, where $n > m$. Assume that in fact any possible sequence of games was possible to reach this result. Let $P_{n,m}$ denote the probability that from the first game until the last, Sampras is always in the lead. Find:

1. $P_{2,1}$; $P_{3,1}$; $P_{n,1}$
2. $P_{3,2}$; $P_{4,2}$; $P_{n,2}$
3. $P_{4,3}$; $P_{5,3}$; $P_{5,4}$
4. Make a conjecture about a formula for $P_{n,m}$.

2.30 A bag contains 3 black, 5 white, and 2 red marbles. A marble is selected at random. It turns out to be black; find the probability that the next marble selected (without replacing the first) is also black.

2.31 My friend Andrei has designed a system to win at the roulette. He likes to bet on red, but he waits until there have been 6 previous black spins and only then he bets on red. He reasons that the chance of winning is quite large since the probability of 7 consecutive back spins is quite small. What do you think of his system? Calculate the probability that he wins using this strategy.

Actually, Andrei plays his strategy 4 times and he actually wins 3 times out of the 4 he played. Calculate the probability of the event that just occurred.

2.32 Toss a nickel and a dime at random and record the result as an ordered pair. A = heads on both nickel and dime, B = having at least one head, C = heads appears on the nickel, and D = tails appears on the nickel Then calculate

$$\mathbf{P}(A|B), \mathbf{P}(C|B), \mathbf{P}(D|B).$$

2.33 Ali Baba is caught by the sultan while stealing his daughter. The sultan is being gentle with him and he offers Ali Baba a chance to regain his liberty. There are 2 urns and m white balls and n black balls. Ali Baba has to put the balls in the 2 urns however he likes with the only condition that no urn is empty. After that the sultan will chose an urn at random and then pick a ball from that urn. If the chosen ball is white, Ali Baba is free to go; otherwise Ali Baba's head will be at the same level as his legs.

How should Ali Baba divide the balls to maximize his chance of survival?

2.34 Consider the probability space $([0, 1], \mathscr{B}[0, 1], \lambda)$ with λ denoting the Lebesque measure. Define

$$A_n = \left[0, \frac{1}{2^n}\right) \bigcup \left(\frac{2}{2^n}, \frac{3}{2^n}\right) \bigcup \cdots \bigcup \left[\frac{2^n - 2}{2^n}, \frac{2^n - 1}{2^n}\right).$$

Show that the sets A_n, $n \geq 1$ are independent.

2.35 Let p$\in (0, 1)$. Construct a probability space $(\Omega, \mathcal{F}, \mathbf{P})$ and a sequence $(A_n)_n$ included in $\mathcal{F}$ such that

$$\sum_n \mathbf{P}(A_n) = \infty$$

and

$$\mathbf{P}(\limsup_n A_n) = p.$$

Hint: Use the probability space $([0, 1], \mathcal{B}[0, 1], \lambda)$.

2.36 A Pap smear is a screening procedure to detect cervical cancer. For women with this type of cancer, 16 percent of tests are false negatives. For women without this cancer, about 19 percents are false positives. In the United States, there are about 8 women in 100,000 who have this type of cancer. A woman tests positive on this screening test. Calculate her probability of having cancer.

2.37 We want to analyze the gender distribution in a family with n children. We consider the probability space

$$\Omega = \{g, b\}^n = \{(x_1, \ldots, x_n) \mid x_i \in \{g, b\}, i = 1, \ldots, n\}$$

("g" is girl and "b" is boy). On Ω we consider the uniform probability. Define the events

$$A = \{\text{the family has children of both gender}\}$$

and

$$B = \{\text{the family has at most one girl}\}.$$

(**a**) Show that for every $n \geq 2$, we have

$$\mathbf{P}(A) = \frac{2^n - 2}{2^n} \quad \text{and} \quad \mathbf{P}(B) = \frac{n+1}{2^n}.$$

(**b**) Deduce that A and B are independent if and only if $n = 3$.

CHAPTER THREE

Random Variables: Generalities

3.1 Introduction/Purpose of the Chapter

Random variables (or stochastic variables) are used in mathematics and many other sciences to understand and model events based on data obtained from scientific experiments. A random variable is typically describing some phenomenon whose value, size, volume, and so on, is not known before it is produced and depends on the particular hazard. Random variables are described completely by their probability distribution. This probability distribution associated with a random variable quantifies the chance that a certain value, size, and so on, occurs for a given random variables.

3.2 Vignette/Historical Notes

The origin of the word stochastic is the word *stokhastikos* (Greek), which means capable of guessing. The word literally comes from *stokhos*, which was a pointed stick set up for archers to shoot at.

Let us point out some important dates in the history of the study of random variables. An important moment is constituted by the publication in 1713 of Bernoulli's work entitled *Ars Conjectandi* (*The Art to Guess*), where, for the first time, sequences of Bernoulli random variables are considered and a first variant of

Handbook of Probability, First Edition. Ionuţ Florescu and Ciprian Tudor.

the Central Limit Theorem is stated and proved. The concept of "independence" is mainly due to De Moivre. Another important moment in defining random variables is the appearance of Laplace's monograph *Théorie Analitique des Probabilités* (1812), where the current state of the art of the theory at the time is described. The fundamental limit theorems are presented in the monograph in a rather complete formulation. Siméon Denis Poisson (1781–1840) later introduced the probability law which was named after him. Carl Friedrich Gauss (1777–1855) is credited with first introducing normally distributed errors, which have the distribution carrying his name.

3.3 Theory and Applications

3.3.1 DEFINITION

The first step is to give the following definition.

Definition 3.1 (Measurable function) *Let $(\Omega, \mathcal{F})$ be a measurable space and let $(\Omega_1, \mathcal{F}_1)$ be another measurable space. An application $X : \Omega \to \Omega_1$ is called measurable if and only if for any $A \in \mathcal{F}_1$ one has*

$$X^{-1}(A) \in \mathcal{F}.$$

Definition 3.2 (Random variable on $\mathbb{R}$) *Let $(\Omega, \mathcal{F}, \mathbf{P})$ be a probability space and consider $\mathbb{R}$ with the Borel sets on $\mathbb{R}$: $\mathscr{B}(\mathbb{R})$. A measurable function $X : \Omega \to \mathbb{R}$ is called a random variable. The set of its values is the image of Ω through X and it is denoted by $X(\Omega)$. Note that a random variable is simply a measurable function with co-domain $(\mathbb{R}, \mathscr{B}(\mathbb{R}))$*

Consequence: Since the Borel sets in $\mathbb{R}$ are generated by $(-\infty, x]$, we can have the definition of a random variable directly by

$$X : \Omega \longrightarrow \mathbb{R} \quad \text{such that} \quad X^{-1}(-\infty, x] \in \mathcal{F} \quad \text{or}$$
$$\{\omega : X(\omega) \leq x\} \in \mathcal{F}, \qquad \forall x \in \mathbb{R}.$$

We shall sometimes use $\{X(\omega) \leq x\}$ or just $X(\omega) \leq x$ to denote the preimage of $(-\infty, x]$: $X^{-1}(-\infty, x)$. Traditionally, the random variables are denoted with capital letters from the end of the alphabet $X, Y, Z, \ldots$, and their values are denoted with corresponding small letters $x, y, z, \ldots$. In practice it is difficult to construct functions which are not measurable, and almost every function considered in this book will be measurable.

If one rolls three dies, one can denote by X, Y, X the result of the first, second, and third die, respectively. The introduction of random variables allows us to study in an easier way the sum $S = X + Y + Z$. It is possible to use only events to study S, but with the help of random variables the sum can be handled much easier.

Remark 3.3 *A sum, product, or composition of measurable functions is a measurable function. Therefore, if* X *and* Y *are random variables,* $XY, X+Y$, *and* $\sin(X), X^2e^Y$ *are examples of random variables. If* $(X_i)_{i\geq 1}$ *denotes a sequence of random variables, then*

$$\inf_{n\geq 1} X_n, \quad \sup_{n\geq 1} X_n, \quad \liminf_{n\geq 1} X_n, \quad \limsup_{n\geq 1} X_n$$

are random variables.

Remark 3.4 *The notation* $X^{-1}(A)$ *simply denotes the preimage of the set* A*. The notation does not imply that the function* X *has an inverse in the usual sense. For example, let us look at the function* $X(\omega) = \sin(\omega)$ *for* $\omega \in \mathbb{R}$ *is measurable. The set*

$$X^{-1}([0, 5]) = \bigcup_{2k\in\mathbb{Z}} \left[\frac{(2k-1)\pi}{2}, \frac{(2k+1)\pi}{2}\right]$$

is simply a union of intervals in $\mathbb{R}$*. Yet the function* $\sin(\omega)$ *is not invertible on* $\mathbb{R}$*.*

3.3.2 THE DISTRIBUTION OF A RANDOM VARIABLE

Definition 3.5 (Distribution of random variable) *Let* $(\Omega, \mathcal{F}, \mathbf{P})$ *be a probability space and let* $X : \Omega \to \mathbb{R}$ *be a random variable.*

The law (or the distribution) of X*, denoted by* $\mathbf{P}_X$*, is an application defined on the Borel sets of* $\mathbb{R}$ *denoted* $\mathscr{B}(\mathbb{R})$*, with values in* $[0, 1]$ *by*

$$\mathbf{P}_X(B) = \mathbf{P}\left(\{\omega : X(\omega) \in B\}\right) = \mathbf{P}\left(X^{-1}(B)\right) = \mathbf{P} \circ X^{-1}(B)$$

for every $B \in \mathscr{B}(\mathbb{R})$*.*

An example will make clear the relationship between random variables and probability distributions. Suppose you flip a coin two times. This simple statistical experiment can have four possible outcomes: HH, HT, TH, and TT. Now, let the variable X represent the number of heads that result from this experiment. The variable X can take on the values 0, 1, or 2. In this example, X is a random variable because its value is determined by the outcome of a statistical experiment.

Proposition 3.6 *The set function* $\mathbf{P}_X$ *defined above is a probability on the measurable space* $(\mathbb{R}, \mathscr{B}(\mathbb{R}))$*. Consequently,*

$$(\mathbb{R}, \mathscr{B}(\mathbb{R}), \mathbf{P}_X)$$

is a probability space.

Proof: We have

$$\mathbf{P}_X(\mathbb{R}) = \mathbf{P} \circ X^{-1}(\mathbb{R}) = \mathbf{P}(\omega \in \Omega, X(\omega) \in \mathbb{R}) = \mathbf{P}(\Omega) = 1.$$

Let $(A_i)_{i\geq 1}$ be a sequence of disjoint sets in $\mathscr{B}(\mathbb{R})$. Then

$$\begin{aligned}\mathbf{P}_X\left(\cup_{i\geq 1} A_i\right) &= (\mathbf{P} \circ X^{-1})\left(\cup_{i\geq 1} A_i\right) = \mathbf{P}\left(X^{-1}(\cup_{i\geq 1} A_i)\right) \\ &= \mathbf{P}\left(\cup_{i\geq 1} X^{-1}(A_i)\right) = \sum_{i\geq 1} \mathbf{P}(X^{-1}(A_i)) \\ &= \sum_{i\geq 1} (\mathbf{P} \circ X^{-1}(A_i) = \sum_{i\geq 1} \mathbf{P}_X(A_i),\end{aligned}$$

where we used that the sets $X^{-1}(A_i)$ are disjoint. If they are not, and there exists an ω common to two such sets, say $X^{-1}(A_i)$ and $X^{-1}(A_j)$, then $X(\omega)$ would be in both A_i and A_j and thus in intersection, which is empty by hypothesis. ■

The probability set function $\mathbf{P}_X$ is called the distribution of the random variable X.

There is one more simplification we can make. If we recall the result of exercises 2.23 and 2.24, we know that all Borel sets are generated by the same category of sets (semiopen, open, closed, etc.). Using the same ideas, it is enough to describe how to calculate $\mathbf{P}_X = \mathbf{P} \circ X^{-1}$ for the generators of the σ-algebra (that is, for one such category of intervals). We can do this for any type of generating sets we wish (open sets, closed sets, etc.); but it turns out that the simplest way is to use sets of the form $(-\infty, x]$, since we only need to specify one end of the interval (the other is always $-\infty$). With this observation, we only need to specify the probability measure $\mathbf{P}_X = \mathbf{P} \circ X^{-1}$ directly on the generators to completely characterize the probability measure.

Remark 3.7 (About random variables)

1. *If A is a Borel set in $\mathbb{R}$, we will sometimes use the short notation $\{X \in A\}$ to indicate the set*

   $$\{\omega \in \Omega \mid X(\omega) \in A)\} = X^{-1}(A).$$

2. *The set $\{X \in A\}$ is included in Ω. To be able to compute its probability, the set should be an element of the sigma $\mathcal{F}$. Since X is measurable, this is true, and so we can calculate the probability for any Borel set.*
3. *Clearly, the law of the random variable X depends on the probability* $\mathbf{P}$.
4. *Please do not confuse a random variable X with its law. For example, if one rolls two dies and denote by X, respectively Y, their outcomes, then clearly X and Y have the same law. On the other hand, $\mathbf{P}(X = Y) < 1$ and therefore the two random variables are different even though their law is identical.*

3.3.3 THE CUMULATIVE DISTRIBUTION FUNCTION OF A RANDOM VARIABLE

Definition 3.8 (The cumulative distribution function of a random variable) *The (cumulative) distribution function (cdf) of a random variable X is a function $F : \mathbb{R} \to [0, 1]$ with*

$$F(x) = \mathbf{P}_X(-\infty, x] = \mathbf{P}\left(\{\omega : X(\omega) \in (-\infty, x]\}\right) = \mathbf{P}\left(\{\omega : X(\omega) \leq x\}\right).$$

Usually one uses the abbreviation "c.d.f." or "CDF" or simply "cdf" for the cumulative distribution function.

As the value x increases, the function $F(x)$ cumulates all the probability related to outcomes less than or equal with the number x.

Let us return to the coin flip experiment. If we flip a coin two times, we might ask: What is the probability that the coin flips would result in one or fewer heads? The answer would be a cumulative probability. It would be the probability that the coin flip experiment results in zero heads plus the probability that the experiment results in one head.

$$\mathbf{P}(X \leq 1) = \mathbf{P}(X = 0) + \mathbf{P}(X = 1) = 0.25 + 0.50 = 0.75.$$

We present the basic properties of a c.d.f.

Proposition 3.9 *The distribution function for any random variable X has the following properties:*

(i) *F is increasing (i.e., if $x < y$, then $F(x) \leq F(y)$).*

(ii) *F is right continuous (i.e., $\lim_{h \downarrow 0} F(x + h) = F(x)$).*

(iii) $\lim_{x \to -\infty} F(x) = 0$ *and* $\lim_{x \to \infty} F(x) = 1$.

Proof: The first property is a consequence of the fact that a probability is a non-decreasing function; that is, if $A \subset B$, then $\mathbf{P}(A) \leq \mathbf{P}(B)$. Clearly if $x \leq y$, then

$$(X \leq x) \subseteq (X \leq y)$$

and thus $F(x) \leq F(y)$.

Denote by $A_h = (X \leq x + h)$. This sequence of sets is decreasing with respect to $h \downarrow 0$ and $\cap_{h \downarrow 0} A_h = (X \leq x)$. Then by the monotone convergence property of the measure, we obtain

$$\begin{aligned} \lim_{h \downarrow 0} F(x + h) &= \lim_{h \downarrow 0} \mathbf{P}(A_h) \\ &= \mathbf{P}(\cap_{h \downarrow 0} A_h) = \mathbf{P}(X \leq x) \\ &= F(x). \end{aligned}$$

Finally

$$\lim_{x \to \infty} F(x) = \mathbf{P}(X \leq \infty) = \mathbf{P}(X \in \mathbb{R}) = \mathbf{P}(\Omega) = 1$$

and

$$\lim_{x \to -\infty} F(x) = \mathbf{P}(X \leq -\infty) = \mathbf{P}(\emptyset) = 0.$$

■

Remark 3.10 *In other mathematical books an increasing function is sometimes called nondecreasing to distinguish it from strictly increasing functions (if $x < y$, then $F(x) < F(y)$).*

Proving the following lemma is elementary using the properties of the probability measure (Proposition 2.2) and is left as an exercise.

Lemma 3.11 *Let F be the distribution function of X. Then:*

(i) $\mathbf{P}(X \geq x) = 1 - F(x)$.

(ii) $\mathbf{P}(x < X \leq y) = F(x) - F(y)$.

(iii) $\mathbf{P}(X = x) = F(x) - F(x-)$, *where* $F(x-) = \lim_{y \nearrow x} F(y)$ *the left limit of* F *at* x.

Remark 3.12 *Suppose that $F(x_0) = 0$ where F is a c.d.f. Then*

$$F(x) = 0 \qquad \textit{for every } x \leq x_0.$$

EXAMPLE 3.1 Using the c.d.f. for a discrete random variable

Suppose we roll a fair six-sided die (the probability of landing on each face is equal). We ask, What is the probability that the die will land on a number that is smaller than 5?

Solution: When a die is tossed, there are 6 possible outcomes represented by $S = \{1, 2, 3, 4, 5, 6\}$. Each possible outcome is equally likely to occur.

This problem involves the cumulative probability function. However, pay attention to whether the outcome is included or not. The probability that the die will land on a number smaller than 5 is equal to

$$\begin{aligned}\mathbf{P}(X < 5) = F(4) &= \mathbf{P}(X = 1) + \mathbf{P}(X = 2) + \mathbf{P}(X = 3) + \mathbf{P}(X = 4)\\ &= 1/6 + 1/6 + 1/6 + 1/6 = 2/3.\end{aligned}$$

■

There exists a "reciprocal argument" to Proposition 3.9.

Theorem 3.13 *Let F be a function satisfying conditions i to iii in Proposition 3.9. Then, there exists a probability space $(\Omega, \mathscr{F}, \mathbf{P})$ and a random variable X defined on this probability space such that F is the c.d.f. of X.*

Proof: We choose the probability space to be $\Omega = (0, 1)$ with $\mathscr{F}$ the Borel σ-algebra of subsets of the unit interval and $\mathbf{P} = \lambda$ the Lebesque measure on the intervals. Define

$$X(\omega) = \sup\{z \mid F(z) < \omega\}.$$

For any number c such that $X(\omega) > c$ the definition implies that $\omega > F(c)$. On the other hand, if $\omega > F(c)$, then, since F is right continuous, for some $\varepsilon > 0$ we obtain

$$\omega > F(c + \varepsilon)$$

and this implies

$$X(\omega) \geq c + \varepsilon > c.$$

It follows that $X(\omega) > c$ if and only if $\omega > F(c)$. Therefore

$$\mathbf{P}[X(\omega) > c] = \mathbf{P}[\omega > F(c)] = 1 - F(c),$$

and thus obviously

$$\mathbf{P}[X(\omega) \leq c] = F(c).$$

Therefore, we have found a random variable X such that F is the cumulative distribution function of X. ■

EXAMPLE 3.2 Indicator Random Variable

Let $\mathbf{1}_A$ be the indicator function of a set $A \subseteq \Omega$. This is a function defined on Ω with values in $\mathbb{R}$. Therefore, it may be a random variable. According to the definition, it is a random variable if the function is measurable. It is simple to show that this happens if and only if $A \in \mathcal{F}$ the σ-algebra associated with the probability space. Assuming that $A \in \mathcal{F}$, what is the distribution function of this random variable?

Solution: To answer this question, we have to calculate $\mathbf{P} \circ \mathbf{1}_A^{-1}((-\infty, x])$ for any x. However, the function 1_A only takes two values 0 and 1. We can calculate immediately:

$$\mathbf{1}_A^{-1}((-\infty, x]) = \begin{cases} \emptyset & \text{if } x < 0, \\ A^c & \text{if } x \in [0, 1), \\ \Omega & \text{if } x > 1. \end{cases}$$

Therefore,

$$F(x) = \begin{cases} 0 & \text{if } x < 0, \\ \mathbf{P}(A^c) & \text{if } x \in [0, 1), \\ 1 & \text{if } x \geq 1. \end{cases}$$

■

We have defined a random variable as any measurable function with image (codomain) $(\mathbb{R}, \mathscr{B}(\mathbb{R}))$. A more specific case is obtained when the random variable also has the domain equal to $(\mathbb{R}, \mathscr{B}(\mathbb{R}))$. In this case the random variable is called a Borel function.

Definition 3.14 (Borel measurable function) *A function $g : \mathbb{R} \to \mathbb{R}$ is called Borel (measurable) function if g is a measurable function from $(\mathbb{R}, \mathscr{B}(\mathbb{R}))$ into $(\mathbb{R}, \mathscr{B}(\mathbb{R}))$.*

EXAMPLE 3.3

Show that any continuous function $g : \mathbb{R} \to \mathbb{R}$ is Borel measurable.

Solution: This is very simple. Recall that the Borel sets are generated by open sets. So it is enough to see what happens to the preimage of an open set B. Since g is a continuous function, $g^{-1}(B)$ is an open set and thus $g^{-1}(B) \in \mathscr{B}(\mathbb{R})$. Therefore by definition g is Borel measurable. ■

Let us define the σ-algebra generated by a random variable.

Definition 3.15 (σ-algebra generated by random variables) *For $X : \Omega \to \mathbb{R}$ a random variable, define*

$$\sigma(X) = \{X^{-1}(B), B \in \mathscr{B}(\mathbb{R}\}.$$

Note that $\sigma(X)$ is the smallest σ-algebra such that X is a measurable function with values in $\mathbb{R}$.

3.3.4 INDEPENDENCE OF RANDOM VARIABLES

We defined in Section 2.3.7 the notion of independence of two σ-algebras. We define now the independence of random variables.

Definition 3.16 *Two random variables $X, Y : \Omega : \mathbb{R}$ are independent if and only if their generated σ-algebras $\sigma(X)$ and $\sigma(Y)$ are independent. That implies, for every A, B Borel sets in $\mathbb{R}$, it holds that*

$$\mathbf{P}(X \in A, Y \in B) = \mathbf{P}(X \in A)\mathbf{P}(Y \in B).$$

Remark 3.17 *Since the Borel σ-algebra is generated by intervals, the above definition is equivalent to:*

For any $x, y \in \mathbb{R}$,

$$\mathbf{P}(X \leq x, Y \leq y) = \mathbf{P}(X \leq x)\mathbf{P}(Y \leq y).$$

As in the case of σ-fields, the above definition can be extended for a finite or countably infinite number of random variables.

Definition 3.18 *The random variables $X_1, X_2, \ldots, X_n$ are (mutually) independent if the fields $\sigma(X_1), \sigma(X_2), \ldots, \sigma(X_n)$ are independent. This is equivalent to*

$$\mathbf{P}\left(\bigcap_{i=1}^{n}(X_i \in B_i)\right) = \prod_{i=1}^{n}\mathbf{P}(X_i \in B_i)$$

for every Borel set $B_1, \ldots, B_n$.

Definition 3.19 *An arbitrary family of random variables $(X_i)_{i \in I}$ is (mutually) independent if every finite subfamily is independent.*

This definition is an analogue of Theorem 2.24.

Remark 3.20 *The random variables $X_1, \ldots, X_n$ are (mutually) independent if and only if*

$$\mathbf{P}(X_1 \leq a_1, \ldots, X_n \leq a_n) = \mathbf{P}(X_1 \leq a_1)\ldots\mathbf{P}(X_n \leq a_n).$$

A weaker notion of independence is that of pairwise independence.

Definition 3.21 *The random variables $X_1, X_2, \ldots, X_n$ are pairwise independent if for every $i, j = 1, .., n$ with $i \neq j$ the random variables X_i and X_j are independent.*

Remark 3.22 *If the random variables $X_1, \ldots, X_n$ are (mutually) independent, then they are pairwise independent. The converse implication is not true (see, e.g., Exercise on 238).*

EXERCISES

Problems with Solution

3.1 We roll a fair die twice. Let X be the random variable equal with the maximum of the results of the first and the second roll.

(a) Calculate the image of Ω through X and the probability distribution of X.

(b) Calculate the c.d.f. F of X.

Solution: We have

$$\Omega = \{w = (j_1, j_2), j_i \in \{1, \dots, 6\} \times \{1, \dots, 6\}\}$$

and $|\Omega| = 6 \times 6 = 36$. Let $\mathbf{P}$ the uniform probability on Ω, that is, for every $\omega \in \Omega$,

$$\mathbf{P}(\omega) = \frac{1}{|\Omega|} = \frac{1}{36}.$$

(a) The random variable X can be expressed as follows:

$$\forall w = (j_1, j_2) \in \Omega, X(\omega) = \max(j_1, j_2)$$

so

$$X : \Omega \to \{1, \dots, 6\}.$$

Let us compute the law of X. First,

$$\mathbf{P}(X = 1) = \mathbf{P}_X(\{1\}) = \mathbf{P}(w = (1, 1)) = \frac{1}{36},$$

and then

$$\mathbf{P}(X = 2) = \mathbf{P}(\{\omega = (1, 2)\} \cup \{\omega = (2, 1)\} \cup \{\omega = (2, 2)\}) = \frac{3}{36}.$$

In this way

$$\mathbf{P}(X = 3) = \mathbf{P}\left(\cup_{k=1}^{3}\{\omega = (3, k)\} \cup \cup_{k=1}^{2}\{w = (k, 3)\}\right) = \frac{5}{36},$$

$$\mathbf{P}(X = 4) = \mathbf{P}\left(\cup_{k=1}^{4}\{w = (4, k)\} \cup \cup_{k=1}^{3}\{w = (k, 4)\}\right) = \frac{7}{36},$$

$$\mathbf{P}(X = 5) = \mathbf{P}\left(\cup_{k=1}^{5}\{w = (5, k)\} \cup \cup_{k=1}^{4}\{w = (k, 5)\}\right) = \frac{9}{36},$$

$$\mathbf{P}(X = 6) = \mathbf{P}\left(\cup_{k=1}^{6}\{w = (6, k)\} \cup \cup_{k=1}^{5}\{w = (k, 6)\}\right) = \frac{11}{36}.$$

(b) The cumulative distribution function of X is

$$\begin{aligned} F(b) &= \mathbf{P}(X \leq b) = \mathbf{P}(X \in (-\infty, b] \cap X(\Omega)) \\ &= \mathbf{P}(X \in (-\infty, b] \cap \{1, \dots, 6\}). \end{aligned}$$

Therefore

$$F(b) = 0 \text{ if } b < 1,$$

$$F(b) = \mathbf{P}(X = 1) = \frac{1}{36} \text{ if } b \in [1, 2),$$

$$F(b) = \mathbf{P}(X \in \{1, 2\} = \mathbf{P}(X = 1) + \mathbf{P}(X = 2) = \frac{4}{36} \text{ if } b \in [2, 3),$$

$$F(b) = \mathbf{P}(X \in \{1, 2, 3\}) = \frac{9}{36} \text{ if } b \in [3, 4),$$

$$F(b) = \mathbf{P}(X \in \{1, 2, 3, 4\} = \frac{16}{36} \text{ if } b \in [4, 5),$$

$$F(b) = \mathbf{P}(X \in \{1, 2, 3, 4, 5\} = \frac{25}{36} \text{ if } b \in [5, 6),$$

$$F(b) = \mathbf{P}(X \in \{1, 2, 3, 4, 5, 6\} = \frac{36}{36} = 1 \text{ if } b \in [5, \infty).$$

■

3.2 A random variable X denotes the number of books bought by a customer in a bookstore. We know that the random variable has the distribution

$$\mathbf{P}(0 \leq X \leq 1) = \frac{8}{12}, \qquad \mathbf{P}(1 \leq X \leq 2) = \frac{7}{12}, \qquad \mathbf{P}(0 \leq X \leq 3) = \frac{10}{12},$$

$$\mathbf{P}(X = 3) = \mathbf{P}(X \geq 4).$$

Compute

$$\mathbf{P}(X = i)$$

for every $i = 0, 1, 2, 3$.

Solution: We have

$$\mathbf{P}(0 \leq X \leq 1) = \mathbf{P}(X = 0) + \mathbf{P}(X = 1) = \frac{8}{12},$$

$$\mathbf{P}(1 \leq X \leq 2) = \mathbf{P}(X = 1) + \mathbf{P}(X = 2) = \frac{7}{12},$$

$$\mathbf{P}(0 \leq X < 3) = \mathbf{P}(X = 0) + \mathbf{P}(X = 1) + \mathbf{P}(X = 2) = \frac{10}{12}.$$

From the last two relations we get

$$\mathbf{P}(X = 0) = \frac{3}{12}.$$

Next, from the first relation we have

$$\mathbf{P}(X = 1) = \frac{5}{12}$$

and replacing this in the second identity we obtain

$$\mathbf{P}(X = 2) = \frac{2}{12}.$$

Moreover,

$$1 = \sum_{k \geq 0} \mathbf{P}(X = k) = \mathbf{P}(0 \leq X < 3) + \mathbf{P}(X = 3) + \mathbf{P}(X \geq 4)$$

so

$$\mathbf{P}(X = 3) + \mathbf{P}(X \geq 4) = \frac{2}{12}$$

and since $\mathbf{P}(X = 3) = \mathbf{P}(X \geq 4)$ it holds that

$$\mathbf{P}(X = 3) = \frac{1}{12}.$$

■

3.3 Let X, Y be two random variables such that

$$X(\omega) \leq Y(\omega) \qquad \text{for every } \omega.$$

Show that

$$F_X(a) \geq F_Y(a) \qquad \text{for every } a \in \mathbb{R}.$$

Note this relationship is called stochastic dominance of the first order.

Solution: Since $X(\omega) \leq Y(\omega)$ for every ω, it holds that if $Y(\omega) \leq a$, then necessarily $X(\omega) \leq a$. Therefore the set $\{Y \leq a\} \subseteq \{X \leq a\}$ for every $a \in \mathbb{R}$. Thus:

$$F_X(a) = \mathbf{P}(X \leq a) \geq \mathbf{P}(Y \leq a) = F_Y(a).$$

■

3.4 Let F be given by

$$\mathbf{F}(t) = \begin{cases} 0 & \text{if } t < 0, \\ b & \text{if } t \in [0, 2), \\ 2c + b & \text{if } t \in [2, 3), \\ d & \text{if } t \geq 3. \end{cases}$$

(a) Under which conditions on a, b, c, d is the function F a cumulative distribution function?

(b) For the rest of the problem we assume that F is the c.d.f. of an r.v. X.

Express in terms of F the probabilities

$$\mathbf{P}(X > 2.5), \qquad \mathbf{P}(-1 < X \leq 1.5)$$

(c) We give

$$\mathbf{P}(X > 2.5) = 0.25, \qquad \mathbf{P}(-1 < X \leq 1.5) = 0.25.$$

Compute a, b, c, d.

(d) Compute the law of X.

Solution: (a)

$$\lim_{x\to\infty} F(x) = 1 \Rightarrow d = 1$$

$$\lim_{x\to-\infty} F(x) = 0 \Rightarrow a = 1.$$

Since F is increasing, we get

$$b \leq 2c + b \Rightarrow c \geq 0.$$

$$\forall x \in \mathbb{R}, F(x) \in [0, 1] \Rightarrow 0 \leq b \leq 1 \quad \text{and} \quad 0 \leq 2c + b \leq 1.$$

(b)

$$\mathbf{P}(X > 2.5) = 1 - \mathbf{P}(X \leq 2.5) = 1 - F(2.5) = 1 - 2c - b$$

and

$$\mathbf{P}(-1 < X \leq 1.5) = F(1.5) - F(-1) = b - a = b.$$

(c) We obtain the following system:

$$\begin{aligned} a &= 0, \\ 1 - 2c - b &= 0.25, \\ b &= 0.25, \\ d &= 1, \end{aligned}$$

so

$$a = 0, \qquad b = 0.25, \qquad c = 0.25, \qquad d = 1.$$

(d) We use

$$\forall t \in \mathbb{R}, \mathbf{P}(X = t) = F(t) - \lim_{h\to 0^+} F(t - h).$$

Then

$$\mathbf{P}(X = 0) = b = 0.25,$$

$$\mathbf{P}(X = 2) = 2c + b - b = 0.5,$$

$$\mathbf{P}(X = 3) = 1 - 2c - b = 0.25,$$

and

$$\mathbf{P}(X = t) = 0 \text{ otherwise.}$$

This implies

$$X(\Omega) = \{0, 2, 3\}.$$

■

3.5 Let $X_1, \ldots, X_n$ be independent identically distributed random variables and define, for $X_i > 0$

$$Y_i = \frac{X_i}{X_1 + \cdots + X_n}, \qquad \forall i = 1, \ldots, n.$$

Prove that $Y_1, \ldots, Y_n$ are identically distributed and calculate $\mathbf{E}(Y_1)$.

Solution: For every $a \in \mathbb{R}$, one has

$$\begin{aligned}
\mathbf{P}(Y_1 \leq a) &= \mathbf{P}(X_1 \leq a(X_1 + \cdots + X_n)) \\
&= \mathbf{P}((a-1)X_1 + \cdots + a(X_2 + \cdots + X_n) > 0) \\
&= \mathbf{P}((a-1)X_2 + a(X_1 + X_3 + \cdots + X_n) > 0) \\
&= \mathbf{P}(Y_2 < a)
\end{aligned}$$

where we used that the X's are i.i.d. Thus Y_1 and Y_2 have the same c.d.f. and therefore the same law. For the second part, use the linearity of the expectation and

$$\mathbf{E}\left[\frac{X_1 + \cdots + X_n}{X_1 + \cdots + X_n}\right] = 1.$$

■

Problems without Solution

3.6 Prove the Proposition 3.9. Specifically, prove that the function F in Definition 3.8 is increasing, right continuous and taking values in the interval $[0, 1]$, using only Proposition 2.2.

3.7 Show that any piecewise constant function is Borel measurable. A function $f : \Omega \to \mathbb{R}$ is called piecewise constant if there exist constants $a_1, a_2, \ldots$ and disjoint sets $A_1, A_2, \ldots$ in $\mathcal{F}$ such that

$$f(x) = \sum_{i=1}^{\infty} a_i \mathbf{1}_{A_i}$$

3.8 Let $g = \sum_{i=1}^{\infty} b_i \mathbf{1}_{B_i}$ where the sets B_i are not necessarily disjoint. Show that this function can be written as a piecewise constant function as in the previous problem.

3.9 Give an example of two distinct random variables with the same distribution function.

3.10 Let X, Y be two continuous random variables such that

$$\mathbf{P}_X(\{x\}) = \mathbf{P}_Y(\{x\}) = 0$$

for every $x \in \mathbb{R}$ (that means that X, Y are continuous r.v.). Show that

$$\mathbf{P}(X = Y) = 0.$$

3.11 Let $X_1, \ldots, X_n$ be independent random variables. Show that there exists $a \in \mathbb{R}$ such that

$$\mathbf{P}(X_1 + \cdots + X_n) = a = 1$$

if and only if for every i there exists $a_i \in \mathbb{R}$ with

$$\mathbf{P}(X_i = a_i) = 1$$

(i.e., all random variables are constants).

3.12 An aircraft engine fails with probability $1 - p$ where $p \in (0, 1)$, independently of the other engines. To complete its flight, the aircraft needs that the majority of its engines works.

(a) Calculate the probability to successfully complete the flight for an aircraft with 3 engines.

(b) Calculate the probability to successfully complete the flight for an aircraft with 5 engines.

(c) Find the value of p for which the aircraft with 5 engines is preferable (i.e., has greater probability of completing its flight).

Hint: Let M_i be the event "the engine i works," for $i = 1, 2, 3$. Clearly

$$\mathbf{P}(M_i) = p, \qquad i = 1, 2, 3$$

and

$$\mathbf{P}(M_i^c) = 1 - p \qquad i = 1, 2, 3.$$

For a 3-engine plane, the probability of a successful flight is

$$\begin{aligned}
\mathbf{P}\ &\big((M_1 \cap M_2 \cap M_3^c) \cup (M_1 \cap M_3 \cap M_2^c) \cup (M_2 \cap M_3 \cap M_1^c) \\
&\quad \cup (M_1 \cap M_2 \cap M_3)\big) \\
&= \mathbf{P}(M_1 \cap M_2 \cap M_3^c) + \mathbf{P}(M_1 \cap M_3 \cap M_2^c) \\
&\quad + \mathbf{P}(M_2 \cap M_3 \cap M_1^c) + \mathbf{P}(M_1 \cap M_2 \cap M_3) \\
&= 3p^2(1-p) + p^3.
\end{aligned}$$

CHAPTER FOUR

Random Variables: The Discrete Case

4.1 Introduction/Purpose of the Chapter

This chapter treats discrete random variables. After having introduced the general notion of a random variable, we discuss specific cases. Discrete random variables are presented next, and continuous random variables are left to the next chapter. In this chapter we learn about calculating simple probabilities using a probability mass function. Several probability functions for discrete random variables warrant special mention because they arise frequently in real-life situations. These are the probability functions for, among others, the so-called geometric, hypergeometric, binomial, and Poisson distributions. We focus on the physical assumptions underlying the application of these functions to real problems. Although we can use computers to calculate probabilities from these distributions, it is often convenient to use special tables, or even use approximate methods in which one probability function can be approximated quite closely by another function. We introduce the concepts of distribution, cumulative distribution function, expectation, and variance for discrete random variables. We also discuss higher-order moments of such variables.

Handbook of Probability, First Edition. Ionuţ Florescu and Ciprian Tudor.

4.2 Vignette/Historical Notes

Historically, the discrete random variables were the first type of random outcomes studied in practice. The documented exchange of letters in 1964 between Pascal and Fermat was prompted by a game of dice which essentially dealt with discrete random outcomes (faces of the dies). Even earlier, the 16th-century Italian mathematician and physician Cardano wrote "On Casting the Die," a study dealing with discrete random variables; however, it was not published until 1663, 87 years after the death of Cardano. He introduced concepts of combinatorics into calculations of probability and defined probability as "the number of favorable outcomes divided by the number of possible outcomes."

In 1655 during his first visit to Paris the Dutch scientist Christian Huygens, learned of the work on probability carried out in this correspondence. On his return to Holland in 1657, Huygens wrote a small work *De Ratiociniis in Ludo Aleae* (hard to translate dative in English—approximate translation *On Reasonings in Dice Games*) the first printed work on the calculus of probabilities. It was a treatise on problems associated with gambling, once again dealing with discrete random variables. Because of the inherent appeal of games of chance, probability theory soon became popular, and the subject developed rapidly during the 18th century.

One of the major contributors during this period was Jacob Bernoulli (1654–1705). Jacob (Jacques) Bernoulli was a Swiss mathematician who was the first to use the term *integral*. He was the first mathematician in the Bernoulli family, a family of famous scientists of the 18th century. Jacob Bernoulli's most original work was *Ars Conjectandi* (*The Art of Conjecturing*—The art of drawing conclusions) published in Basel in 1713, eight years after his death. The work was incomplete at the time of his death, but it still was a work of the greatest significance in the development of the Theory of Probability.

In the late 18th century, it became increasingly evident that analogies exist between games of chance and random phenomena in physical, biological, and social sciences. Thus the theory of probability and discrete variables, in particular, exploded in the following two centuries; and even today, particle interaction models coagulation and fragmentation models are initially described in a discrete setting to make them easier to understand.

4.3 Theory and Applications

4.3.1 DEFINITION AND BASIC FACTS

After defining the discrete random variables, we list their basic properties. In fact, we will translate the general notion introduced in the previous chapter to the particular case of random variables with a finite or countable possible number of outcomes.

Definition 4.1 *We will say that a random variable is discrete if its image $X(\Omega)$ is a finite or countable subset of $\mathbb{R}$.*

As a simple example, if a coin is tossed three times, the number of heads obtained can be 0, 1, 2, or 3. In this example, the number of heads can only take 4 distinct values $\{0, 1, 2, 3\}$, and so the variable is discrete.

Remark 4.2 *In the case of discrete random variables the law of X is completely determined by the set $X(\Omega)$ and the probabilities $\mathbf{P}_X(\{x\}) = \mathbf{P}(X = x)$ for every $x \in X(\Omega)$. Further, we have*

$$\sum_{x \in X(\Omega)} \mathbf{P}(X = x) = 1.$$

Remark 4.3 *The cumulative distribution function (c.d.f.) is given by*

$$F(t) = \mathbf{P}(X \le t) = \sum_{\{x \in X(\Omega) | x \le t\}} \mathbf{P}(X = x).$$

A c.d.f. of a discrete random variable is a piecewise constant function (e.g., Figure 4.1).

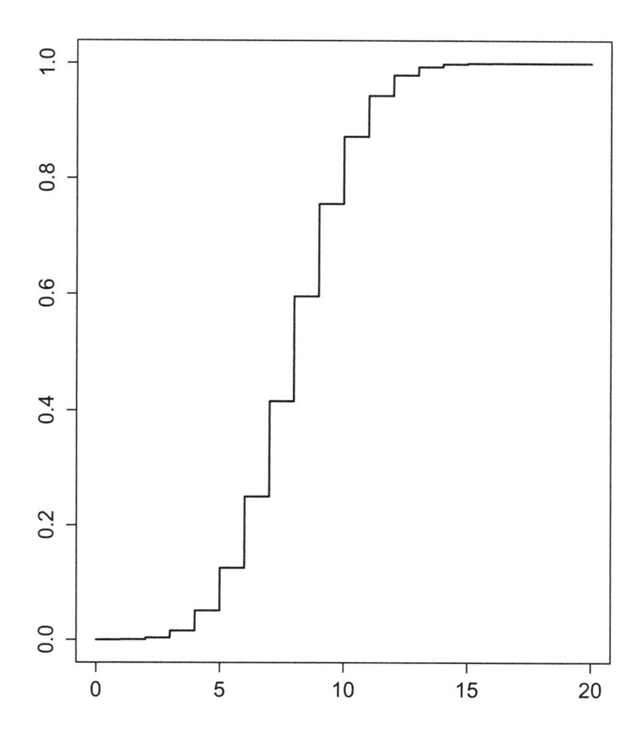

FIGURE 4.1 A general discrete c.d.f. The figure depicts a ***Binomial***(20,0.4).

EXAMPLE 4.1 Calculating the c.d.f. of a discrete random variable

A die is rolled repeatedly until a 6 is obtained. Let X be the random variable representing the number of times we roll the die.

$$\mathbf{P}(X = 1) = \frac{1}{6}.$$

Clearly, if we only roll the die once and we get a 6 on our first roll, this is the only case when $X = 1$. The probability of this is $1/6$. Furthermore,

$$\mathbf{P}(X = 2) = \frac{5}{6} \times \frac{1}{6} = \frac{5}{36}$$

If we roll the die twice before getting the first 6, we must have rolled something that isn't a 6 with our first roll, the probability of which is $5/6$, and we must roll a 6 on our second roll, the probability of which is $1/6$.

Compute the c.d.f. of X.

Solution: Let us compute the cumulative distribution function $F(t)$ for some values of t.

$$F(1) = \mathbf{P}(X \leq 1)$$

is the probability that the number of rolls until we get a 6 is less than or equal to 1. Thus, this number must be either 0 or 1. Since, $\mathbf{P}(X = 0) = 0$ and $\mathbf{P}(X = 1) = 1/6.$, we obtain $\mathbf{P}(X \leq 1) = 1/6$.

Similarly,

$$\begin{aligned}\mathbf{P}(X \leq 2) &= \mathbf{P}(X = 0) + \mathbf{P}(X = 1) + \mathbf{P}(X = 2)\\ &= 0 + 1/6 + 5/36 = 11/36.\end{aligned}$$

This calculation can proceed like this, and we readily obtain the c.d.f. $F(t)$ for all t. In fact, we shall encounter this particular distribution again when we talk about geometric random variables. The c.d.f. may be calculated in general using geometric sums. As an exercise, the reader is left with calculating $\mathbf{P}(X \leq 6)$. ■

The cumulative distribution function has particular properties in the case of discrete random variables.

Proposition 4.4 *Let $X : \Omega \to \mathbb{R}$ be a discrete random variable on $(\Omega, \mathcal{F}, \mathbf{P})$. Then the cumulative distribution function F_X is piecewise constant, and it has a finite or countable number of jumps. That is, the function is constant almost everywhere; in other words, there exists at most a countable sequence $\{a_1, a_2, \dots\}$ such that the function F is constant for every point except these a_i's.*

Proof: Let us show that the number of jumps is countable. Recall that a c.d.f. is right continuous. Let us denote by

$$D = \{a \in \mathbb{R} \mid F_X(a-) < F_X(a)\}.$$

Let an $a \in D$, thus

$$F_X(a-) < F_X(a).$$

Between any two reals we can find a rational; therefore there exists an $n_0 \geq 1$ integer such that

$$F_X(a) - F_X(a-) > \frac{1}{n_0}.$$

Denote by

$$D_n = \left\{a \in \mathbb{R} \mid F_X(a) - F_X(a-) > \frac{1}{n}\right\}.$$

We just saw that

$$D \subseteq \bigcup_{n \geq n_0} D_n \subseteq \bigcup_{n \geq 1} D_n.$$

To show that D is countable, it suffices to check that every set D_n is countable (a countable union of countable sets is countable). Since the cumulative distribution function F_X is increasing and takes values in $[0, 1]$, it cannot have more than n jumps with length bigger than $\frac{1}{n}$. Therefore

$$|D_n| \leq n,$$

where we denoted $|D_n| = Card(D_n)$ the cardinality (number of elements) of D_n. ■

Remark 4.5 *Since a discrete random variable X only takes a finite or countable number of values, its distribution is entirely determined by $X(\Omega)$. In fact, the law of X is determined by*

$$\mathbf{P}(X = x) = F_X(x) - F_X(x-)$$

for every $x \in X(\Omega)$.

These discrete values are formalized next.

Definition 4.6 (Probability mass function) *A discrete random variable X with $X(\Omega) = \{a_1, a_2, \dots\}$ is completely characterized by the probabilities:*

$$p_i = \mathbf{P}(X = a_i).$$

The collection of these numbers is called the probability mass function (or p.m.f.) of the random variable X.

4.3.2 MOMENTS

We made in this handbook the choice to treat independently the moments of discrete and continuous random variables. In this section we present the moments of discrete random variables.

Definition 4.7 *Let X be a discrete random variable on a probability space $(\Omega, \mathcal{F}, \mathbf{P})$. We will say that X is integrable (or admits an expectation) if*

$$\sum_{x \in X(\Omega)} |x| \; \mathbf{P}(X = x) < \infty.$$

In this case, its expectation (or mean) is defined by

$$\mathbf{E}(X) = \sum_{x \in X(\Omega)} x \, \mathbf{P}(X = x). \tag{4.1}$$

Using the p.m.f. notation, this is expressed as

$$\mathbf{E}(X) = \sum_i a_i p_i.$$

If X is a random variable with a finite number of values, then it will always have an expectation (a finite sum of real numbers is finite). If X has a countable (nonfinite) number of values, then its expectation is given by a series and we naturally need to assume that the series is convergent.

The series

$$\sum_{x \in X(\Omega)} |x| \; \mathbf{P}(X = x)$$

is a series of real numbers with positive terms. Its convergence implies the convergence of the series $\sum_{x \in X(\Omega)} x \, \mathbf{P}(X = x)$.

The expectation of X can be interpreted as a weighted average of all the values X takes with the weights given by the probabilities that X take these values.

In a geometrical interpretation, $\mathbf{E}(X)$ is the position of the center of mass for the particles with positions at the values of the random variable and mass the corresponding weights.

Proposition 4.8 (Properties of the expectation) *The following hold.*

a. *The expectation is linear. That is, if X, Y are two discrete integrable random variables on the same probability space $(\Omega, \mathcal{F}, \mathbf{P})$, then*

$$\mathbf{E}(aX + bY) = a\mathbf{E}(X) + b\mathbf{E}(Y)$$

for every $a, b \in \mathbb{R}$. In particular, if a is a constant, then

$$\mathbf{E}(a) = a.$$

b. *If $X \geq 0$ (i.e., if $X(\omega) \geq 0$ for every $\omega \in \Omega$), then*

$$\mathbf{E}(X) \geq 0.$$

In particular, if $X \geq Y$, then

$$\mathbf{E}(X) \geq \mathbf{E}(Y).$$

Proof: 1. Let us first show that, if $a \in \mathbb{R}$ and X is a discrete random variable admitting an expectation, then aX has an expectation and $\mathbf{E}(ax) = a\mathbf{E}(X)$. This is clear for $a = 0$. Suppose $a \neq 0$ and let $Z = aX$. It is clear that Z is a discrete random variable since $Z(\Omega)$ is at most countable. We have

$$\begin{aligned}\mathbf{E}(|Z|) &= \sum_{x \in X(\Omega)} |ax|\mathbf{P}(Z = ax) \\ &= |a| \sum_{x \in X(\Omega)} |x|\mathbf{P}(aX = ax) = |a| \sum_{x \in X(\Omega)} |x|\mathbf{P}(X = x)\end{aligned}$$

using $a \neq 0$. Hence Z has an expectation and a similar calculation shows that

$$\mathbf{E}(aX) = a\mathbf{E}(X).$$

Consider now two random variables X and Y and let us show that $Z = X + Y$ admits expectation. It is an easy exercise to see that Z is discrete. Now,

$$\begin{aligned}\sum_{z \in Z(\Omega)} |z|\mathbf{P}(Z = z) &= \sum_{z \in Z(\Omega)} |z|\mathbf{P}\left(\bigcup_{(x,y) \in X(\Omega) \times Y(\Omega), x+y=z} (X = x, Y = y)\right) \\ &= \sum_{z \in Z(\Omega)} |z| \sum_{(x,y) \in X(\Omega) \times Y(\Omega), x+y=z} \mathbf{P}(X = x, Y = y) \\ &= \sum_{z \in Z(\Omega)} \sum_{(x,y) \in X(\Omega) \times Y(\Omega), x+y=z} |z|\mathbf{P}(X = x, Y = y) \\ &= \sum_{z \in Z(\Omega)} \sum_{(x,y) \in X(\Omega) \times Y(\Omega), x+y=z} |x + y|\mathbf{P}(X = x, Y = y) \\ &\leq \sum_{z \in Z(\Omega)} \sum_{(x,y) \in X(\Omega) \times Y(\Omega), x+y=z} (|x| + |y|)\mathbf{P}(X = x, Y = y).\end{aligned}$$

Consequently,

$$\begin{aligned}\sum_{z \in Z(\Omega)} |z|\mathbf{P}(Z = z) \leq &\sum_{z \in Z(\Omega)} \sum_{(x,y) \in X(\Omega) \times Y(\Omega), x+y=z} |x|\mathbf{P}(X = x, Y = y) \\ &+ \sum_{z \in Z(\Omega)} \sum_{(x,y) \in X(\Omega) \times Y(\Omega), x+y=z} |y|\mathbf{P}(X = x, Y = y)\end{aligned}$$

Let us see why the first sum above is finite (the second could be treated analogously). We have

$$
\begin{aligned}
&\sum_{z\in Z(\Omega)} \sum_{(x,y)\in X(\Omega)\times Y(\Omega), x+y=z} |x|\mathbf{P}(X = x, Y = y) \\
&= \sum_{z\in Z(\Omega)} \sum_{x\in X(\Omega)} |x|\mathbf{P}(X = x, Y = z - x) \\
&= \sum_{z\in Z(\Omega)} \sum_{x\in X(\Omega)} |x|\mathbf{P}(X = x, Z = z) \\
&= \sum_{x\in X(\Omega)} |x|\mathbf{P}(X = x)
\end{aligned}
$$

using the total probability formula (Proposition 2.4) and the fact that the sets $\{Z = z\}_{z\in Z(\Omega)}$ form a partition of the Ω. Again, repeating the same computation, we will obtain $\mathbf{E}(X + Y) = \mathbf{E}(X) + \mathbf{E}(Y)$.

Finally, combining the two results ($\mathbf{E}(X + Y) = \mathbf{E}(X) + \mathbf{E}(Y)$ and $\mathbf{E}(aX) = a\mathbf{E}(X)$) gives the general linear formula.

2. $X \geq 0$ means $x \geq 0$ for every $x \in X(\Omega)$. The conclusion is obtained using just the definition of the expectation (a sum of positive numbers is positive).

Of course, if $X(\omega) \geq Y(\omega)$ for all ω, then by defining $Z(\omega) = X(\omega) - Y(\omega)$ we see that $\mathbf{E}(Z) \geq 0$ and using linearity from the first part the conclusion easily follows. ∎

The expectation of a function of a random variable may be computed in the following way.

Theorem 4.9 (Transfer Formula) *Let X be a discrete random variable on $(\Omega, \mathcal{F}, \mathbf{P})$. Assume that $\varphi : \mathbb{R} \to \mathbb{R}$ is a measurable function. Then $\varphi(X) : \Omega \to \mathbb{R}$ is a random variable and this random variable is integrable if and only if*

$$\sum_{x\in X(\Omega)} |\varphi(x)|\ \mathbf{P}(X = x) < \infty.$$

In this case

$$\mathbf{E}(\varphi(X)) = \sum_{x\in X(\Omega)} \varphi(x)\ \mathbf{P}(X = x).$$

Proof: Let $\varphi : \mathbb{R} \to \mathbb{R}$ be a measurable function. Let us denote by $Z = \varphi(X)$. Since only the restriction: $\varphi : X(\Omega) \to Z(\Omega)$ takes nonzero values and since $X(\Omega)$

is discrete, it follows that Z is discrete. Thus we obtain

$$\begin{aligned}
\sum_{z\in Z(\Omega)} |z|\mathbf{P}(Z=z) &= \sum_{z\in Z(\Omega)} \mathbf{P}(\varphi(X)=z)\\
&= \sum_{z\in Z(\Omega)} |z|\mathbf{P}\left(\bigcup_{x\in\varphi^{-1}(\{z\})}(X=x)\right)\\
&= \sum_{z\in Z(\Omega)} |z| \sum_{x\in\varphi^{-1}(\{z\})} \mathbf{P}(X=x)\\
&= \sum_{z\in Z(\Omega)} \sum_{x\in\varphi^{-1}(\{z\})} |\varphi(x)|\mathbf{P}(X=x)\\
&= \sum_{x\in X(\Omega)} |\varphi(x)|\mathbf{P}(X=x).
\end{aligned}$$

For the last equality we use the fact that the restriction $\varphi : X(\Omega) \to Z(\Omega)$ is surjective. Therefore, $(\varphi^{-1}(\{z\}))_{z\in Z(\Omega)}$ is a partition of $X(\Omega)$. Analogously,

$$\begin{aligned}
\mathbf{E}(Z) &= \sum_{z\in Z(\Omega)} z\mathbf{P}(Z=z)\\
&= \sum_{z\in Z(\Omega)} \mathbf{P}(\varphi(X)=z)\\
&= \sum_{z\in Z(\Omega)} z\mathbf{P}\left(\bigcup_{x\in\varphi^{-1}(\{z\})}(X=x)\right)\\
&= \sum_{z\in Z(\Omega)} z \sum_{x\in\varphi^{-1}(\{z\})} \mathbf{P}(X=x)\\
&= \sum_{z\in Z(\Omega)} \sum_{x\in\varphi^{-1}(\{z\})} \varphi(x)\mathbf{P}(X=x)\\
&= \sum_{x\in X(\Omega)} \varphi(x)\mathbf{P}(X=x).
\end{aligned}$$

■

Remark 4.10 *The following are notations.*

1. *Note that the integrability condition from Definition 4.7 can be written in the equivalent way*

$$\mathbf{E}\,|X| < \infty.$$

This condition is identical with the one assumed in the case of continuous random variables (see the next chapter).

2. *Take* $\varphi(x) = x^2$. *Then* X^2 *is integrable if and only if*

$$\sum_{x \in X(\Omega)} x^2 \mathbf{P}(X = x) < \infty$$

and, in fact,

$$\mathbf{E}(X^2) = \sum_{x \in X(\Omega)} x^2 \mathbf{P}(X = x).$$

EXAMPLE 4.2 A basic example

Let X be a random variable with the law

X	1	2	3
$p(x)$	0.2	0.3	0.5

Then

$$\mathbf{E}(X) = 1 \times 0.2 + 2 \times 0.3 + 3 \times 0.5$$

and

$$\mathbf{E}(X^2) = 1^2 \times 0.2 + 2^2 \times 0.3 + 5^2 \times 0.5.$$

Generally, we can compute the expectation of any function of X using the transport formula. For example,

$$\mathbf{E}(\sin X) = \sin(1) \times 0.2 + \sin(2) \times 0.3 + \sin(3) \times 0.5.$$

Definition 4.11 (Variance of a random variable) *Let* X *be a discrete random variable with outcomes* $\{a_1, a_2, \dots\}$ *and p.m.f.* $p_1, p_2, \dots$. *The variance of* X *is defined as*

$$Var(X) = \mathbf{E}\left[(X - \mathbf{E}X)^2\right] = \mathbf{E}(X^2) - (\mathbf{E}X)^2 .$$

Using the p.m.f notation and denoting $\mu = \mathbf{E}X = \sum_i a_i p_i$, *we obtain*

$$\begin{aligned} Var(X) &= \sum_i (a_i - \mu)^2 p_i \\ &= \sum_i a_i^2 p_i - \mu^2 \\ &= \mathbf{E}[X^2] - (\mathbf{E}X)^2. \end{aligned}$$

Definition 4.12 (Standard deviation) *We define the standard deviation of a random variable as*

$$StdDev(X) = \sqrt{V(X)} = \sqrt{\sum_i (a_i - \mu)^2 p_i}.$$

4.4 Examples of Discrete Random Variables

4.4.1 THE (DISCRETE) UNIFORM DISTRIBUTION

This distribution arises in any situation where outcomes are discrete and equally likely. Let us present an example.

EXAMPLE 4.3

Roll a six-sided fair die. Say $X(\omega) = 1$ if the die shows 1 ($\omega = 1$), $X = 2$ if the die shows 2, and so on. Find $F(x) = \mathbf{P}(X \leq x)$.

Solution: If $x < 1$, then $\mathbf{P}(X \leq x) = 0$.

If $x \in [1, 2)$, then $\mathbf{P}(X \leq x) = \mathbf{P}(X = 1) = 1/6$.

If $x \in [2, 3)$, then $\mathbf{P}(X \leq x) = \mathbf{P}(X(\omega) \in \{1, 2\}) = 2/6$.

We continue this way to get

$$\mathbf{F}(x) = \begin{cases} 0 & \text{if } x < 1, \\ i/6 & \text{if } x \in [i, i+1) \text{ with } i = 1, \dots, 5, \\ 1 & \text{if } x \geq 6. \end{cases}$$

■

For discrete random variables it is generally simple to give the probability mass function instead of the c.d.f. since it will describe completely the distribution. Recall that the distribution function is piecewise linear.

Definition 4.13 (Discrete uniform distribution) *A random variable is said to have the Discrete Uniform distribution if it takes values in a discrete set of numbers* $X(\Omega) = \{x_1, x_2, \dots, x_n\}$ *and the probabilities for these numbers are equal to*

$$\mathbf{P}(X = x_i) = \frac{1}{n}, \quad \forall i.$$

We denote a random variable X *with this distribution with* $X \sim DU(n)$.

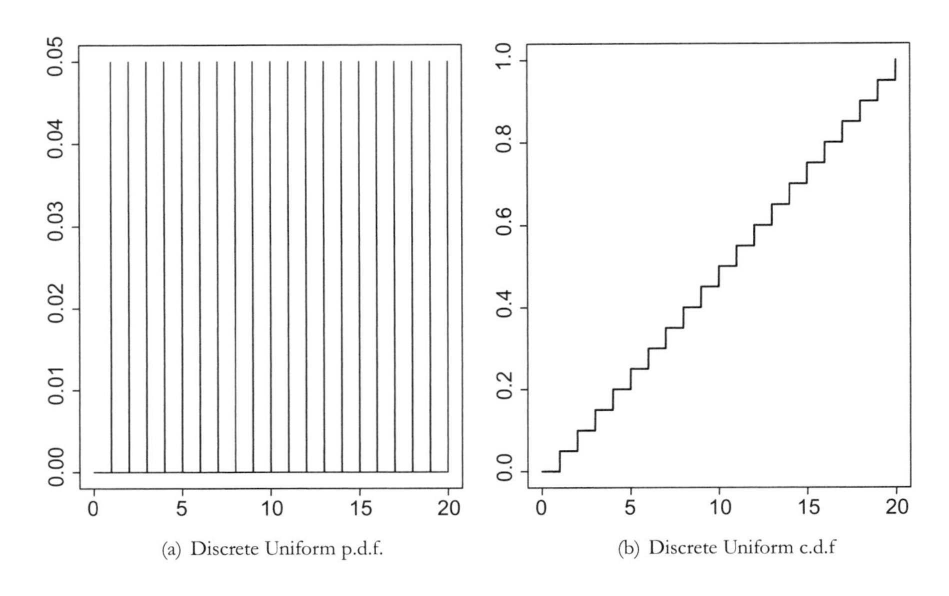

(a) Discrete Uniform p.d.f.

(b) Discrete Uniform c.d.f

FIGURE 4.2 The p.d.f. and c.d.f. of the discrete uniform distribution.

Remark 4.14 *We name this uniform distribution discrete to differentiate from the continuous case when the random variable takes values in an interval (this case will be presented in the next chapter). Traditionally, that distribution is simply called the uniform distribution. Figure 4.2 plots the density and distribution of this random variable.*

Remark 4.15 *The outcomes* $\{x_1, x_2, \dots, x_n\}$ *are not important, and generally we use* $\{1, 2, \dots, n\}$ *to denote them. This is a general trend for all discrete random variables where the outcomes may be put in one-to-one correspondence with discrete sets of the type* $\{1, 2, \dots, n\}$.

The probabilities clearly define a probability density ($\sum_{i=1}^{n} \frac{1}{n} = 1$), and in the case when the outcomes are $\{1, 2, \dots, n\}$ we may calculate the expectation and variance as

$$\mathbf{E}X = \sum_{i=1}^{n} i\frac{1}{n} = \frac{1}{n}\sum_{i=1}^{n} i = \frac{1}{n}\frac{n(n+1)}{2} = \frac{n+1}{2}$$

and

$$\begin{aligned} V(X) = \mathbf{E}(X^2) - (\mathbf{E}(X))^2 &= \sum_{i=1}^{n} i^2\frac{1}{n} - \left(\frac{n+1}{2}\right)^2 \\ &= \frac{1}{n}\frac{n(n+1)(2n+1)}{6} - \left(\frac{n+1}{2}\right)^2 \end{aligned}$$

$$
\begin{aligned}
&= \frac{n+1}{2}\left(\frac{2n+1}{3} - \frac{n+1}{2}\right) \\
&= \frac{n+1}{2}\frac{n-1}{6} \\
&= \frac{n^2-1}{12}.
\end{aligned}
$$

Please note that in the general case when the outcomes are $\{x_1, x_2, \ldots, x_n\}$ the formulas need to be recalculated every time.

4.4.2 BERNOULLI DISTRIBUTION

Before we give the formal definition let us present a couple of examples of situations where this distribution was encountered previously.

EXAMPLE 4.4 Indicator r.v. (continued)

This indicator variable is also called the Bernoulli random variable. Notice that the variable only takes values 0 and 1, and the probability that the variable takes the value 1 may be easily calculated using the previous definitions:

$$\mathbf{P} \circ \mathbf{1}_A^{-1}(\{1\}) = \mathbf{P}\{\omega : \mathbf{1}_A(\omega) = 1\} = \mathbf{P}(A).$$

Therefore the variable is distributed as a Bernoulli random variable with parameter $p = \mathbf{P}(A)$. Alternately, we may obtain this probability using the previously computed distribution function:

$$\mathbf{P}\{\omega : \mathbf{1}_A(\omega) = 1\} = F(1) - F(1-) = 1 - \mathbf{P}(A^c) = \mathbf{P}(A).$$

Let us formalize this example.

Definition 4.16 (Bernoulli distribution) *A random variable with a Bernoulli distribution with parameter* $p \in (0, 1)$ *takes only two values:*

$$\mathbf{X} = \begin{cases} 1 & \textit{with } \mathbf{P}(X = 1) = p, \\ 0 & \textit{with } \mathbf{P}(X = 0) = 1 - p. \end{cases}$$

We denote a random variable X *with this distribution with* $X \sim$ *Bernoulli*(p).

Clearly the sequence p, $1 - p$ defines a probability mass function since

$$p + (1 - p) = 1.$$

A random variable with *Bernoulli*(p) distribution has mean p since

$$\mathbf{E}(X) = 0\mathbf{P}(X = 0) + 1\mathbf{P}(X = 1) = p$$

and the variance is given by

$$Var(X) = \mathbf{E}(X^2) - (\mathbf{E}(X))^2 = p - p^2 = p(1-p).$$

4.4.3 BINOMIAL (n, p) DISTRIBUTION

In probability theory and statistics, the *Binomial* distribution is the discrete probability distribution of the number of successes in a sequence of n independent yes/no experiments, each of which yields success with probability $p \in (0, 1)$. Such a success/failure experiment is also called a *Bernoulli* experiment or *Bernoulli* trial. In fact, when $n = 1$, the *Binomial* distribution reduces to a *Bernoulli* distribution.

Definition 4.17 (Binomial distribution) *A binomial random variable with parameters n and p takes values in* $\mathbb{N}$ *with*

$$\mathbf{P}(X = k) = \begin{cases} \binom{n}{k} p^k (1-p)^{n-k} & \text{for any } k \in \{0, 1, 2, \ldots, n\}, \\ 0 & \text{otherwise.} \end{cases}$$

We denote a random variable X with this distribution as follows: $X \sim Binom(n, p)$.

Remark 4.18 *A* $Binomial(n, p)$ *random variable X has the same distribution as* $Y_1 + \cdots Y_n$ *where* $Y_i \sim Bernoulli(p)$.

Recall that for $a, b \in \mathbb{R}$ we have

$$(a+b)^n = \sum_{k=0}^{n} \binom{n}{k} a^k b^{n-k}, \tag{4.2}$$

the binomial expansion.

Proposition 4.19 *The sequence* $p_k = \mathbf{P}(X = k)$, $k = 0, 1, \ldots$ *defines a discrete probability distribution.*

Proof: Since $p_k \geq 0$ for every k, it suffices to check that $\sum_k p_k = 1$. Applying (4.2) to $a = p$ and $b = 1 - p$, we get

$$\sum_k p_k = \sum_{k=0}^{n} \binom{n}{k} p^k (1-p)^{n-k} = (p + (1-p))^n = 1.$$

■

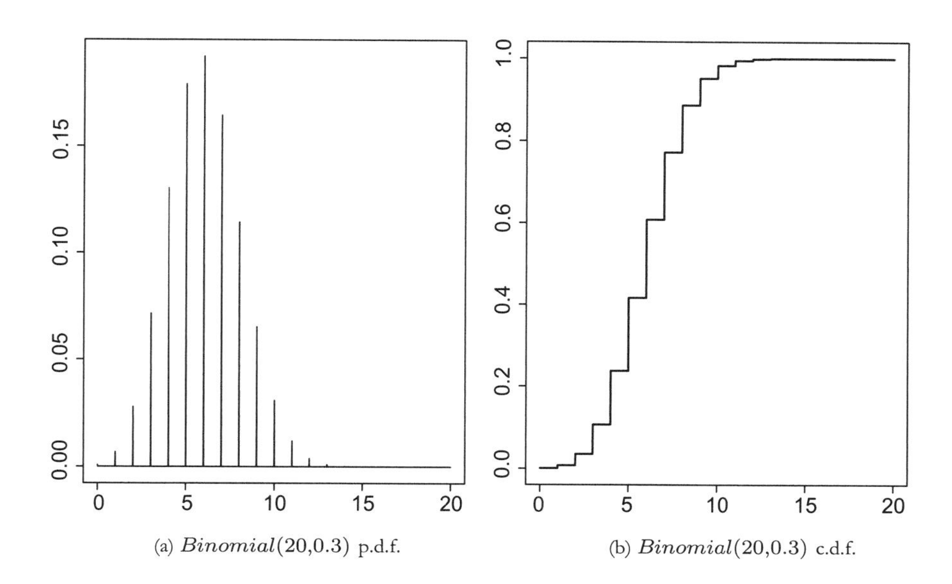

(a) *Binomial*(20,0.3) p.d.f.

(b) *Binomial*(20,0.3) c.d.f.

FIGURE 4.3 **The p.d.f. and c.d.f. of the binomial distribution.**

Figure 4.3 presents the density and distribution for a *Binomial*(20, 0.3) random variable.

Let us calculate the expectation and variance of the binomial law.

Proposition 4.20 *Suppose $X \sim Binom(n, p)$. Then*

$$\mathbf{E}(X) = np, \quad Var(X) = np(1-p).$$

Proof: Using formula (4.1), we can write

$$\begin{aligned}
\mathbf{E}X &= \sum_{k=0}^{n} k\binom{n}{k}p^k(1-p)^{n-k} = \sum_{k\geq 1} k\binom{n}{k}p^k(1-p)^{n-k} \\
&= \sum_{k=1}^{n} \frac{n!}{(k-1)!(n-k)!}p^k(1-p)^{n-k} \\
&= np\sum_{k=1}^{n} \frac{(n-1)!}{(k-1)!(n-k)!}p^{k-1}(1-p)^{n-k} \\
&= np\sum_{k=0}^{n-1} \frac{(n-1)!}{k!(n-k-1)!}p^k(1-p)^{n-k-1} \\
&= np(p+(1-p))^{n-1} = np
\end{aligned}$$

using again (4.2). Next,

$$\begin{aligned}\mathbf{E}[X(X-1)] &= \sum_{k=0}^{n} k(k-1)\binom{n}{k}p^k(1-p)^{n-k}\\ &= \sum_{k=2}^{n} k(k-1)\binom{n}{k}p^k(1-p)^{n-k}\end{aligned}$$

because the first two terms of the sum vanish. Thus,

$$\begin{aligned}\mathbf{E}[X(X-1)] &= \sum_{k=2}^{n} n(n-1)\binom{n-2}{k-2}p^k(1-p)^{(n-2)-(k-2)}\\ &= n(n-1)p^2\sum_{k=2}^{n}\binom{n-2}{k-2}p^{k-2}(1-p)^{(n-2)-(k-2)}\\ &= n(n-1)p^2\sum_{k=0}^{n-2}\binom{n-2}{k}p^k(1-p)^{(n-2)-k}\\ &= n(n-1)p^2\end{aligned}$$

again from (4.2). Finally,

$$\begin{aligned}Var(X) = \mathbf{E}X^2 - (\mathbf{E}X)^2 &= \mathbf{E}[X(X-1)] + \mathbf{E}X - (\mathbf{E}X)^2\\ &= n(n-1)p^2 + np - n^2p^2 = np(1-p).\end{aligned}$$

■

If $X \sim Binom(n, p)$, then its cumulative distribution function is given by

$$F(x) = \mathbf{P}(X \leq x) = \sum_{k=0}^{[x]}\binom{n}{k}p^k(1-p)^{n-k}.$$

Remark 4.21 *The c.d.f. of a binomial r.v. $F(x)$ can be also expressed in terms of the incomplete beta function as follows:*

$$F(k) = (n-k)\binom{n}{k}\int_0^{1-p} t^{n-k-1}(1-t)^k.$$

We will prove that the sum of two independent binomially distributed random variable with the same parameter p is a binomially distributed random variable. Intuitively, this is easy to understand taking into account the experiment described by the binomial law (number of successes in k trials plus number of successes in other, independent n trials should be the number of successes in the total $n + k$ trials).

Proposition 4.22 *If $X \sim Binom(n, p)$ and $Y \sim Binom(m, p)$ and X, Y are independent, then*

$$X + Y \sim Binom(n + m, p).$$

Proof: Note that

$$(X + Y)(\Omega) = \{0, 1, 2, \ldots, m + n\}.$$

For every k between 0 and $m + n$ we have

$$\begin{aligned}\mathbf{P}(X + Y = k) &= \sum_{i=0}^{k} \mathbf{P}(X = i)\mathbf{P}(Y = k - i) \\ &= \sum_{i=0}^{k} \binom{n}{i} p^i (1 - p)^{n-i} \binom{m}{k - i} p^{k-i} (1 - p)^{m-k+i} \\ &= p^k (1 - p)^{m+n-k} \sum_{i=0}^{k} \binom{n}{i} \binom{m}{k - i}.\end{aligned}$$

We will use the relation

$$\sum_{i=0}^{k} \frac{\binom{n}{i}\binom{m}{k-i}}{\binom{n+m}{k}} = 1$$

and we will get

$$\mathbf{P}(X + Y = k) = \binom{n + m}{k} p^k (1 - p)^{m+n-k}$$

for every $k = 0, \ldots, n + m$, which means that $X + Y \sim Binom(n + m, p)$.

The relation in terms of combinatorial terms used will be proven a bit later, when we show that the hypergeometric distribution is a probability distribution. ∎

4.4.4 GEOMETRIC (p) DISTRIBUTION

Definition 4.23 (Geometric distribution) *Let the law of a random variable X be defined by*

$$\mathbf{P}(X = k) = \begin{cases} (1 - p)^{k-1} p & \text{for any } k \in \{1, 2 \cdots\}, \\ 0 & \text{otherwise.} \end{cases}$$

We will write $X \sim Geometric(p)$ to denote that the random variable X has this distribution.

Remark 4.24 *The geometric distribution is the law of the total number of tosses of a biased coin needed to obtain the first head (counting the toss of the head). The parameter p of this law is a real number between* 0 *and* 1 *and represents the probability of success.*

Since the number X is exactly the number of trials to get the first success, its distribution is sometimes called the geometric "number of trials" distribution. We can also talk about the geometric "number of failures distribution" distribution, defined as

$$\mathbf{P}(Y = k) = \begin{cases} (1-p)^k p & \text{for any } k \in \{0, 1, 2, \ldots\}, \\ 0 & \text{otherwise,} \end{cases}$$

and also talk about modeling just the number of tosses leading to the first head.

Figure 4.4 presents the density and distribution for a *Geometric*(0.2) random variable.

In this book when we write $X \sim \textit{Geometric}(p)$ we will mean that X has a geometric number of trials distribution with the probability of success p. In the rare cases when we use the number of failures distribution, we will specify very clearly.

Proposition 4.25 *The sequence* $\mathbf{P}(X = k)$ *defines a probability distribution.*

Proof: Indeed, since for every $q \in (0, 1)$ we have

$$\sum_{k \geq 0} q^k = \frac{1}{1-q},$$

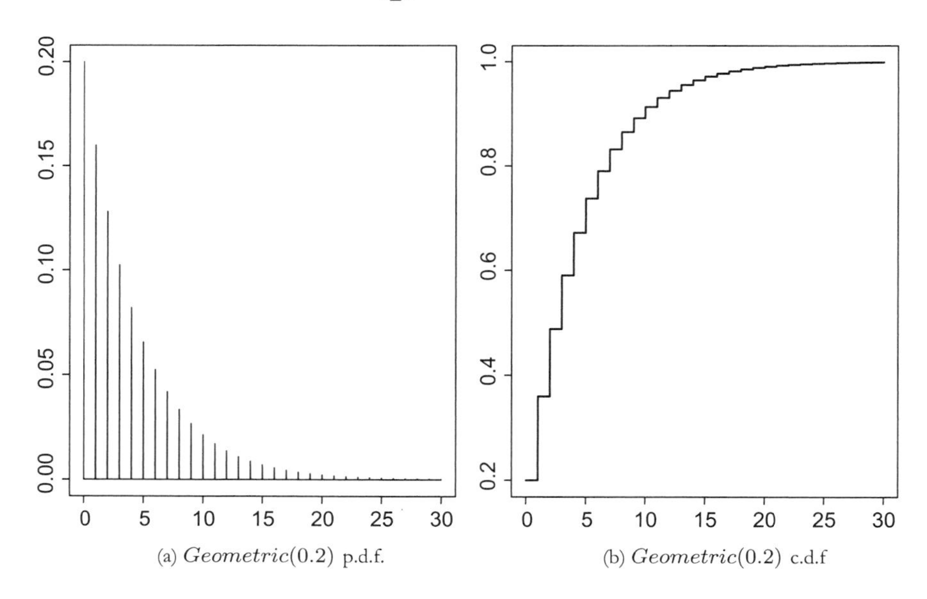

(a) *Geometric*(0.2) p.d.f.

(b) *Geometric*(0.2) c.d.f

FIGURE 4.4 The p.d.f. and c.d.f. of the geometric distribution.

we get

$$\sum_{k\geq 1}\mathbf{P}(X=k)=\sum_{k=1}^{\infty}p(1-p)^{k-1}=p\sum_{k=0}^{\infty}(1-p)^{k}$$
$$=p\frac{1}{1-(1-p)}=1.$$

■

To compute the expectation of a random variable $X \sim \mathit{Geometric}(p)$, we need the following result. It is a general result for discrete random variables with positive values.

Proposition 4.26 *Let X be a random variable with values in $\mathbb{N}$. Then X is integrable if and only if the series*

$$\sum_{n=0}^{\infty}\mathbf{P}(X>n)$$

is convergent, and in this case

$$\mathbf{E}X=\sum_{k=0}^{\infty}k\mathbf{P}(X=k)=\sum_{n=0}^{\infty}\mathbf{P}(X>n).$$

Proof: By definition we have

$$\begin{aligned}\mathbf{E}X&=\sum_{k=0}^{\infty}k\mathbf{P}(X=k)=\sum_{k=0}^{\infty}\left(\sum_{n=0}^{k-1}1\right)\mathbf{P}(X=k)\\&=\sum_{k=0}^{\infty}\sum_{n<k}\mathbf{P}(X=k)=\sum_{n=0}^{\infty}\sum_{k>n}\mathbf{P}(X=k)\\&=\sum_{n=0}^{\infty}\mathbf{P}(X>n),\end{aligned}$$

where we changed the order of summation which is valid since the sum exists (the random variable has finite expectation). ■

Let us recall some elements concerning power series. These are useful in order to compute the characteristics of the geometric distribution. If $x \in (0, 1)$, the power series

$$\sum_{n\geq 0}x^{n}$$

is convergent and its sum is $\frac{1}{1-x}$. It is possible to "differentiate under the sum," which means

$$\sum_{n\geq 1} nx^{n-1} = \frac{d}{dx}\left(\frac{1}{1-x}\right) = \frac{1}{(1-x)^2}. \tag{4.3}$$

By differentiating once more, we obtain

$$\sum_{n\geq 2} n(n-1)x^{n-2} = \frac{2}{(1-x)^3}. \tag{4.4}$$

It is also possible to change the order of summation and integration.

$$\int_0^a \frac{1}{1-x}\,dx = \int_0^a \sum_{n\geq 0} x^n dx = \sum_{n\geq 0}\int_0^a x^n dx = \sum_{n\geq 0}\frac{1}{n+1}a^{n+1}. \tag{4.5}$$

Consequently,

$$-\ln(1-a) = \sum_{n\geq 0}\frac{1}{n+1}a^{n+1}$$

for every $a \in (0, 1)$.

Proposition 4.27 *Let $X \sim Geometric(p)$. Then*

$$\mathbf{E}X = \frac{1}{p}, \qquad Var(X) = \frac{1-p}{p^2}.$$

Proof: We use Proposition 4.26. We first compute $\mathbf{P}(X > n)$.

$$\begin{aligned}\mathbf{P}(X > n) &= \sum_{k=n+1}^{\infty}\mathbf{P}(X = k) = \sum_{k=n+1}^{\infty} p(1-p)^{k-1}\\ &= p(1-p)^n \sum_{k=n+1}^{\infty} p(1-p)^{k-n-1} = (1-p)^n\end{aligned}$$

by (4.2). Now,

$$\mathbf{E}X = \sum_{nn\geq 0}\mathbf{P}(X > n) = \sum_{n\geq 0}(1-p)^n = \frac{1}{p}.$$

Alternative proof for the expectation:

$$\begin{aligned}\mathbf{E}X &= \sum_{k\geq 1} kp(1-p)^{k-1}\\ &= p\sum_{k\geq 1} k(1-p)^{k-1}\\ &= p\frac{1}{(1-(1-p))^2} = \frac{1}{p},\end{aligned}$$

by relation (4.3) above. Concerning the variance, we compute, as usual, the expectation of $X(X-1)$.

$$\begin{aligned}\mathbf{E}X(X-1) &= \sum_{k\geq 1} k(k-1)p(1-p)^{k-1}\\ &= \sum_{k\geq 2} k(k-1)p(1-p)^{k-1}\\ &= p(1-p)\sum_{k\geq 2} k(k-1)(1-p)^{k-2}\\ &= p(1-p)\frac{2}{p^3} = \frac{2(1-p)}{p^2},\end{aligned}$$

using the identity (4.4). Consequently,

$$\mathbf{E}X^2 = \frac{2(1-p)}{p^2} + \mathbf{E}X = \frac{2-p}{p^2}$$

which implies

$$Var(X) = \mathbf{E}X^2 - (\mathbf{E}X)^2 = \frac{1-p}{p^2}.$$

■

The cumulative distribution of the *Geometric*(p) can be computed as follows.

Proposition 4.28 *Let* $X \sim$ *Geometric*(p). *Then*

$$F_X(n) = \mathbf{P}(X \leq n) = 1-(1-p)^n$$

for every $n \geq 1$ *integer.*

Proof: Indeed, for every $n \geq 1$ integer we have

$$\begin{aligned}\mathbf{P}(X \leq n) &= \sum_{k=1}^{n} p(1-p)^{k-1} = p \sum_{k=0}^{n-1} (1-p)^k \\ &= p \frac{1-(1-p)^n}{1-(1-p)} = 1-(1-p)^n.\end{aligned}$$

■

EXAMPLE 4.5 Calculations using the Geometric distribution

A New Zealand Herald data report quoted obstetrician Dr. Freddie Graham as stating that the chances of a successful pregnancy resulting from implanting a frozen embryo are about 1 in 10. Suppose a couple who are desperate to have children will continue to try this procedure until the woman carries a successful pregnancy. We will assume that each individual attempt is independent of any other. The probability of "a successful pregnancy" at any attempt is $p = 0.1$. Let X be the number of times the couple tries the procedure up to and including the successful attempt. Then X has a geometric distribution.

(a) The probability of first becoming pregnant on the 4th try is

$$\mathbf{P}(X = 4) = 0.9^3 \times 0.1 = 0.0729.$$

(b) The probability of becoming pregnant before the 4th try is

$$\begin{aligned}\mathbf{P}(X \leq 3) &= \mathbf{P}(X = 1) + \mathbf{P}(X = 2) + \mathbf{P}(X = 3) \\ &= 0.1 + 0.9 \times 0.1 + 0.9^2 \times 0.1 = 0.271.\end{aligned}$$

(c) The probability of a successful attempt occurring at either the second, third, or fourth attempt is

$$\begin{aligned}\mathbf{P}(2 \leq X \leq 4) &= \mathbf{P}(X = 2) + \mathbf{P}(X = 3) + \mathbf{P}(X = 4) \\ &= 0.9 \times 0.1 + 0.9^2 \times 0.1 + 0.9^3 \times 0.1 = 0.2439.\end{aligned}$$

The important thing in the example above is to realize the utility of probability distributions and probability in general. The probability distribution is useful regardless of the nature of the area, be it gambling or biology. In the subsections which follow, we will meet several simple physical models which have widespread practical applications. Each physical model has an associated probability distribution.

4.4.5 NEGATIVE BINOMIAL (r, p) DISTRIBUTION

Definition 4.29 (Negative binomial distribution) *For $p \in (0, 1)$ define the law of X as follows:*

$$\mathbf{P}(X = k) = \begin{cases} \binom{k-1}{r-1}(1-p)^{r-k}p^r & \text{for any } k \in \{r, r+1, \dots\}, \\ 0 & \text{otherwise.} \end{cases} \tag{4.6}$$

We will denote a random variable with this distribution $X \sim$ Negative Binomial (r, p).

Similarly with the *Geometric*(p) distribution we can talk about "number of failures" distribution, and the formulation is similar.

Figure 4.5 presents the density and distribution for a *Geometric*(0.2) random variable.

Remark 4.30 *Let us stop for a moment and see how these distributions are related. Suppose we do a simple experiment, and we repeat this experiment many times. The experiment has only two possible outcomes: "success" with probability p and "failure" with probability $1 - p$.*

- *The variable X that takes value 1 if the experiment is a success and 0 if it is a failure has a Bernoulli(p) distribution.*

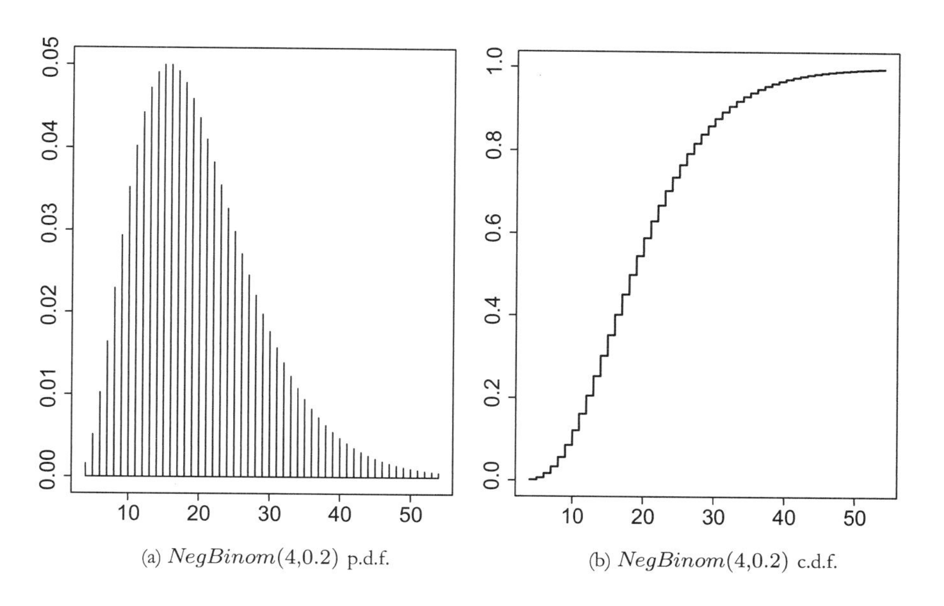

(a) $NegBinom(4,0.2)$ p.d.f. (b) $NegBinom(4,0.2)$ c.d.f.

FIGURE 4.5 The p.d.f. and c.d.f. of the NBinom distribution.

- *Repeat the experiment n times in such a way that no experiment influences the outcome of any other experiment*[1] *and we count how many of the n repetition actually resulted in successes. Let Y be the variable counting this number. Then $Y \sim Binom(n, p)$.*
- *Suppose that instead of repeating the experiment a fixed number of times n, we repeat the experiment as many times as are needed to see the first success. The number of trials needed is going to be distributed as a Geometric(p) random variable. If we count failures until the first success, we obtain the Geometric(p) "number of failures" distribution.*
- *If we repeat the experiment until we see r successes, the number of trials needed is a NegativeBinomial(r, p) random variable.*

Here are two examples for your practice.

Exercise 1. *Prove that equation (4.6) defines a probability distribution.*

Exercise 2. *Let X be a random variable with distribution given by (4.6). Prove that*

$$\mathbf{E}X = \frac{p(r-k)}{1-p} \quad \text{and} \quad VarX = \frac{p(r-k)}{(1-p)^2}.$$

4.4.6 HYPERGEOMETRIC DISTRIBUTION (N, m, n)

The Polya urn scheme named after George Pólya is any model where we represent objects of real interest (people, cars, atoms, etc.) by balls of different colors. The way the balls are drawn create many interesting models applicable to real problems. For example, a single ball (or multiple balls) may be drawn and then recolored and put back, multiple urns may be used for drawing the balls, and so on.

In the simplest such urn model we have an urn containing N balls split between two colors; call them black and white. There are a total of m black balls (where obviously $0 \le m \le N$) and of course $N - m$ white ones. We pick a random sample of a fixed size n and we count how many black balls (or equivalently white ones) we have chosen. It turns out that X has a distribution that is worth knowing.

Definition 4.31 (Hypergeometric distribution) *This distribution is defined by the probabilities*

$$\mathbf{P}(X = k) = \frac{\binom{m}{k}\binom{N-m}{n-k}}{\binom{N}{n}}, \; k \in \{0, 1, \dots, m\}. \tag{4.7}$$

We denote a random variable X with this distribution as $X \sim$ Hypergeometric(N, m, n).

[1] This is the idea of independence which we already discussed in Chapter 3.

Proposition 4.32 *The probabilities in* (4.7) *define a valid probability distribution.*

Proof: The proof comes from the Vandermonde identity:

$$\binom{m+n}{r} = \sum_{k=0}^{r} \binom{m}{k}\binom{n}{r-k}$$

for $r, n, m \in \mathbb{N}$. To prove this identity, one considers the identity

$$(a+b)^{m+n} = (a+b)^m (a+b)^n,$$

which uses the binomial expansion for all the binoms in the expression then identifies the coefficients of the power terms. ■

Figure 4.6 presents the density and distribution for a *Geometric*(0.2) random variable.

Remark 4.33 *Despite its simplicity, this particular urn model is appropriate for modeling any population of size N where we want to know the count m of individuals who have a characteristic of interest by looking at the random number X obtained in a random sample. For example, in a sample survey in San Francisco the black balls and white balls may correspond to people who are (black balls) or are not (white balls) HIV positive, people who do or do not smoke, or people who will or will not vote for a particular presidential candidate. Using the previous notation, N is the size of the*

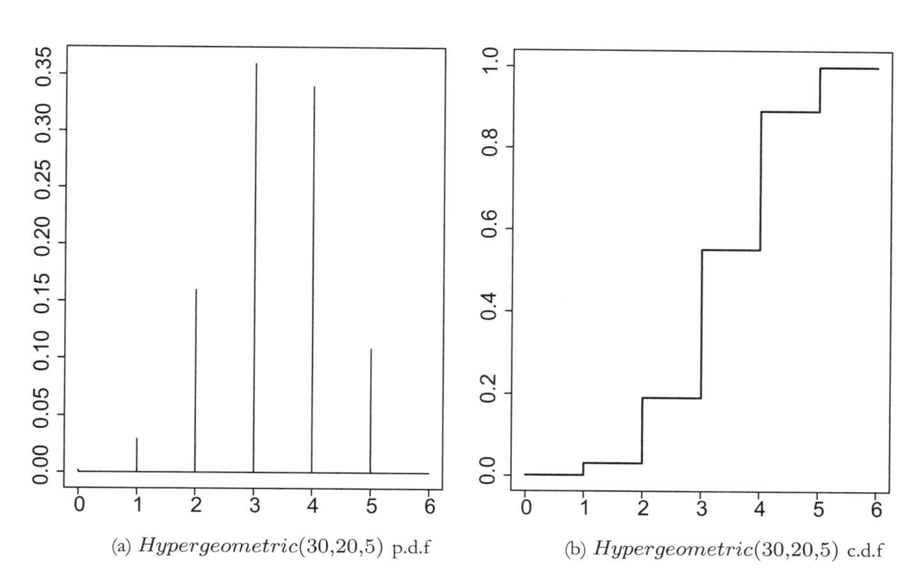

(a) *Hypergeometric*(30,20,5) p.d.f

(b) *Hypergeometric*(30,20,5) c.d.f

FIGURE 4.6 The p.d.f. and c.d.f. of the hypergeometric distribution for 30 total balls of which 20 are white and we pick a sample of 5.

population, m is the number of individuals in the population with the characteristic of interest, and X counts the number with that characteristic in the chosen sample of size n. In all such cases the distribution of X is the Hypergeometric(N, m, n).

Naturally, the reason for conducting such surveys is to estimate m, or equivalently the proportion of "black" balls $p = m/N$, from an observed value of X. However, before we can do this we need to be able to calculate probabilities associated with the hypergeometric distribution. This method is practical and relevant when the numbers N and m are relatively small.

Remark 4.34 *In general, the situation where the hypergeometric distribution is applicable (and useful) is either when the population N is small or when the sample size n is comparable with the population size (e.g., $n = N/10$).*

However, if the population is large enough and the sampling of individuals is random enough and the sample size n is small relative to the population, one may use a binomial distribution as an good approximate for the distribution. To understand why, consider the change in probability of drawing a black ball from an urn containing $1,000,000$ black and $1,000,000$ white balls. The first ball is black with probability 0.5. Depending on whether the first ball was black or white, the chance that the second is black is $999,999/1,999,999 = 0.49999975$ or $1,000,000/1,999,999 = 0.50000025$, both extremely close to 0.5. The binomial distribution is much easier to work with; and it may be further approximated using the Poisson distribution or the Gaussian distribution; both of these distributions are also much easier to work with.

Proposition 4.35 *Suppose X follows a hypergeometric distribution with parameters (N, m, n) as in the definition. Then*

$$\mathbf{E}X = n\frac{m}{N} \quad \text{and} \quad VarX = n\frac{m}{N}\frac{N-m}{N}\frac{N-n}{N-1}.$$

Proof: Left as an exercise (see exercise 4.14). ∎

4.4.7 POISSON DISTRIBUTION

Definition 4.36 (Poisson distribution) *Suppose that the random variable X takes values in $\mathbb{N}$ and*

$$\mathbf{P}(X = k) = \frac{\lambda^k}{k!}e^{-\lambda}, \qquad k = 0, 1, 2, \ldots \tag{4.8}$$

A random variable with values in $\mathbb{N}$ and the probability distribution given by (4.8) is called a Poisson distributed random variable with rate (or intensity) $\lambda > 0$. We will use the notation $X \sim Poisson(\lambda)$.

Proposition 4.37 *The sequence $p_k = \mathbf{P}(X = k)$, $k = 0, 1, 2, \ldots$ defines a discrete probability distribution.*

Proof: $p_k \geq 0$ for every $k \geq 0$ and

$$\sum_{k\geq 0} p_k = e^{-\lambda} \sum_{k\geq 0} \frac{\lambda^k}{k!}$$
$$= e^{-\lambda} e^{\lambda} = 1.$$

■

The Poisson distribution is a good model for many situations where we count the number of events occurring in a fixed time interval—for example, counting the number of cars passing by a particular point on a highway in an hour, counting the number of telephone calls arriving to a central switchboard in a minute, and so on.

In real applications, X is bounded; for instance, it is hard to conceive having 10 trillion cars passing by in an hour, given that the total earth population is a fraction of that. Although the Poisson distribution gives positive probabilities to all values of X going to infinity, it is still useful in practice as an approximation. Due to the exponential decay in the formula, the Poisson probabilities rapidly become extremely small.

For example, if $\lambda = 1$, then $\mathbf{P}(X = 50) = 1.1 \times 10^{-65}$ and $\mathbf{P}(X > 50) = 1.7 \times 10^{-16}$ (which are essentially zero numbers for all practical purposes).

Figure 4.7 presents the density and distribution for a *Geometric*(0.2) random variable.

Exercise 3. *Using the Poisson probability formula, verify the following: If* $\lambda = 1$, *then* $\mathbf{P}(X = 0) = 0.36788$ *and* $\mathbf{P}(X = 3) = 0.061313$.

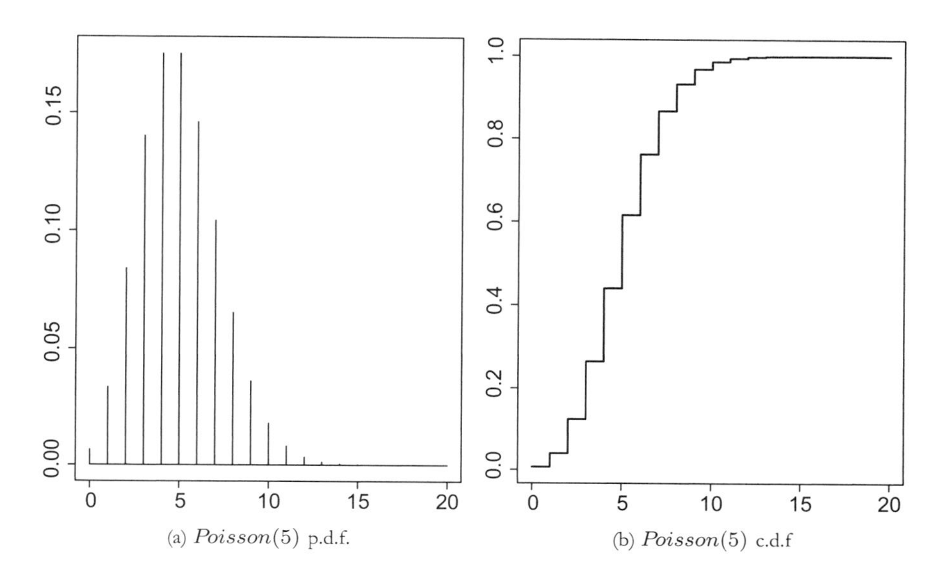

FIGURE 4.7 The p.d.f. and c.d.f. of the Poisson distribution with mean 5.

EXAMPLE 4.6 A practical application of the Poisson distribution

Consider a type of event occurring randomly through time—say, for example, earthquakes. Let X count the number of events occurring in a specific unit interval of time—for example, yearly number of earthquakes in California.

Under the following conditions, X may be shown mathematically to have a Poisson(λ) distribution.

1. The events occur at a constant average rate of λ per unit time.
2. Occurrences are independent of one another.
3. More than one occurrence cannot happen at the same time.

With earthquakes, condition 1 would not hold if there is an increasing or decreasing trend in the underlying levels of seismic activity. We would have to be able to distinguish "primary" quakes from the aftershocks they cause and only count primary shakes; otherwise condition 2 would not hold. Condition 3 is probably true for this example. However, if we were looking at car accidents, we could not count the number of damaged cars without condition 3 being violated since most accidents involve collisions between cars and several cars are often damaged at the same instant. Instead, we would have to count accidents as whole entities no matter how many vehicles or people were involved. It should be noted that except for the rate, the above three conditions required for a process to be Poisson do not depend on the unit of time. If λ is the rate per second, then 60λ is the rate per minute. The choice of time unit will depend on the questions asked about the process.

The Poisson distribution often provides a good description of many situations where points are randomly distributed in time or space—for example, the number of microorganisms in a given volume of liquid, the number of errors in collections of accounts, the number of errors per page of a book manuscript, the number of stars in a quadrant of space, cosmic rays at a Geiger counter, telephone calls in a given time interval at an exchange, mistakes in calculations, arrivals at a queue, faults over time in electronic equipment, weeds in a lawn, and so on and so forth. In biology, a common question is whether a spatial pattern of where plants grow, or the location of bacterial colonies, is in fact random with a constant rate (and therefore Poisson). If the data do not support a randomness hypothesis, then in which way is the pattern nonrandom? Do the organisms tend to cluster (attraction) or be further apart than one would expect from randomness (repulsion)?

Let us compute now the law and the variance of Poisson's law.

Proposition 4.38 *Suppose* $X \sim$ *Poisson*(λ) *with* $\lambda > 0$. *Then*

$$\mathbf{E}X = \lambda \text{ and } VarX = \lambda.$$

Proof: According to Definition 4.7, we have

$$\begin{aligned}
\mathbf{E}X &= \sum_{k=0}^{\infty} k e^{-\lambda} \frac{\lambda^k}{k!} = \sum_{k=1}^{\infty} k e^{-\lambda} \frac{\lambda^k}{k!} \\
&= \sum_{k=1}^{\infty} e^{-\lambda} \frac{\lambda^k}{(k-1)!} = \lambda e^{-\lambda} \sum_{k=1}^{\infty} \frac{\lambda^{k-1}}{(k-1)!} \\
&= \lambda e^{-\lambda} \sum_{k=0}^{\infty} \frac{\lambda^k}{k!} \\
&= \lambda.
\end{aligned}$$

To obtain the variance, we first compute $\mathbf{E}X(X-1)$. We have

$$\begin{aligned}
\mathbf{E}X(X-1) &= \sum_{k=2}^{\infty} k(k-1) e^{-\lambda} \frac{\lambda^k}{k!} \\
&= e^{-\lambda} \lambda^2 \sum_{k=2}^{\infty} \frac{\lambda^{k-2}}{(k-2)!} = e^{-\lambda} \lambda^2 \sum_{k=0}^{\infty} \frac{\lambda^k}{k!} \\
&= \lambda^2.
\end{aligned}$$

So

$$\begin{aligned}
VarX &= \mathbf{E}X(X-1) + \mathbf{E}X - (\mathbf{E}X)^2 \\
&= \lambda^2 + \lambda - \lambda^2 = \lambda.
\end{aligned}$$

■

The Poisson distribution satisfies the following "additivity" property.

Proposition 4.39 *Let $X \sim Poisson(\lambda_1)$ and $Y \sim Poisson(\lambda_2)$. Suppose that X and Y are independent. Then*

$$X + Y \sim Poisson(\lambda_1 + \lambda_2).$$

Proof: For every $k \geq 0$ integer we have

$$\begin{aligned}
\mathbf{P}(X + Y = k) &= \sum_{i=0}^{k} \mathbf{P}(X = i)\mathbf{P}(Y = k - i) \\
&= \sum_{i=0}^{k} e^{-\lambda_1} \frac{\lambda_1^i}{i!} e^{-\lambda_2} \frac{\lambda_2^{k-i}}{(k-i)!}
\end{aligned}$$

$$= e^{-(\lambda_1+\lambda_2)} \sum_{i=0}^{k} \frac{\lambda_1^i}{i!} \frac{\lambda_2^{k-i}}{(k-i)!}$$

$$= e^{-(\lambda_1+\lambda_2)} \sum_{i=0}^{k} \frac{\lambda_1^i}{i!} \frac{\lambda_2^{k-i}}{(k-i)!} \frac{k!}{k!}$$

$$= \frac{e^{-(\lambda_1+\lambda_2)}}{k!} \sum_{i=0}^{k} \binom{k}{i} \lambda_1^i \lambda_2^{k-i}$$

$$= \frac{e^{-(\lambda_1+\lambda_2)}}{k!} (\lambda_1 + \lambda_2)^k,$$

where we used the Newton formula (4.2). ■

An alternative proof will be given later with the help of the characteristic function of the Poisson distribution.

Remark 4.40 *Another property to be noticed for Poisson random variables with parameter $\lambda > 0$ is that for every $k > 2\lambda - 1$, one has*

$$\mathbf{P}(X > k) < \mathbf{P}(X = k).$$

Thus the probability of the entire tail of the distribution gets extremely small. For the proof we refer to exercise 4.13.

EXERCISES

Problems with Solution

4.1 For the experiment in exercise 3.1, compute the expectation and the variance of the random variable X

Solution: We have

$$\mathbf{E}X = \sum_{k=1}^{6} k \frac{2k-1}{36} = \frac{161}{36},$$

$$\mathbf{E}X^2 = \sum_{k=1}^{6} k^2 \frac{2k-1}{36} = \frac{791}{36},$$

and

$$Var(X) = \frac{2555}{1296}.$$

■

4.2 Let X be a discrete random variable. Prove that for every $a \in \mathbb{R}$,

$$\mathbf{E}(X-a)^2 = Var(X) + (\mathbf{E}X - a)^2.$$

Deduce from this result the minimum of the mapping

$$a \to \mathbf{E}((X-a)^2).$$

Solution: The first part is easy since

$$\mathbf{E}(X-a)^2 = \mathbf{E}X^2 - 2a\mathbf{E}X + a^2$$

and

$$\begin{aligned} Var(X) + (\mathbf{E}X - a)^2 &= Var(X) + (\mathbf{E}X)^2 - 2a\mathbf{E}X + a^2 \\ &= \mathbf{E}X^2 - 2a\mathbf{E}X + a^2 \end{aligned}$$

from the formula for the variance. Then

$$\begin{aligned} \inf_{a\in\mathbb{R}} \mathbf{E}(X-a)^2 &= Var(X) + \inf_{a\in\mathbb{R}} (\mathbf{E}X - a)^2 \\ &= VarX \end{aligned}$$

since the above infimum is obtained for $a = \mathbf{E}X$. Another way to interpret this result is to say that $\mathbf{E}X$ is the constant that approximates the random variable X best in the L^2 sense. ■

4.3 Let X, Y be uniform random variables on $\{0, 1, \ldots, \}$, that is,

$$\mathbf{P}(X=0) = \mathbf{P}(X=1) = \cdots = \mathbf{P}(X=n) = \frac{1}{n+1}$$

and the same for Y. Assume that X and Y are independent. Compute

$$\mathbf{P}(X=Y)$$

and

$$\mathbf{P}(X \leq Y).$$

Solution: First

$$\begin{aligned}
\mathbf{P}(X = Y) &= \sum_{k=0}^{n} \mathbf{P}(X = k, Y = k) \\
&= \sum_{k=0}^{n} \mathbf{P}(X = k)\mathbf{P}(Y = k) \\
&= \sum_{k=0}^{n} \frac{1}{n+1}\frac{1}{n+1} \\
&= \frac{n+1}{(n+1)^2} = \frac{1}{n+1}.
\end{aligned}$$

Next,

$$\begin{aligned}
\mathbf{P}(X \leq Y) &= \mathbf{P}(X = 0, Y \in \{0, \ldots, n\}) + \mathbf{P}(X = 1, Y \in \{1, \ldots, n\}) \\
&\quad + \cdots + \mathbf{P}(X = n, Y = n) \\
&= \sum_{k=0}^{n} \mathbf{P}(X = k) \sum_{j=k}^{n} \mathbf{P}(Y = j) \\
&= \sum_{k=0}^{n} \frac{1}{n+1} \sum_{j=k}^{n} \frac{1}{n+1} \\
&= \frac{1}{(n+1)^2} \sum_{k=0}^{n} \sum_{j=k}^{n} 1 \\
&= \frac{1}{(n+1)^2} \sum_{k=0}^{n} (n - k + 1) \\
&= \frac{1}{(n+1)^2} \left(\frac{n(n+1)}{2} + n + 1 \right).
\end{aligned}$$

■

4.4 Let $X \sim \textit{Geometric}(p)$ with $p \in (0, 1)$. Prove that for every $a, b \in \mathbb{N}$ we have

$$\mathbf{P}(X = a + b \mid X > a) = \mathbf{P}(X = b).$$

Solution: Since a, b are positive, we obtain

$$\mathbf{P}(X = a + b \mid X > a) = \frac{\mathbf{P}(X = a + b)}{\mathbf{P}(X > a)};$$

and using Proposition 4.28, we will have

$$\mathbf{P}(X = a + b) = p(1 - p)^{a+b-1} \quad \text{and} \quad \mathbf{P}(X > a) = (1 - p)^a$$

so

$$\mathbf{P}(X = a + b \mid X \geq a) = p(1-p)^b = \mathbf{P}(X = b).$$

■

Remark 4.41 *The above exercise shows that the geometric distribution is a distribution "without memory." It is like the variable restarts itself at a if a is the current time. The same property is satisfied by the exponential law (see Chapter 5).*

4.5 Let X be a *Geometric*(b) random variable, with $0 < b < 1$. Let $m \geq 1$ integer and define

$$Y = \max(X, m)$$

and

$$Z = \min(X, m).$$

(a) Find the probability distribution of Y.
(b) Prove that $Y + Z = X + m$.
(c) Find $\mathbf{E}(Y)$ and $\mathbf{E}(Z)$.

Solution: Note that for every $k \geq 1$ integer

$$(Y = k) = ((X \leq m, m = k) \cup (X \geq m, X = k)).$$

It suffices to compute $\mathbf{P}(Y = k)$ for every $k \geq 1$ integer. Suppose $k < m$. Then

$$\begin{aligned}\mathbf{P}(Y = k) &= \mathbf{P}(X \leq m, m = k) + \mathbf{P}(X \geq m, X = k)\\ &= \mathbf{P}(\emptyset) + \mathbf{P}(\emptyset) = 0\end{aligned}$$

and for $k > m$

$$\mathbf{P}(Y = k) = \mathbf{P}(X \geq m, X = k) = \mathbf{P}(X = k) = p(1-p)^{k-1}.$$

Finally, when $m = k$ we obtain

$$\mathbf{P}(Y = k) = \mathbf{P}(X = m) = p(1-p)^{m-1}.$$

■

4.6 Let $X_1, \ldots, X_n$ be independent random variables with the same distribution given by

$$\mathbf{P}(X_i = 0) = \mathbf{P}(X_i = 2) = \frac{1}{4}$$

and

$$\mathbf{P}(X_i = 1) = \frac{1}{2}.$$

Let

$$S_n = X_1 + \cdots + X_n.$$

(a) Find $\mathbf{E}(S_n)$ and $Var(S_n)$.
(b) Find $Var(S_n)$

Solution: Clearly for every $i = 1, \ldots, n$ we have

$$X_i(\Omega) = \{0, 1, 2\}$$

and

$$\begin{aligned}\mathbf{E}X_i &= 0\mathbf{P}(X_i = 0) + 1\mathbf{P}(X_i = 1) + 2\mathbf{P}(X_i = 2)\\ &= \frac{1}{2} + 2\frac{1}{4} = 1.\end{aligned}$$

Thus

$$\mathbf{E}S_n = \sum_{i=1}^{n} \mathbf{E}X_i = n.$$

Also, since

$$X_i^2(\Omega) = \{0, 1, 4\},$$

$$\begin{aligned}\mathbf{E}X_i^2 &= 0\mathbf{P}(X_i^2 = 0) + 1\mathbf{P}(X_i^2 = 1) + 4\mathbf{P}(X_i^2 = 4)\\ &= 0\mathbf{P}(X_i = 0) + 1\mathbf{P}(X_i = 1) + \mathbf{P}(X_i = 2)\\ &= \frac{1}{2} + 4\frac{1}{4} = \frac{3}{2}.\end{aligned}$$

Therefore,

$$Var(X_i) = \frac{3}{2} - 1 = \frac{1}{2}$$

and

$$Var(S_n) = \sum_{i=1}^{n} Var(X_i) = \frac{n}{2}$$

using the fact that the r.v. X_i are independent. ■

4.7 Let X be a random variable with distribution *Poisson*(λ). Define the r.v. Y by

$$Y = 0 \qquad \text{if } X \text{ is zero or even}$$

and

$$Y = \frac{X+1}{2} \qquad \text{if } X \text{ is odd.}$$

Find the law of Y.

Solution: We have

$$\begin{aligned}\mathbf{P}(Y=0) &= \mathbf{P}(Y \text{ is zero or even }) \\ &= \mathbf{P}(X=0) + \sum_{k=2;k \text{ is even}} \mathbf{P}(X=k) \\ &= e^{-\lambda} + \sum_{k=2;k \text{ even}} e^{-\lambda}\frac{\lambda^k}{k!} \\ &= e^{-\lambda} + \sum_{j=1}^{\infty} e^{-\lambda}\frac{\lambda^{2j}}{(2j)!}.\end{aligned}$$

and for every $k \geq 1$,

$$\mathbf{P}(Y=k) = \mathbf{P}(X=2k-1) = e^{-\lambda}\frac{\lambda^{2k-1}}{(2k-1)!}.$$

Clearly $Y(\Omega) = 0, 1, 2, \ldots$. ■

4.8 Let X be a random variable with a geometric distribution with parameter p. Calculate $\mathbf{E}\left(\frac{1}{1+X}\right)$.

Solution: By the definition of the expectation and using (4.5) we obtain

$$\begin{aligned}\mathbf{E}\left(\frac{1}{1+X}\right) &= \sum_{k\geq 1}\frac{1}{k+1}p(1-p)^{k-1} \\ &= \frac{p}{(1-p)^2}\sum_{k\geq 1}\frac{1}{k+1}(1-p)^{k+1} \\ &= \frac{p}{(1-p)^2}\left(\sum_{k\geq 0}\frac{1}{k+1}(1-p)^{k+1} - (1-p)\right) \\ &= \frac{p}{(1-p)^2}\left(-\ln(1-p) - (1-p)\right).\end{aligned}$$

■

4.9 Let X be a discrete r.v. with the following distribution:

$$\mathbf{P}(X=-3)=a,\ \mathbf{P}(X=-2)=\frac{5}{32},\ \mathbf{P}(X=-1)=b,\ \mathbf{P}(X=0)=\frac{5}{16}$$

and

$$\mathbf{P}(X=1)=c,\ \mathbf{P}(X=2)=\frac{1}{32}.$$

We know that

$$\mathbf{E}(X)=-\frac{1}{2} \quad \text{and} \quad Var(X)=\frac{5}{44}.$$

(a) Find a, b, c.
(b) Calculate

$$\begin{aligned}&\mathbf{E}(3+2X),\\ &\mathbf{E}(3+2X)^2,\\ &Var(3+2X).\end{aligned}$$

Solution: (a) From

$$\sum_{k=-3}^{2}\mathbf{P}(X=k)=1, \qquad \mathbf{E}X=-\frac{1}{2}, \qquad \mathbf{E}X^2=\frac{3}{2}$$

we get

$$\begin{aligned}a+b+c &= \tfrac{1}{2},\\ -3a-b+c &= -\tfrac{1}{4},\\ 9a=b+c &= \tfrac{3}{4},\end{aligned}$$

so

$$a=\frac{1}{32}, \qquad b=\frac{5}{16}, \qquad c=\frac{5}{32}.$$

(b)

$$\mathbf{E}(3+2X)=3+2\mathbf{E}X=2$$

and

$$\mathbf{E}(3+2X)^2=\mathbf{E}(9+12X+4X^2)=9.$$

Then

$$Var(X)=5.$$

■

4.10 Let X be a discrete random variable distributed as a Bernoulli with parameter $\frac{1}{3}$ and let

$$Y = 1 - X.$$

Give the law of Y. Calculate $\mathbf{E}Y$ and $Var(Y)$.

Solution: Note that $Y = \varphi(X)$ with $\varphi(u) = 1 - u$. Let us find the law of Y. Since $X(\Omega) = \{0, 1\}$ we see that

$$Y(\Omega) = \{\varphi(0), \varphi(1)\} = \{0, 1\}.$$

Therefore

$$\mathbf{P}(Y = 1) = \mathbf{P}(\varphi(X) = 1) = \mathbf{P}(X \in \varphi^{-}(\{1\})).$$

Since

$$\varphi^{-1}(\{1\}) = \{x \in \mathbb{R}, 1 - x = 1\} = \{0\}$$

we have

$$\mathbf{P}(Y = 1) = \mathbf{P}(X = 0) = \frac{2}{3}.$$

In the same way, we obtain

$$\mathbf{P}(Y = 0) = \mathbf{P}(1 - X = 0) = \mathbf{P}(X = 1) = \frac{1}{3}.$$

Consequently, Y follows a Bernoulli law with parameter $\frac{1}{3}$. Moreover,

$$\mathbf{E}Y = \frac{2}{9} \quad \text{and} \quad Var(Y) = \frac{2}{9}.$$

■

4.11 Let X be a discrete random variable with

$$\mathbf{P}(X = -1) = \frac{1}{8}, \quad \mathbf{P}(X = 0) = \frac{1}{4}, \quad \mathbf{P}(X = 1) = \frac{5}{8}.$$

Show that the r.v.s. $Y = X^2$ and $Z = |X|$ have the same distribution and find this distribution.

Solution: We have

$$X(\Omega) = \{-1, 0, 1\} \quad \text{and thus} \quad Y(\Omega) = \{0, 1\}$$

and

$$\mathbf{P}(Y = 1) = \mathbf{P}(X^2 = 1) = \mathbf{P}(X = -1) + \mathbf{P}(X = 1) = \frac{3}{4}$$

$$\mathbf{P}(Y = 0) = \mathbf{P}(X^2 = 0) = \mathbf{P}(X = 0) = \frac{1}{4}.$$

We deduce that $Y \sim Bernoulli(\frac{3}{4})$. Similarly, Z follows the same law. ■

4.12 Let X be a random variable with Poisson distribution with parameter $\lambda > 0$. Let $\varphi : \mathbb{R} \to \mathbb{R}$ be a bounded function. Show that

$$\mathbf{E}X\varphi(X) = \lambda \mathbf{E}\varphi(X + 1)$$

Solution: Since φ is bounded, the expectation of $\varphi(X)$ exists and

$$\begin{aligned}
\mathbf{E}X\varphi(X) &= \sum_{k\geq 0} k\varphi(k)\mathbf{P}(X = k) \\
&= \sum_{k\geq 1} k\varphi(k)\mathbf{P}(X = k) \\
&= e^{-\lambda} \sum_{k\geq 1} \frac{\lambda^k}{(k-1)!}\varphi(k) \\
&= \sum_{k\geq 0} \varphi(k + 1)\mathbf{P}(X = k).
\end{aligned}$$

■

4.13 Suppose that X follows a Poisson distribution with parameter $\lambda > 0$.

(a) Show that

$$\mathbf{P}(X \geq k) < \mathbf{P}(X = k)\frac{k+1}{k+1-\lambda} \qquad \text{for every } k > \lambda - 1.$$

(b) Deduce that for every $k > 2\lambda - 1$, one has

$$\mathbf{P}(X > k) < \mathbf{P}(X = k).$$

Solution: (a) Let $k > \lambda - 1$. We have

$$\begin{aligned}
\mathbf{P}(X \geq k) &= e^{-\lambda} \sum_{j=k}^{\infty} \frac{\lambda^j}{j!} \\
&= e^{-\lambda}\frac{\lambda^k}{k!}\left(1 + \frac{\lambda}{k+1} + \frac{\lambda^2}{(k+1)(k+2)} + \cdots\right) \\
&\leq e^{-\lambda}\frac{\lambda^k}{k!}\sum_{j=0}^{\infty}\frac{\lambda^j}{(k+1)^j}
\end{aligned}$$

Using the fact that for $k > \lambda - 1$ the series $\sum_{j=0}^{\infty} \frac{\lambda^j}{(k+1)^j}$ is convergent and

$$\sum_{j=0}^{\infty} \frac{\lambda^j}{(k+1)^j} = \frac{1}{1 - \frac{\lambda}{k+1}} = \frac{k+1}{k+1-\lambda}$$

we get the conclusion of part (a).

Concerning point (b), we have

$$\begin{aligned}\mathbf{P}(X > k) &= \mathbf{P}(X \geq k) - \mathbf{P}(X = k)\\ &\leq \mathbf{P}(X = k)\frac{k+1}{k+1-\lambda} - \mathbf{P}(X = k)\\ &= \mathbf{P}(X = k)\frac{\lambda}{k+1-\lambda}\end{aligned}$$

and we conclude the result since

$$\frac{\lambda}{k+1-\lambda} < 1 \qquad \text{for } k > 2\lambda - 1.$$

■

Problems without Solution

4.14 If X follows a hypergeometric distribution with parameters (N, m, n), show that

$$\mathbf{E}X = n\frac{m}{N} \quad \text{and} \quad VarX = n\frac{m}{N}\frac{N-m}{N}\frac{N-n}{N-1}.$$

4.15 Let Y be a random variable with law $Poisson(\lambda)$. Define the r.v. Z by

$$Z = \frac{Y}{2} \qquad \text{if Y is even}$$

and

$$Z = \frac{1-Y}{2} \qquad \text{if Y is odd.}$$

Find the law of Z.

4.16 Let $X \sim Geometric(p)$. Define

$$U = 4\left[\frac{X}{2}\right] - 2X + 1,$$

where $[x]$ is the integer part of x (the biggest integer less or equal than x). Find the law of U.

4.17 Please answer the following questions:

(a) What is the probability to obtain a "double" when we roll two fair dies (both faces show the same outcome)?

(b) We roll two dies repeatedly until we get the first double. We denote by X the random variables counting the number of rolls. What is the law of X?

(c) Let $n \geq 1$. Compute *directly* the probability to have no double in the first n rolls. Deduce $\mathbf{P}(X > n)$.

Hint: For question (a), the answer is $p = \frac{6}{36} = \frac{1}{6}$ because we have 6 favorable cases and 36 possible outcomes. Therefore, X follows a geometric law with $p = \frac{1}{6}$.

CHAPTER FIVE

Random Variables: The Continuous Case

5.1 Introduction/Purpose of the Chapter

In the previous chapter we discussed the properties of discrete random variables which map events to values in a countable set. In many cases, however, we need to consider variables which take values in an interval. Think about the following experiment: Choose a random point on a segment from the origin to some point A and let be X the abscissa of the chosen point. Then $X(\Omega) = [0, |A|_x]$, where $|A|_x$ is the x-coordinate of the point A and this set is not countable. A continuous random variable is not defined at specific values. Instead, it is defined over an interval of values. Informally, a random variable X is called continuous if its values x form a "continuum," with $\mathbf{P}(X = x) = 0$ for each x.

5.2 Vignette/Historical Notes

Historically the continuous random variables appeared as approximations of the discrete random variables. The 1756 edition of *The Doctrine of Chance* contained what is probably de Moivre's most significant contribution to probability, namely the approximation of the binomial distribution with the normal distribution in the case of a large number of trials—which is now known by most probability textbooks as "The First Central Limit Theorem." Pierre-Simon de Laplace

Handbook of Probability, First Edition. Ionuţ Florescu and Ciprian Tudor.

(1749–1827) published *Théorie Analytique des Probabilités* in 1812. In this book he introduces what is now known as the Laplace transform in applied mathematics, and we will know it as the moment-generating function in this book. This function provides a universal tool to work with distributions of variables—and is most useful for continuous variables.

More recently (1933), Kolmogorov published *Grundbegriffe der Wahrscheinlichkeitsrechnung* his most fundamental book. In it he builds up probability theory in a rigorous way from fundamental axioms in a way comparable with Euclid's treatment of geometry. This theory can deal with discrete or continuous or in fact any kind of random variables. However, in practical (numerical) applications we typically assume some density for random variables since this makes it simpler. The Markov processes and the whole advent in connecting PDE's and stochastic processed has started with continuous random variables.

5.3 Theory and Applications

A crucial notion in the theory of continuous random variables is the concept of probability density function.

5.3.1 PROBABILITY DENSITY FUNCTION (p.d.f.)

Definition 5.1 *A function $f : \mathbb{R} \to \mathbb{R}$ is called a probability density if it is integrable, positive and*

$$\int_{\mathbb{R}} f(x)\, dx = 1.$$

We mention that a function which is continuous with the possible exception of a countable number of points is integrable. This will be the case of all the examples treated in this book.

In the remainder of this section we give examples of density functions.

EXAMPLE 5.1

Let $f \to \mathbb{R}$ be a function defined by

$$f(x) = \left(x + \frac{1}{2}\right) 1_{[0,1]}(x).$$

Then f is a probability density (see Figure 5.1).

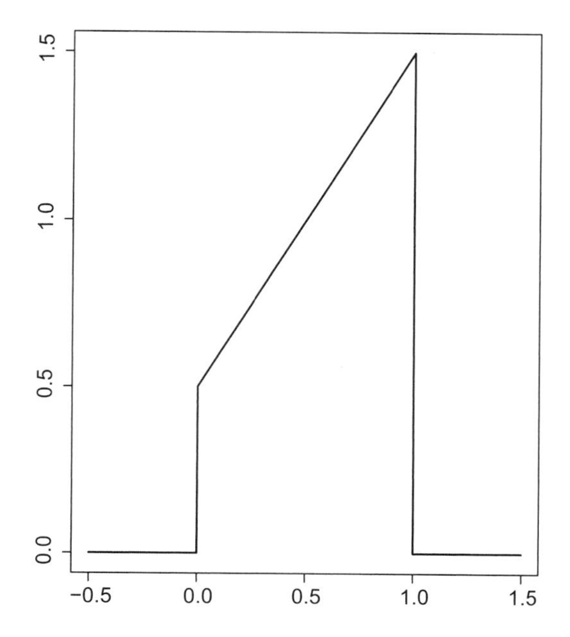

FIGURE 5.1 Density of the function in Example 5.1.

Solution: Note that $f(x) \geq 0$ for every $x \in \mathbb{R}$ and f is continuous (except the two points 0 and 1). Moreover,

$$\int_{\mathbb{R}} f(x)\, dx = \int_0^1 \left(x + \frac{1}{2}\right) = \left[\frac{1}{2}x^2 + \frac{1}{x}\right]_{x=0}^{x=1} = 1.$$

■

EXAMPLE 5.2

Let $c > 0$ be a constant and define

$$f(x) = cx1_{[0,1]}(x) + (2 - x)1_{[1,2]}(x).$$

Find c that makes f a probability density function (see Figure 5.2).

Solution: We have that f is clearly positive and continuous almost everywhere (it may not be continuous at 1 and 2). We compute

$$\int_{\mathbb{R}} f(x)\, dx = c\int_0^1 x\, dx + \int_1^2 (2 - x)\, dx = \frac{c}{2} + \frac{1}{2},$$

and this implies $c = 1$.

■

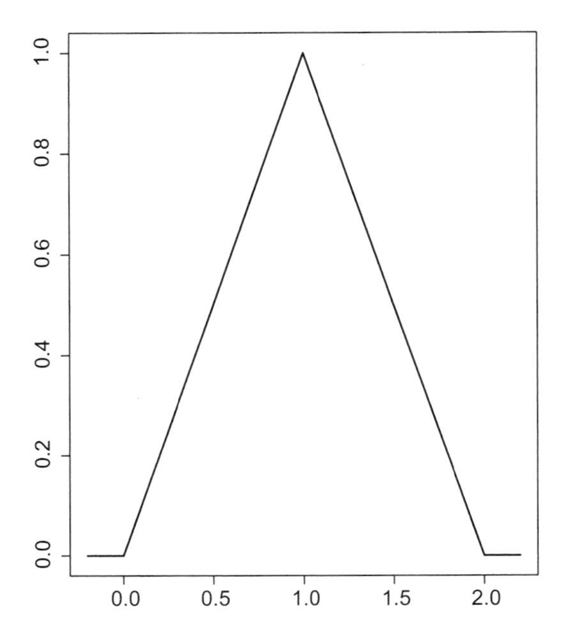

FIGURE 5.2 Density of the function in Example 5.2.

EXAMPLE 5.3

Define $f : \mathbb{R} \to \mathbb{R}$ by

$$f(x) = \frac{1}{x^2} 1_{[1,\infty)}(x).$$

Then f is a probability density (see Figure 5.3).

Solution: f is positive and

$$\int_{\mathbb{R}} f(x)\, dx = \int_1^\infty \frac{1}{x^2}\, dx = \left[-\frac{1}{x} \right]_{x=1}^{x=\infty} = 1.$$

■

So what is the connection between random variables and these functions?

Definition 5.2 (Continuous random variable) *Let $X : \Omega \to \mathbb{R}$ be a random variable on the probability space $(\Omega, \mathscr{F}, \mathbf{P})$. We say that the random variable*

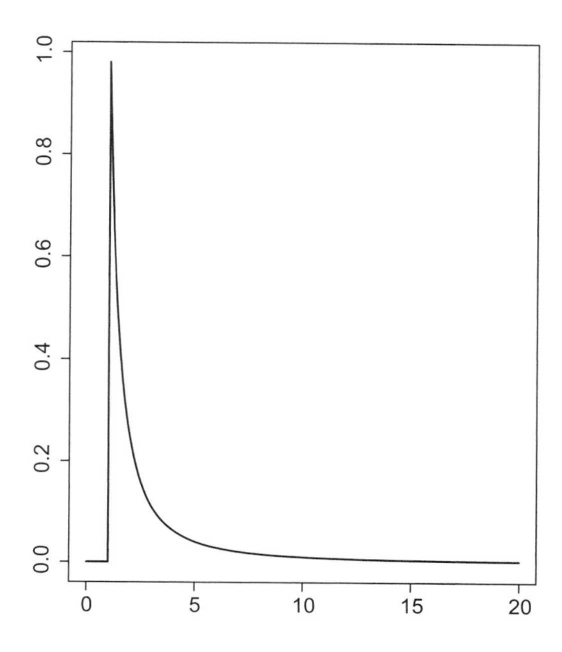

FIGURE 5.3 Density of the function in Example 5.3.

X is continuous (or X ***has a density***) *if there exists a probability density function f such that*

$$\mathbf{P}(a \leq X \leq b) = \int_a^b f(x)\,dx$$

for every $a, b \in \mathbb{R}$, $a \leq b$.

Remark 5.3 *The probability that the continuous random variable, say X, has any exact value a value a is 0. Indeed, formally*

$$\begin{aligned}\mathbf{P}(X = a) &= \lim_{\Delta x \to 0} \mathbf{P}(a \leq X \leq a + \Delta x)\\ &= \lim_{\Delta x \to 0} \int_a^{a+\Delta x} f_X(x)\,dx\\ &= 0.\end{aligned}$$

In general, for continuous random variables, we have

$$\mathbf{P}(X = a) \neq f_X(a).$$

Let us state a proposition which will be used to identify distributions of functions of random variables.

Proposition 5.4 *Let X and Y two random variables with respective probability densities f_X and f_Y. The two random variables have the same distribution (i.e., $f_X(z) = f_Y(z)$, $\forall z$) if and only if for any measurable bounded function F we have*

$$\mathbf{E}_X[F(X)] = \mathbf{E}_Y[F(Y)],$$

where each expectation is calculated with respect to the the subscripted random variable or written in integral terms:

$$\int F(z) f_X(z)\,dz = \int F(z) f_Y(z)\,dz.$$

Proof: Exercise. Use the standard construction in one direction and the delta function in the other direction. ■

5.3.2 CUMULATIVE DISTRIBUTION FUNCTION (c.d.f.)

The cumulative distribution function of a continuous random variable X is still defined by $F_X(x) = \mathbf{P}(X \le x)$. But in the case of continuous r.v., it can be expressed in terms of the density as follows.

Proposition 5.5 *Let X be a random variable with density f. Then*

$$F_X(x) = \mathbf{P}(X \le x) = \int_{-\infty}^{x} f(y)\,dy$$

for every $x \in \mathbb{R}$.

Proof: It is an immediate consequence of Definition 5.2. ■

You may read this as follows: The density (p.d.f.) f is the derivative of the distribution (c.d.f.) F_X (however, not everywhere). The exact details will be formalized in the next proposition.

EXAMPLE 5.4

Let X be a random variable with density function

$$f(x) = x 1_{[0,1]}(x) + (2 - x) 1_{[1,2]}(x).$$

Calculate $F_X(\frac{1}{2}) = \mathbf{P}(X \le \frac{1}{2})$.

Solution: We have seen in the Example given on page 121 that f is indeed a density. Then, by Proposition 5.5, we have

$$\mathbf{P}\left(X \le \frac{1}{2}\right) = \int_{\infty}^{\frac{1}{2}} f(x)\,dx = \int_{0}^{\frac{1}{2}} x\,dx = \frac{1}{2}.$$

■

Proposition 5.6 (Properties of the c.d.f. of the continuous random variables) *The probability distribution of a continuous random variable X has the following properties.*

i. *F_X is a continuous and increasing function, $F_X : \mathbb{R} \to [0, 1]$.*
ii. *$\lim_{x\to-\infty} F_X(x) = 0$ and $\lim_{x\to\infty} F_X(x) = 1$.*
iii. *For every $a \leq b$, $a, b \in \mathbb{R}$,*

$$F(b) - F(a) = \int_a^b f(u)\, du.$$

iv. *F is differentiable in any point a where f is continuous, and in this case*

$$F'(a) = f(a).$$

Proof: The first two properties (parts i and ii in the proposition above) are valid for any general random variable. We refer to the proof of Proposition 3.9. Part iii is an easy consequence of Proposition 5.5. Part iv is left as an exercise. ■

Remark 5.7 *The first point of the above proposition explains where the name is coming from: A continuous random variable is a random variable whose distribution is continuous (as opposed to right continuous in the general case). In fact the precise terminology is absolutely continuous with respect to the Lebesque measure and admits a Radon–Nykodim derivative (which is the density). We mention that several books on probability theory define the density as the derivative of the cumulative distribution function.*

Remark 5.8 *If f is a probability density function, note that*

$$\lim_{x\to-\infty} f(x) = \lim_{x\to\infty} f(x) = 0.$$

If this is not the case, then the p.d.f. cannot possibly integrate to a finite value.

EXAMPLE 5.5

Let X be a continuous random variable with density function

$$f(x) = 2x\mathbf{1}_{(0,1)}(x).$$

Calculate

$$\mathbf{P}\left(\frac{1}{4} \leq X \leq \frac{1}{2}\right).$$

Solution: By Proposition 5.6, we have

$$\begin{aligned}
\mathbf{P}\left(\frac{1}{4} \leq X \leq \frac{1}{2}\right) &= \int_{\frac{1}{2}}^{\frac{1}{4}} f(x)\,dx \\
&= \left[x^2\right]_{x=\frac{1}{4}}^{x=\frac{1}{2}} \\
&= \frac{3}{16}.
\end{aligned}$$

■

EXAMPLE 5.6

Let

$$f(x) = \frac{a}{x^5}$$

if $x \in (1, 2)$ and assume

$$f(x) = 0$$

otherwise. Here a denotes a real constant.

1. Find the parameter a in order to obtain a probability density function.
2. Find the probability

$$\mathbf{P}\left(1 \leq X \leq \frac{3}{2}\right).$$

Solution: From the definition of the density, we need

$$a \geq 0.$$

Since

$$\int_{\mathbb{R}} f(x)\,dx = \frac{15}{64}a$$

we obtain $a = \frac{64}{15}$. Using the value of a we just found, we obtain

$$\begin{aligned}
\mathbf{P}(1 \leq X \leq 3/2) &= \frac{64}{15}\int_1^{\frac{3}{2}} x^{-5}\,dx \\
&= \frac{64}{15}\left[-\frac{1}{4}x^{-4}\right]_{x=1}^{x=\frac{3}{2}} \\
&= \frac{208}{243}.
\end{aligned}$$

■

5.3.3 MOMENTS

Remember that for a discrete random variable we defined

$$\mathbf{E}X = \sum_{x \in X(\Omega)} x\mathbf{P}(X = x).$$

For a continuous random variable we substitute the sum by an integral and $\mathbf{P}(X = x)$ by the density $f(x)$. That is,

$$\mathbf{E}X = \int_{\mathbb{R}} xf(x)\, dx$$

whenever the above integral is well-defined. In fact, the definition of the moments of a continuous random variable is a particular case of the general notion developed in Appendix B. The general approach is relegated there since, while important, it requires general work with measures and Lebesque integration.

In order to give an unitary, more intuitive approach in the context of continuous random variables, we will adopt the following definition.

Definition 5.9 *Let X be a random variable on $(\Omega, \mathscr{F}, \mathbf{P})$ having a density f. Let $h : \mathbb{R} \to \mathbb{R}$ be a measurable function such that*

$$\int_{\mathbb{R}} |h(x)| f(x)\, dx < \infty.$$

Then

$$\mathbf{E}[h(X)] = \int_{\mathbb{R}} h(x)f(x)\, dx.$$

In particular, if $X \in L^1(\Omega)$ (see Appendix B for the definition), then X admits an expectation and

$$\mathbf{E}(X) = \int_{\mathbb{R}} x f(x)\, dx.$$

If $X \in L^p(\Omega)$, then the moment of order p of X exists and

$$\mathbf{E}(X^p) = \int_{\mathbb{R}} x^p f(x)\, dx.$$

For $X \in L^2(\Omega)$ the variance of X exists and it is given by

$$V(X) = \int_{\mathbb{R}} (x - \mathbf{E}(X))^2 f(x)\, dx = \mathbf{E}[X^2] - (\mathbf{E}X)^2.$$

Furthermore, as in the discrete case the standard deviation of X also exists and is the square root of the variance.

$$StdDev(X) = \sqrt{V(X)}.$$

EXAMPLE 5.7 Continuing the Example on page 120

Let X be a r.v. with density f as in Example 5.1. Let us compute the expectation of X.

$$\begin{aligned}\mathbf{E}X &= \int_0^{\frac{1}{2}} x\left(x+\frac{1}{2}\right)dx\\ &= \left[\frac{1}{3}x^3\right]_{x=0}^{x=\frac{1}{2}} + \left[\frac{1}{4}x^2\right]_{x=0}^{x=\frac{1}{2}}\\ &= \frac{5}{48}.\end{aligned}$$

We will compute the moments for several classical random variables later in this chapter.

5.3.4 DISTRIBUTION OF A FUNCTION OF THE RANDOM VARIABLE

The purpose of this paragraph is to discuss the following question: How do you find the density of a random variable Y constructed as a function $h(X)$, where X is a random variable with a given density function? An explicit and useful formula is given below.

Theorem 5.10 *Let $(\Omega, \mathcal{F}, \mathbf{P})$ be a probability space and let $X : \Omega \to \mathbb{R}$ be a random variable with density f. Let $h : D \subset \mathbb{R} \to \mathbb{R}$ (D is an open set) satisfying the following assumptions:*

1. *h is differentiable on its domain D.*
2. *h is injective; that is, $h(x) = h(y)$ implies $x = y$.*
3. *The support of f (or of X),*

$$Supp(f) = \{x; f(x) > 0\}$$

is contained in D.

Then the density of the random variable $Y = h(X)$ is given by

$$f_Y(y) = \left|\frac{1}{h'(h^{-1}(y))}\right| f(h^{-1})(y) 1_{\{y \in h(D)\}}$$

for all points y such that f is continuous at $h^{-1}(y)$.

Proof: Since h is injective and continuous (differentiable), then either h is increasing or is decreasing. Let us assume that it is increasing (the decreasing case is similar). We compute the distribution function of Y:

$$\begin{aligned}\mathbf{P}(Y \leq y) &= \mathbf{P}(h(X) \leq y) = \mathbf{P}(X \leq h^{-1}(y)) \\ &= \int_{-\infty}^{h^{-1}(y)} f(x)\, dx;\end{aligned}$$

and now using the change of variables $x = h^{-1}(z)$, we obtain

$$\begin{aligned}\mathbf{P}(Y \leq y) &= \int_{-\infty}^{y} f(h^{-1}(z) \left(h^{-1}\right)'(z)\, dz \\ &= \int_{-\infty}^{y} f(h^{-1}(z) \left|\left(h^{-1}\right)'(z)\right| dz\end{aligned}$$

because h is increasing. Now suppose that $h^{-1}(y)$ is a continuity point for f. Then F_Y is differentiable at y with derivative

$$\left|\frac{1}{h'(h^{-1}(y))}\right| f(h^{-1})(y),$$

and the formula is thus proven. ∎

In the particular case when h is an affine function ($h(x) = ax + b$), we have the following.

Corollary 5.11 *Let X be a r.v. with density f and let*

$$Y = aX + b$$

with $a \neq 0$. Then the density of Y is

$$f_Y(y) = \frac{1}{|a|} f\left(\frac{x-b}{a}\right).$$

Proof: We apply the result in the previous theorem to the function

$$h(x) = ax + b.$$

∎

Remark 5.12 *Let X be a random variable with density*

$$f(x) = \frac{1}{\sqrt{2\pi}} e^{-\frac{x^2}{2}}.$$

Using Corollary 5.11, show that the density of $Y = \sigma X + \mu$ (with $\sigma > 0$) is given by

$$f_Y(y) = \frac{1}{\sqrt{2\pi\sigma^2}} e^{-\frac{(x-\mu)^2}{2\sigma^2}}.$$

As we shall see in the next section, X follows a standard normal distribution $N(0, 1)$ and the resulting random variable $Y = \sigma X + \mu$ has the same normal law but with mean μ and variance σ^2 ($N(\mu, \sigma^2)$).

5.4 Examples

In this section we will present some of most used continuous probability distributions and discuss their basic properties.

5.4.1 UNIFORM DISTRIBUTION ON AN INTERVAL [*a*,*b*]

The uniform distribution models continuous random variables which take values in an interval without any particular preference within that interval. It is the equivalent of the discrete uniform distribution presented in Chapter 4. As an example, suppose that buses arrive at a given bus stop every 15 minutes. If you arrive at that bus stop at a particular random time, the time you have to wait for the next bus to arrive could be described by a uniform distribution over the interval from 0 to 15. It is different from a discrete uniform because the time can be anywhere in the interval. However, if we only consider the minutes—to the nearest integer—we obtain a discrete set of outcomes and the corresponding distribution is the discrete uniform.[1] We hope you can see the advantage of using the whole number and not just the rounded value.

In general, for an arbitrary interval $[a, b]$ the random variable with a uniform distribution represents the position of a point taken at random (without any preference) within the interval $[a, b]$.

Definition 5.13 *We say that a random variable X follows the uniform distribution on the interval $[a, b]$ if its probability density function is given by*

$$f(x) = \begin{cases} \frac{1}{b-a}, & \text{if } x \in [a, b], \\ 0, & \text{otherwise.} \end{cases} \tag{5.1}$$

We will use the notation

$$X \sim U[a, b].$$

[1] Technically, in the example we should round to the nearest middle of the minute, i.e., 0.5, 1.5, ..., 14.5 to obtain the discrete uniform—otherwise the ends 0 and 15 will have less probability than the rest.

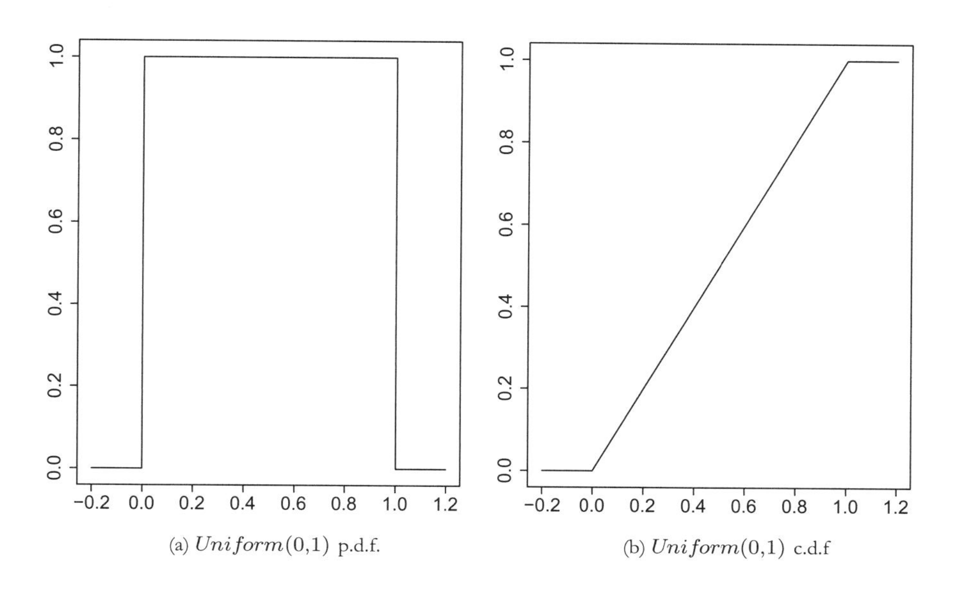

(a) $Uniform(0,1)$ p.d.f. (b) $Uniform(0,1)$ c.d.f

FIGURE 5.4 **The p.d.f. and c.d.f. of the uniform distribution on the interval [0, 1].**

Figure 5.4 presents the density and distribution for a *Uniform*(0, 1) random variable.

Proposition 5.14 *The function defined by (5.1) is a probability density function.*

Proof: It is immediate that f is positive and

$$\int_{\mathbb{R}} f(x)\,dx = \frac{1}{b-a}\int_a^b dx = 1.$$

■

Proposition 5.15 *Let X be a random variable with uniform distribution; that is, its density is given by (5.1). Then*

$$\mathbf{E}(X) = \frac{a+b}{2} \quad \textit{and} \quad V(X) = \frac{(b-a)^2}{12}.$$

Proof: We have

$$\mathbf{E}(X) = \int_{\mathbb{R}} x f(x)\,dx = \frac{1}{b-a}\int_a^b x\,dx = \frac{1}{b-a}\left[\frac{1}{2}x^2\right]_{x=a}^{x=b} = \frac{a+b}{2}$$

and

$$\mathbf{E}(X^2) = \int_{\mathbb{R}} x^2 f(x)dx = \frac{1}{b-a}\int_a^b x^2 dx$$
$$= \frac{1}{b-a}\left[\frac{1}{3}x^3\right]_{x=a}^{x=b} = \frac{b^3-a^3}{3(b-a)} = \frac{a^2+ab+b^2}{3}.$$

Thus

$$V(X) = \mathbf{E}(X^2) - (\mathbf{E}(X))^2 = \frac{a^2+ab+b^2}{3} - \frac{(a+b)^2}{4} = \frac{(b-a)^2}{12}.$$

■

Concerning the c.d.f. of the uniform law, we have the following proposition.

Proposition 5.16 *Let $X \sim U[a, b]$. Then the cumulative distribution function of X is*

$$F_X(t) = \begin{cases} 0 & \text{if } t < a, \\ \frac{t-a}{b-a} & \text{if } t \in [a, b], \\ 1 & \text{if } t > b. \end{cases}$$

Proof: Using the formula of the density of X, we obtain

$$F_X(t) = \mathbf{P}(X \le t)$$
$$= \frac{1}{b-a}\int_{-\infty}^{t} 1_{[a,b]}(x)\,dx = \frac{1}{b-a}\int_a^{t\wedge b} dx$$

and the conclusion follows. ■

EXAMPLE 5.8

Let $X \sim U[0, 1]$. Compute

$$\mathbf{P}\left(\frac{1}{4} \le X \le \frac{3}{4}\right).$$

Solution: This is an immediate application of the result above. Indeed,

$$\mathbf{P}\left(\frac{1}{4} \le X \le \frac{3}{4}\right) = F_X\left(\frac{3}{4}\right) - F_X\left(\frac{1}{4}\right)$$
$$= \frac{3}{4} - \frac{1}{4} = \frac{1}{2}.$$

■

5.4.2 EXPONENTIAL DISTRIBUTION

Definition 5.17 *Let $\lambda > 0$. Then the density of the exponential distribution is given by*

$$f(x) = \lambda e^{-\lambda x}, \qquad x \geq 0. \tag{5.2}$$

We will use the notation

$$X \sim Exp(\lambda)$$

to indicate that the r.v. has an exponential distribution with parameter $\lambda > 0$.

Proposition 5.18 *The function given by (5.2) is a probability density.*

Proof: Indeed, $f(x) > 0$ for every $x \in \mathbb{R}$ and

$$\int_{\mathbb{R}} f(x)\,dx = \lambda \int_0^{\infty} e^{-\lambda x}\,dx = \left[-e^{-\lambda x}\right]_{x=0}^{x=\infty} = 1.$$

■

The exponential distribution has been used extensively to model lifetimes of organisms. As we see in the next proposition, the average lifetime of an organism thus modeled is $1/\lambda$.

Figure 5.5 presents the density and distribution for an *Exponential*($\lambda = 0.2$) random variable.

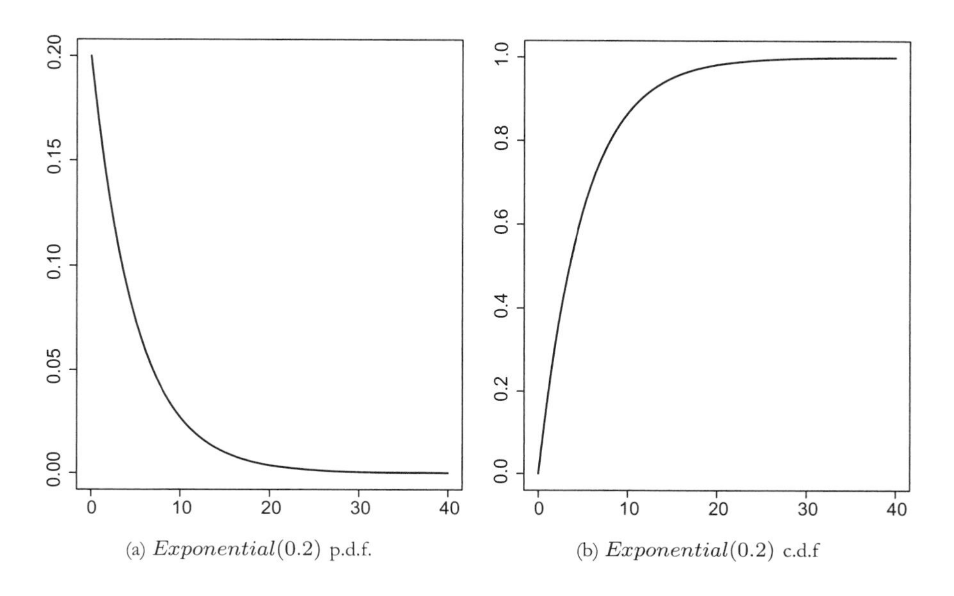

(a) $Exponential(0.2)$ p.d.f. (b) $Exponential(0.2)$ c.d.f

FIGURE 5.5 ***Exponential*** **distribution.**

Proposition 5.19 *Let X be a random variable with exponential distribution; that is, its density is given by (5.2). Then*

$$\mathbf{E}(X) = \frac{1}{\lambda} \quad \text{and} \quad V(X) = \frac{1}{\lambda^2}.$$

Proof: Let us compute first the expectation. We have

$$\mathbf{E}(X) = \int_0^\infty \lambda t \lambda e^{-\lambda t}\, dt.$$

We will integrate by parts with $u = t$, $v' = \lambda\lambda e^{-\lambda t}$, so $u' = 1$, $v = -e^{-\lambda t}$, and we get

$$\mathbf{E}(X) = \left[-te^{-\lambda t}\right]_{t=0}^{t=\infty} + \int_0^\infty e^{-\lambda t}\, dt = \frac{1}{\lambda}.$$

Let us compute now $\mathbf{E}(X^2)$. It is given by

$$\mathbf{E}(X^2) = \lambda \int_0^\infty x^2 e^{-\lambda x}\, dx$$

and by integrating by parts ($u = x^2$ and $v' = \lambda e^{-\lambda x}$ with $u' = 2x$, $v = -e^{-\lambda x}$), we obtain

$$\begin{aligned}\mathbf{E}(X^2) &= \left[-x^2 e^{-\lambda x}\right]_{x=0}^{x=\infty} + 2\int_0^\infty xe^{-\lambda x}\, dx \\ &= 2\int_0^\infty xe^{-\lambda x}\, dx = \frac{2}{\lambda^2}\end{aligned}$$

using the calculation done for $\mathbf{E}(X)$. Then

$$V(X) = \mathbf{E}(X^2) - (\mathbf{E}(X))^2 = \frac{2}{\lambda^2} - \frac{1}{\lambda^2} = \frac{1}{\lambda^2}.$$

■

In fact, as we will mention later, the exponential distribution is a particular case of the gamma distribution. The results in this paragraph can be deduced from the general derivations in Section 5.4.4 (which follows shortly).

Let us compute the c.d.f. of the exponential law.

Proposition 5.20 *Suppose $X \sim Exp(\lambda)$, $\lambda > 0$. Then*

$$F_X(x) = 0 \qquad \text{if } x < 0$$

and

$$F_X(x) = 1 - e^{-\lambda x}, \qquad \text{if } x \geq 0.$$

Proof: The first part is obvious because the support of the density of X is the set of positive real numbers. If $x \geq 0$, then

$$F_X(x) = \int_0^x \lambda e^{-\lambda t} dt = 1 - e^{-\lambda x}.$$

■

Remark 5.21 *The exponential distribution satisfies the following "memoryless property." If we call the random variable lifetime, this property may be interpreted in the following way: Given that the lifetime of the organism is greater than a number t, the probability that the organism will survive s more time units is the same as if the organism is being reborn at s.*

Mathematically, for any $s > 0$ and $t > 0$,

$$\mathbf{P}(X \geq s + t | X \geq s) = \mathbf{P}(X \geq t).$$

For a proof of this result, see exercise 5.11. There is also a converse result: If X is a random variable such that

$$\mathbf{P}(X > 0) = 1 \quad \text{and} \quad \mathbf{P}(X > t) > 0$$

for every $t > 0$ and

$$\mathbf{P}(X > t + s | X > t) = \mathbf{P}(X > s) \qquad \text{for every } s, t \geq 0,$$

then X has an exponential distribution. See exercise 5.12.

EXAMPLE 5.9

The lifetime of a radioactive element is a continuous random variable with the following p.d.f.:

$$f_X(t) = \frac{1}{100} e^{-\frac{t}{100}}$$

for $t > 0$ and zero otherwise. Recognizing the exponential distribution, we see immediately that the lifetime has an average of 100 years.

The probability that an atom of this element will decay within 50 years is

$$\begin{aligned}\mathbf{P}(0 \leq X \leq 50) &= \int_0^{50} \frac{1}{100} e^{-\frac{t}{100}} \\ &= 1 - e^{-0.5} \\ &= 0.39.\end{aligned}$$

5.4.3 NORMAL DISTRIBUTION (μ, σ^2)

The normal distribution is the most widely used distribution in practice. There are several reasons for this. The errors associated with any repeated measurements are normally distributed, so the distribution appears quite often in practice. Furthermore, the result that will be presented in the limit theorems section (the central limit theorem) shows that the distribution appears quite naturally as the limit of averages of random variables with arbitrary distribution. Finally, despite its complicated form, it has certain properties that allow analytical work with this distribution.

Definition 5.22 (Normal distribution) *A random variable normally distributed has the density*

$$f(x) = \frac{1}{\sqrt{2\pi\sigma^2}} e^{\frac{-(x-\mu)^2}{2\sigma^2}}, \qquad x \in \mathbb{R}, \tag{5.3}$$

where $\mu \in \mathbb{R}$ is called the mean parameter and $\sigma > 0$ is called the standard deviation parameter. We use the notation $X \sim N(\mu, \sigma^2)$ to denote this distribution. The distribution is also known as the Gaussian distribution.

Remark 5.23 *Please note that we shall always use the variance and not the standard deviation in the notation. For example, if we write $X \sim N(0, 2)$, we mean that the variance of X is equal to 2 and correspondingly the standard deviation is $\sqrt{2}$.*

Figure 5.6 presents the density and distribution for a *Normal*(0, 1) random variable.

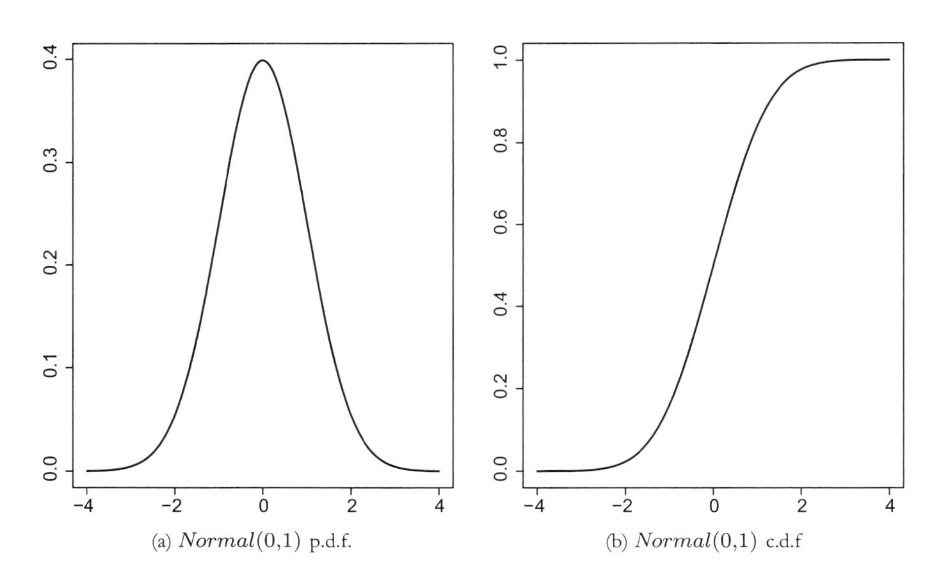

(a) *Normal*(0,1) p.d.f. (b) *Normal*(0,1) c.d.f

FIGURE 5.6 ***Normal*** **distribution.**

The normal distribution is commonly encountered in practice, and it is used throughout statistics, natural sciences, and social sciences as a simple model for complex phenomena. For example, the observational error in an experiment is usually assumed to follow a normal distribution, and the propagation of uncertainty is computed using this assumption. Note that a normally distributed variable has a symmetric distribution about its mean. Quantities that grow exponentially, such as prices, incomes, or populations, are often skewed to the right, and hence may be better described by other distributions, such as the log-normal distribution or Pareto distribution. In addition, the probability of seeing a normally distributed value that is far from the mean (i.e., extreme values) drops off extremely rapidly. As a result, statistical inference using a normal distribution is not robust to the presence of outliers (data that is unexpectedly far from the mean). When data contains outliers, a heavy-tailed distribution such as the Student's t distribution may be a better fit for the data.

Proposition 5.24 *The function given by (5.3) is a probability density function.*

Proof: Let us show that

$$\int_{\mathbb{R}} \frac{1}{\sqrt{2\pi\sigma^2}} e^{\frac{-(x-\mu)^2}{2\sigma^2}} dx = 1.$$

We first write

$$\begin{aligned}
\int_{\mathbb{R}} \frac{1}{\sqrt{2\pi\sigma^2}} e^{\frac{-(x-\mu)^2}{2\sigma^2}} dx &= \left[\left(\int_{\mathbb{R}} \frac{1}{\sqrt{2\pi\sigma^2}} e^{\frac{-(x-\mu)^2}{2\sigma^2}} dx \right)^2 \right]^{\frac{1}{2}} \\
&= \left[\frac{1}{2\pi\sigma^2} \int_{\mathbb{R}} e^{\frac{-(x-\mu)^2}{2\sigma^2}} dx \int_{\mathbb{R}} e^{\frac{-(y-\mu)^2}{2\sigma^2}} dy \right]^{\frac{1}{2}} \\
&= \frac{1}{\sqrt{2\pi\sigma^2}} \left[\int_{\mathbb{R}} \int_{\mathbb{R}} e^{\frac{-x^2}{2\sigma^2}} e^{\frac{-y^2}{2\sigma^2}} dxdy \right]^{\frac{1}{2}},
\end{aligned}$$

where we used the change of variables $x' = x - \mu$ and $y' = y - u$. Now, using polar coordinates $x = r\cos\theta$ and $y = r\sin\theta$, we obtain

$$\begin{aligned}
\int_{\mathbb{R}} \frac{1}{\sqrt{2\pi\sigma^2}} e^{\frac{-(x-\mu)^2}{2\sigma^2}} dx &= \frac{1}{\sqrt{2\pi\sigma^2}} \left[\int_0^{\infty} \int_0^{2\pi} re^{-\frac{r^2}{2\sigma^2}} d\theta dr \right]^{\frac{1}{2}} \\
&= \frac{1}{\sigma} \left[\int_0^{\infty} re^{-\frac{r^2}{2\sigma^2}} dr \right]^{\frac{1}{2}} \\
&= \left(\left[-e^{-\frac{r^2}{2\sigma^2}} \right]_{r=0}^{r=\infty} \right)^{\frac{1}{2}} = 1.
\end{aligned}$$

■

Proposition 5.25 *Let X be a random variable with density given by (5.3). Then*

$$\mathbf{E}(X) = \mu \quad \text{and} \quad V(X) = \sigma^2.$$

Proof: Using the change of variables $x - \mu = y$,

$$\begin{aligned}\mathbf{E}(X) &= \frac{1}{\sqrt{2\pi\sigma^2}} \int_{\mathbb{R}} x e^{\frac{-(x-\mu)^2}{2\sigma^2}} dx = \frac{1}{\sqrt{2\pi\sigma^2}} \int_{-\infty}^{\infty} (y+\mu) e^{-\frac{y^2}{2\sigma^2}} dy \\ &= \frac{1}{\sqrt{2\pi\sigma^2}} \int_{-\infty}^{\infty} y e^{-\frac{y^2}{2\sigma^2}} dy + \mu \frac{1}{\sqrt{2\pi\sigma^2}} \int_{-\infty}^{\infty} e^{-\frac{y^2}{2\sigma^2}} dy.\end{aligned}$$

The first integral is zero because we integrate over $\mathbb{R}$ an odd function $ye^{-\frac{y^2}{2\sigma^2}}$. Thus

$$\mathbf{E}(X) = \mu \frac{1}{\sqrt{2\pi\sigma^2}} \int_{-\infty}^{\infty} e^{-\frac{y^2}{2\sigma^2}} dy = \mu$$

by using Proposition 5.24. Now,

$$\begin{aligned}V(X) &= \frac{1}{\sqrt{2\pi\sigma^2}} \int_{\mathbb{R}} (x - \mathbf{E}(X))^2 e^{\frac{-(x-\mu)^2}{2\sigma^2}} dx \\ &= \frac{1}{\sqrt{2\pi\sigma^2}} \int_{\mathbb{R}} (x-\mu)^2 e^{\frac{-(x-\mu)^2}{2\sigma^2}} dx \\ &= \frac{1}{\sqrt{2\pi\sigma^2}} \int_{\mathbb{R}} y^2 e^{-\frac{y^2}{2\sigma^2}} dy,\end{aligned}$$

again by the change of variables $x - \mu = y$. We will integrate by parts with $u = y$ and $v' = ye^{-\frac{y^2}{2\sigma^2}} dy$ (thus $u' = 1$ and $v = -\sigma^2 e^{-\frac{y^2}{2\sigma^2}}$) and we obtain

$$V(X) = \frac{1}{\sqrt{2\pi\sigma^2}} \left[-\sigma^2 y e^{-\frac{y^2}{2\sigma^2}} \right]_{y=-\infty}^{y=\infty} + \sigma^2 \frac{1}{\sqrt{2\pi\sigma^2}} \int_{-\infty}^{\infty} e^{-\frac{y^2}{2\sigma^2}} dy = \sigma^2.$$

■

Remark 5.26 *If $X \sim N(0,1)$, then $-X \sim N(0,1)$. Indeed, for any odd, bounded measurable function $F : \mathbb{R} \to \mathbb{R}$, the following holds:*

$$\begin{aligned}\mathbf{E}F(-X) &= \frac{1}{\sqrt{2\pi}} \int_{\mathbb{R}} F(-x) e^{-\frac{x^2}{2}} dx \\ &= \frac{1}{\sqrt{2\pi}} \int_{\mathbb{R}} F(x) e^{-\frac{x^2}{2}} dx\end{aligned}$$

with the change of variables $-x = y$.

Recall that a function $F : \mathbb{R} \to \mathbb{R}$ is odd if and only if $F(-x) = -F(x)$ for all $x \in \mathbb{R}$.

Remark 5.27 *Remark 5.12 says that if $X \sim N(0,1)$, then $\sigma X + \mu \sim N(\mu, \sigma^2)$. Conversely, if $Y \sim N(\mu, \sigma^2)$, then*

$$\frac{Y-\mu}{\sigma} \sim N(0,1).$$

This is the so-called standardization of a normal random variables. That is, any Gaussian r.v. with arbitrary parameters can be transformed into a standard normal random variable.

We will prove later (see Proposition 7.29) that the sum of independent normal random variables is still a normal random variables with parameters equal to the sum of parameters. That is, let $X \sim N(\mu_1, \sigma_1^2)$ and $Y \sim N(\mu_2, \sigma_2^2)$. Assume X and Y are independent. Then

$$X + Y \sim N(\mu_1 + \mu_2, \sigma_1^2 + \sigma_2^2).$$

Remark 5.28 *The above property extends by induction to a finite sum of independent Gaussian random variables. Specifically, $X_i \sim N(\mu, \sigma_i^2)$ for $i = 1, \ldots, n$:*

$$\sum_{i=1}^{n} a_i X_i \sim N\left(\sum_{i=1}^{n} a_i X_i, \sum_{i=1}^{n} a_i^2 \sigma_i^2\right).$$

In Chapter 8 a different proof, sensibly easier, of the above property will be given by using the characteristic function of the normal law (see exercise 8.2).

5.4.4 GAMMA DISTRIBUTION

We start by introducing the gamma integral. For every $x > 0$ the gamma integral (or the gamma function) is defined by

$$\Gamma(x) = \int_0^\infty t^{x-1} e^{-t} dt. \tag{5.4}$$

Remark 5.29 *It is not trivial to see why the function inside the integral above is integrable for $x > 1$. We include this derivation in the paragraph containing the exercises.*

Proposition 5.30 *The gamma function satisfies the following properties:*

(a) *For every $x > 0$ we have $\Gamma(x+1) = x\Gamma(x)$.*

(b) *For every $n \in \mathbb{N}^\star$ we have $\Gamma(n+1) = n!$.*

(c) $\Gamma\left(\frac{1}{2}\right) = \sqrt{\pi}$.

Proof: (a) We have

$$\Gamma(x+1) = \int_0^\infty t^x e^{-t}\, dt$$

Integrating by parts with $u = t^x$ and $v' = e^{-t}$, which gives $u' = xt^{x-1}$, $v = -e^{-t}$, we obtain

$$\begin{aligned}\Gamma(x+1) &= \left[-e^{-t}t^x\right]_{t=0}^{t=\infty} + \int_0^\infty xt^{x-1}e^{-t}\, dt \\ &= x\Gamma(x).\end{aligned}$$

(b) We will use induction to prove this part. First let us verify the relation for $n = 1$. We need to show that $\Gamma(2) = 1$. Repeating the integration by parts from above (with $x = 1$) gives

$$\Gamma(2) = \int_0^\infty te^{-t}\, dt = \left[-te^{-t}\right]_{t=0}^{t=\infty} + \int_0^\infty e^{-t}\, dt = 1.$$

Thus $n = 1$ is verified. Assume as the induction hypothesis $\Gamma(n+1) = n!$. Then from part (a) we have

$$\Gamma(n+2) = (n+1)\Gamma(n+1) = (n+1)n! = (n+1)!.$$

(c) This part follows from Proposition 5.24 with a change of variables $\sqrt{t} = y$. ■

Definition 5.31 *We say that a random variable X has gamma distribution with parameters $a > 0$ and $\lambda > 0$ if its density is given by*

$$f_{a,\lambda}(x) = \frac{\lambda^a}{\Gamma(a)} e^{-\lambda x} x^{a-1} 1_{(0,\infty)}(x) \tag{5.5}$$

with $a, \lambda > 0$. We will denote $X \sim \Gamma(a, \lambda)$, meaning that the random variables X follows a gamma distribution with density (5.5).

Figure 5.7 presents the density and distribution for a $Gamma(10, 0.2)$ random variable.

Proposition 5.32 *The function (5.5) is a probability density.*

Proof: Clearly, $f_{a,\lambda}(x) \geq 0$ for every $x \in \mathbb{R}$ and

$$\int_{\mathbb{R}} f_{a,\lambda}(x)\, dx = \frac{\lambda^a}{\Gamma(a)} \int_0^\infty e^{-\lambda x} x^{a-1} dx = \frac{1}{\Gamma(a)} \int_0^\infty e^{-y} y^{a-1} dy = 1,$$

where we used the change of variables $\lambda x = y$. ■

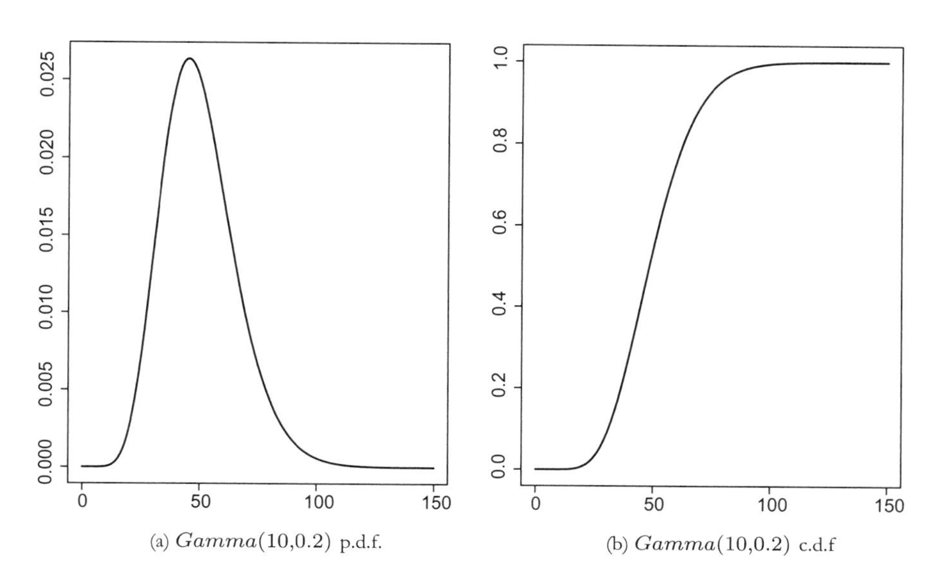

(a) $Gamma(10,0.2)$ p.d.f. (b) $Gamma(10,0.2)$ c.d.f

FIGURE 5.7 *Gamma* distribution.

Proposition 5.33 *Let $X \sim \Gamma(a, \lambda)$. Then*

$$\mathbf{E}X = \frac{a}{\lambda} \quad \textit{and} \quad V(X) = \frac{a}{\lambda^2}.$$

Proof: We have

$$\begin{aligned} \mathbf{E}X &= \frac{\lambda^a}{\Gamma(a)} \int_0^\infty x^a e^{-\lambda x} dx = \frac{\lambda^a}{\Gamma(a)} \lambda^{-a-1} \int_0^\infty y^a e^{-y} dy \\ &= \frac{1}{\lambda \Gamma(a)} \Gamma(a+1) = \frac{a}{\lambda} \end{aligned}$$

using Proposition 5.30, point (a). Next,

$$\mathbf{E}X^2 = \frac{\lambda^a}{\Gamma(a)} \int_0^\infty x^{a+1} e^{-\lambda x} dx = \frac{1}{\lambda^2 \Gamma(a)} \Gamma(a+2) = \frac{(a+1)a}{\lambda^2}$$

again by Proposition 5.30, point a. Finally,

$$V(X) = \frac{(a+1)a}{\lambda^2} - \frac{a^2}{\lambda^2} = \frac{a}{\lambda^2}.$$

■

Remark 5.34 *If $a = 1$, then the law $\Gamma(1, \lambda)$ is the exponential law with parameter λ (with expectation $\frac{1}{\lambda}$).*

An important particular case of the gamma distribution is when the parameters are $a = \lambda = \frac{1}{2}$. In this situation, we have the law of the square of a standard normal random variable.

Proposition 5.35 *Suppose that X is a standard normal random variable, that is, $X \sim N(0,1)$. Then*

$$X^2 \sim \Gamma\left(\frac{1}{2}, \frac{1}{2}\right).$$

Proof: Let F_{X^2} be the distribution function of the r.v. X^2. Then for any $t \leq 0$ we have

$$F_{X^2}(t) = \mathbf{P}(X^2 \leq t) = 0,$$

while for $t > 0$ we obtain

$$F_{X^2}(t) = \mathbf{P}(-\sqrt{t} \leq X \leq \sqrt{t}) = 2\mathbf{P}(X < \sqrt{t}) = 2F_X(\sqrt{t}).$$

By Proposition 5.6, the density function f_{X^2} of the r.v. X^2 is given by

$$f_{X^2}(t) = 0 \qquad \text{if } t \leq 0$$

and for $t > 0$ we have

$$\begin{aligned} f_{X^2}(t) &= (F_{X^2}(t))' = (2F_X(\sqrt{t}))' = 2f_X(\sqrt{t})\frac{1}{2\sqrt{t}} \\ &= \frac{1}{\sqrt{2\pi}} e^{-\frac{t}{2}} t^{-\frac{1}{2}}, \end{aligned}$$

which implies

$$f_{X^2}(t) = f_{\frac{1}{2},\frac{1}{2}}(t)$$

for every $t \in \mathbb{R}$. ∎

The law $\Gamma\left(\frac{1}{2}, \frac{1}{2}\right)$ is also called *chi-square distribution* with 1 degree of freedom. In general the sum of squares of n Gaussian random variables with mean 0 and variance 1 is a chi-square distribution with $n - 1$ degree of freedom. Let us formalize this distribution in a definition (which is also a $\Gamma\left(\frac{n}{2}, \frac{1}{2}\right)$).

Definition 5.36 *We say that a random variables X follows a chi-squared distribution with n degrees of freedom if its density function is:*

$$\frac{1}{2^{n/2}\Gamma(\frac{n}{2})} x^{\frac{n}{2}-1} e^{-\frac{x}{2}}.$$

We will use the notation $X \sim \chi^2_n$.

Figure 5.8 presents the density and distribution for a χ^2_{10} random variable.

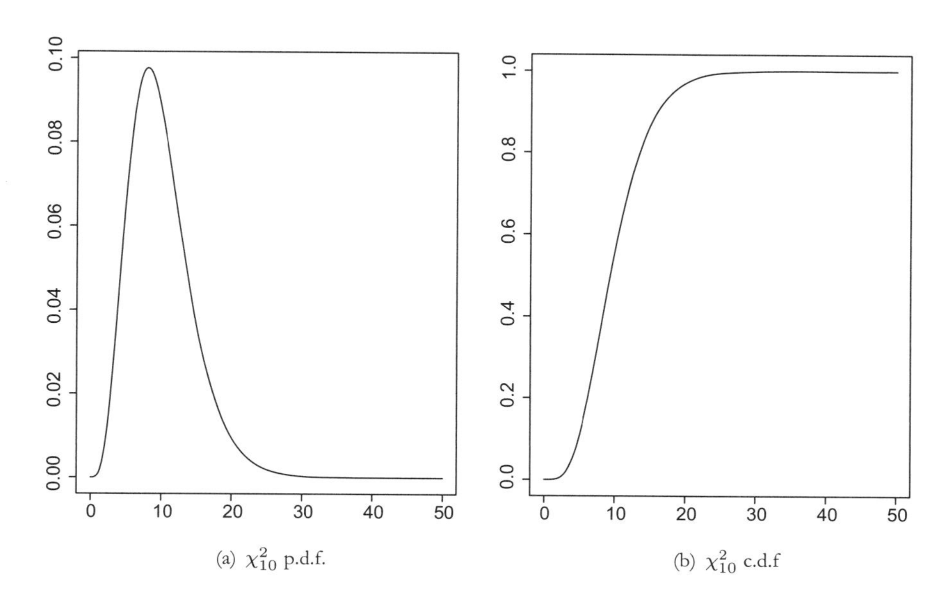

FIGURE 5.8 χ^2_n distribution.

Remark 5.37 *Note that if $X \sim \Gamma(a, \lambda)$ and $Y \sim \Gamma(b, \lambda)$ are independent, then*

$$X + Y \sim \Gamma(a + b, \lambda).$$

We will give a proof of this result in Chapter 8.

If $X \sim \Gamma(a, \lambda)$, then the random variable

$$Y = \frac{1}{X}$$

follows an inverse gamma distribution. This is a probability law appearing in several practical applications. The density of the inverse gamma distribution with parameters $a, \lambda > 0$ is defined by

$$f(x) = \frac{\lambda^a}{\Gamma(a)} \left(\frac{1}{x}\right)^{a+1} e^{-\lambda/x}. \tag{5.6}$$

Figure 5.9 presents the density and distribution for an $InverseGamma(10, 0.2)$ random variable.

The relation between the laws gamma and inverse gamma is proven in the following result.

Proposition 5.38 *Let $X \sim \Gamma(a, \lambda)$. Then $Y = \frac{1}{X}$ is a random variable with inverse gamma distribution with parameters a and $\frac{1}{\lambda}$.*

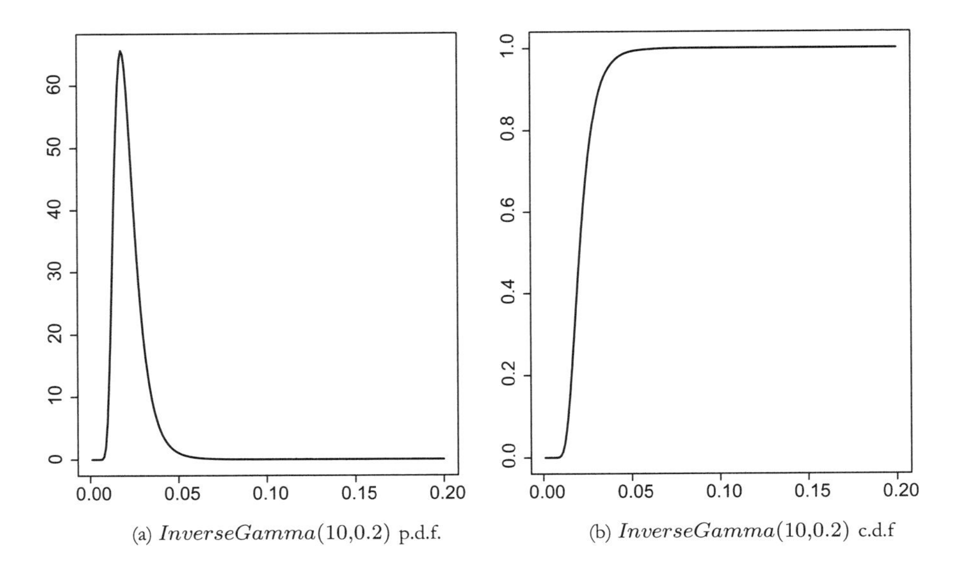

(a) $InverseGamma(10,0.2)$ p.d.f. (b) $InverseGamma(10,0.2)$ c.d.f

FIGURE 5.9 Inverse gamma distribution.

Proof: For any bounded measurable function F we have

$$\begin{aligned}\mathbf{E}F(Y) &= \mathbf{E}\left[F\left(\frac{1}{X}\right)\right] \\ &= \frac{\lambda^a}{\Gamma(a)}\int_0^\infty F\left(\frac{1}{x}\right)e^{-\lambda x}x^{a-1}dx,\end{aligned}$$

and by making a change of variables—namely, $y = \frac{1}{x}$ with $dx = -\frac{1}{y^2}\,dy$—we obtain

$$\begin{aligned}\mathbf{E}F(Y) &= \frac{\lambda^a}{\Gamma(a)}\int_0^\infty F(y)e^{-\frac{\lambda}{y}}y^{1-a}\frac{1}{y^2}dy, \\ &= \frac{\lambda^a}{\Gamma(a)}\int_0^\infty F(y)e^{-\frac{\lambda}{y}}y^{-1-a}dy.\end{aligned}$$

Since this is true for any bounded measurable function using the Proposition 5.4, we obtain the density of Y as given by (5.6). ■

5.4.5 BETA DISTRIBUTION

We introduce first the beta integral or the beta function. For every $a, b > 0$ we define

$$\beta(a, b) = \int_0^1 t^{a-1}(1-x)^{b-1}dx.$$

Proposition 5.39 *Let $a, b > 0$. Then*

$$(a) \quad \beta(a, b) = \beta(b, a).$$

$$(b) \quad \beta(a, b) = \frac{\Gamma(a)\Gamma(b)}{\Gamma(a+b)}.$$

Proof: Point (a) is immediate from the definition of the beta integral and the change of variables $1 - x = y$. For part (b), just substitute the definitions. ■

Definition 5.40 *A random variable X follows a beta distribution with parameters $a, b > 0$ if its density is given by*

$$f_{a,b}(x) = \frac{x^{a-1}(1-x)^{b-1}}{\beta(a, b)} 1_{[0,1]}(x). \tag{5.7}$$

with $a, b > 0$. We use the notation $X \sim Beta(a, b)$ to distinguish from the beta function.

Figure 5.10 presents the density and distributions for various *Beta* random variables.

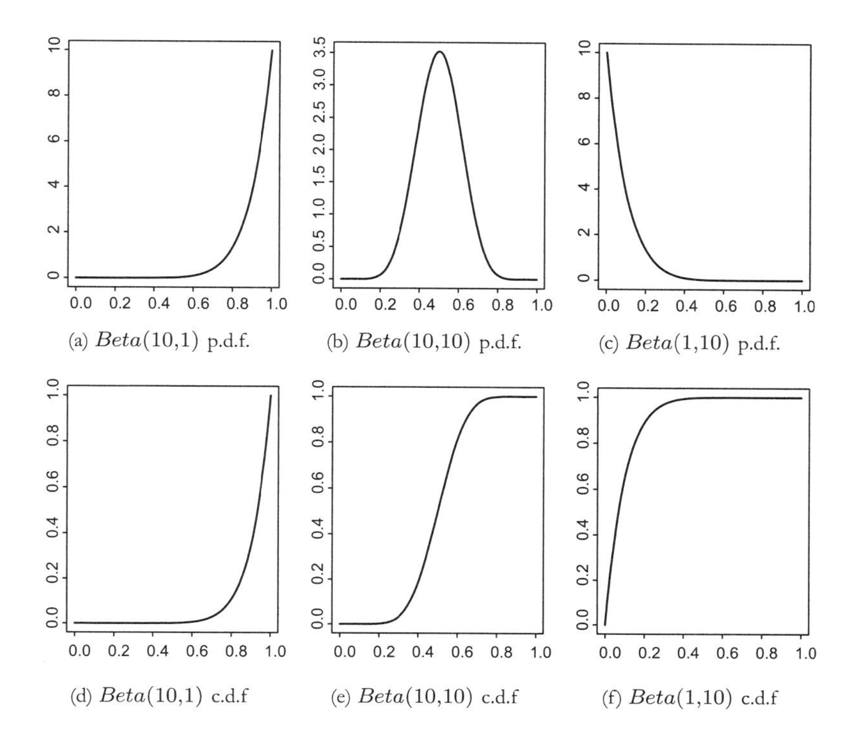

(a) $Beta(10,1)$ p.d.f. (b) $Beta(10,10)$ p.d.f. (c) $Beta(1,10)$ p.d.f.

(d) $Beta(10,1)$ c.d.f (e) $Beta(10,10)$ c.d.f (f) $Beta(1,10)$ c.d.f

FIGURE 5.10 **Beta distributions for various parameter values.**

Proposition 5.41 *The function given by (5.7) is a probability density function.*

Proof: Since the density $f_{a,b}$ is a positive function, we need only to verify that it integrates to 1. But this integral is

$$\frac{1}{\beta(a, b)} \int_0^1 x^{a-1}(1-x)^{b-1} dx = \frac{\beta(a, b)}{\beta(a, b)} = 1.$$

■

Proposition 5.42 *Let $X \sim Beta(a, b)$ with $a, b > 0$. Then*

$$\mathbf{E}X = \frac{a}{a+b} \quad \textit{and} \quad V(X) = \frac{ab}{(a+b)^2(a+b+1)}.$$

Proof: We have

$$\begin{aligned} \mathbf{E}X &= \frac{1}{\beta(a, b)} \int_0^1 x^{a}(1-x)^{b-1} dx = \frac{1}{\beta(a, b)} \beta(a+1, b) \\ &= \frac{\Gamma(a+b)}{\Gamma(a)\Gamma(b)} \frac{\Gamma(a+1)\Gamma(b)}{\Gamma(a+b+1)} = \frac{a}{a+b}, \end{aligned}$$

where we used Proposition 5.39, part b, and Proposition 5.30, part a. To compute the variance, let us compute first the expectation of X^2. We have

$$\begin{aligned} \mathbf{E}X^2 &= \frac{1}{\beta(a, b)} \int_0^1 x^{a+1}(1-x)^{b-1} dx \\ &= \frac{\beta(a+2, b)}{\beta(a, b)} = \frac{a(a+1)}{(a+b+1)(a+b)} \end{aligned}$$

using the properties of the beta and gamma integrals. Then

$$V(X) = \frac{a(a+1)}{(a+b+1)(a+b)} - \left(\frac{a}{a+b}\right)^2 = \frac{ab}{(a+b)^2(a+b+1)}.$$

■

Remark 5.43 *If $a = b = 1$, then $\beta(1, 1)$ is the uniform distribution on the interval $[0, 1]$ (just substitute in the formula). Because of this generalization and the fact that the beta distributed random variable takes values in the interval $[0, 1]$, the distribution is widely used as a distribution for probabilities p.*

Remark 5.44 *If X, Y are independent random variables such that $X \sim \Gamma(a, \lambda)$ and $Y \sim \Gamma(b, \lambda)$, then $\frac{X}{X+Y} \sim Beta(a, b)$. We shall give a proof of this result at a later time.*

5.4.6 STUDENT'S t DISTRIBUTION

Definition 5.45 *A random variable with p.d.f.:*

This distribution has a very interesting history. To start, we remark that if one takes $X_1, \ldots, X_n$ standard normal random variables and forms the expression

$$\frac{\bar{X}}{\sqrt{S^2(X)}},$$

then this expression has a t distribution with $n - 1$ degrees of freedom.

In the previous formulation, we have

$$\bar{X} = \frac{\sum_{i=1}^{n} X_i}{n},$$

the sample average, and

$$S^2(X) = \frac{1}{n-1}\sum_{i=1}^{n}(X_i - \bar{X})^2,$$

the sample variance.

In 1908 William Sealy Gosset published a derivation of the distribution of this expression $\left(\frac{\bar{X}}{\sqrt{S^2(X)}}\right)$ when the original variables $X_1, \ldots, X_n$ are standard normals Student (1908). He was working for the Guinness Brewery in Dublin, Ireland at the time. Guinness forbade members of its staff from publishing scientific papers to not allow the competition to acquire secrets of its famous brew manufacturing. Gosset realized the importance of his discovery and decided such a result deserved to be known even under a pseudonym. The distribution became popular when applied by Sir Ronald Aylmer Fisher (Fisher, 1925) who calls it Student's t distribution.

Definition 5.46 *Let*

$$f(x) = \frac{\Gamma(\frac{n+1}{2})}{\sqrt{\pi n}\Gamma\frac{n}{2}}\left(1 + \frac{t^2}{n}\right)^{-\frac{n+1}{2}}, \tag{5.8}$$

where $n > 0$ is a parameter called degree of freedom and $\Gamma(x)$ is the usual gamma function. A random variable X with probability density given by (5.8) is called a t distributed random variable with n degrees of freedom. We shall use the notation $X \sim t_n$.

Figure 5.11 presents the density and distribution for a t_{10} random variable.

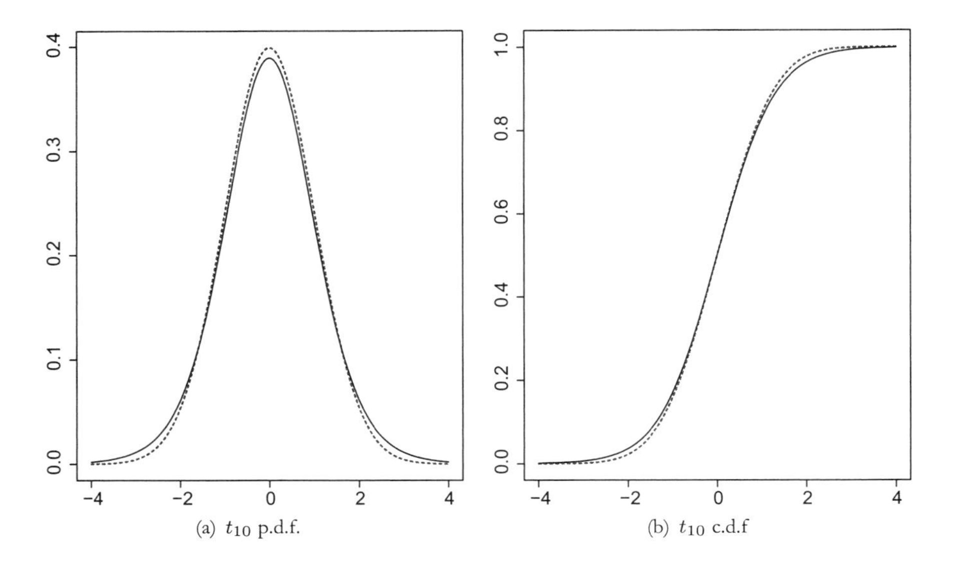

FIGURE 5.11 t_{10} **distribution (continuous line) overlapping a standard normal (dashed line) to note the heavier tails.**

Remark 5.47 *The p.d.f. may be written in terms of the beta function as well:*

$$f(x) = \frac{1}{\sqrt{n}\beta\left(\frac{1}{2}, \frac{n}{2}\right)} \left(1 + \frac{t^2}{n}\right)^{-\frac{n+1}{2}}.$$

The shape of this distribution resembles the bell shape of the normal; however, it has fatter tails. In fact a t_n distribution has no moments greater than n. One more interesting fact is that the t_n p.d.f. converges to the normal p.d.f. as $n \to \infty$.

Proposition 5.48 *The mean and variance of a t_n distributed random variable are*

$$\mathbf{E}[X] = \begin{cases} 0 & \textit{if } n > 1, \\ \infty & \textit{otherwise} \end{cases}$$

and

$$V(X) = \begin{cases} \frac{n}{n-2} & \textit{if } n > 2, \\ \infty & \textit{otherwise.} \end{cases}$$

The proofs of these facts are left as exercises (exercise 5.39).

5.4.7 PARETO DISTRIBUTION

Definition 5.49 *A random variable is distributed as a Pareto random variable with parameters a, $b > 0$ if its probability density is given by*

$$f(x) = \frac{ab^a}{x^{a+1}} 1_{[b,\infty)}(x). \qquad (5.9)$$

We shall use the notation $X \sim P(a, b)$ to denote a random variable with this distribution.

Figure 5.12 presents the density and distribution for a $P(10, 1)$ random variable.

Proposition 5.50 *Function (5.9) is a probability density function.*

Proof: We have

$$\int_{\mathbb{R}} f(x)\, dx = ab^a \int_b^\infty x^{-a-1} dx$$
$$= ab^a \frac{-1}{a} \left[x^{-a}\right]_{x=b}^{x=\infty} = 1.$$

■

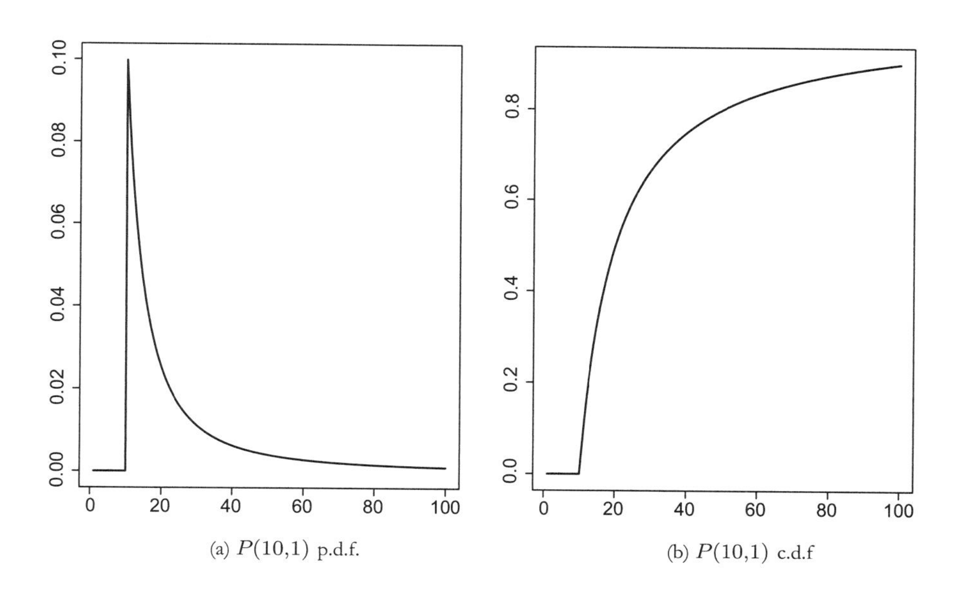

(a) $P(10,1)$ p.d.f.

(b) $P(10,1)$ c.d.f

FIGURE 5.12 ***Pareto*** **distribution.**

Proposition 5.51 *Let $X \sim P(a, b)$. Then*

1. *If $a > 1$, then X is integrable and*

$$\mathbf{E}X = \frac{ab}{a-1}.$$

2. *If $a > 2$, then X is square integrable and in this case*

$$V(X) = \frac{ab^2}{(a-1)^2(a-2)}.$$

Proof: Let us first calculate the expectation. We have

$$\mathbf{E}X = ab^a \int_b^\infty x^{-a} dx$$
$$= \frac{ab^a}{(a-1)b^{a-1}} = \frac{ab}{a-1}.$$

Note that the function x^{-a} is integrable at infinity if and only if $a > 1$. We compute now $\mathbf{E}X^2$. It holds

$$\mathbf{E}X^2 = ab^a \int_b^\infty x^{-a+1} dx$$
$$= \frac{ab^a}{(a-2)b^{a-2}} = \frac{ab^2}{a-2}$$

with the same remark about the second moment existing only if $a > 2$. In this case we can calculate

$$V(X) = \frac{ab^2}{a-2} - \left(\frac{ab}{a-1}\right) == \frac{ab^2}{(a-1)^2(a-2)}.$$

■

Remark 5.52 *If $a \leq 1$ in a Pareto density (5.9), the corresponding random variable has infinite expectation and indeed no moment of any order.*

The following result shows a link between the Pareto and the exponential distributions.

Proposition 5.53 *Let Z be an r.v. with law $Exp(a)$, $a > 0$ and let $b > 0$. Define*

$$X = be^Z.$$

Then X has Pareto distribution $P(a, b)$.

Proof: Recall that the density of Z is given by

$$f_Z(x) = ae^{-ax} 1_{(x>0)}.$$

Consider $F : \mathbb{R} \to \mathbb{R}$ a bounded measurable function. Then

$$\begin{aligned} \mathbf{E}F(X) = \mathbf{E}F(be^Z) &= \int_{\mathbb{R}} F(be^x) f_Z(x)\, dx \\ &= \int_0^\infty F(be^x) ae^{-ax}\, dx, \end{aligned}$$

and by using the change of variables

$$be^x = y \qquad \text{with } x = \log y - \log b$$

we obtain

$$\mathbf{E}F(X) = ab^a \int_b^\infty F(y) y^{-a-1} dy.$$

Finally, using Proposition 5.4, we obtain that the density of X is given by (5.9). ■

5.4.8 THE LOG-NORMAL DISTRIBUTION

In probability theory, a log-normal distribution is a probability distribution of a random variable whose logarithm is normally distributed. If X is a random variable with a normal distribution, then $Y = e^X$ has a log-normal distribution. Likewise, if Y is log-normally distributed, then $X = log(Y)$ is normally distributed. By formalizing, we obtain the following:

Definition 5.54 *A random variable X with density function*

$$f(x) = \frac{1}{x\sigma\sqrt{2\pi}} e^{-\frac{(\log x - \mu)^2}{2\sigma^2}} 1_{(0,\infty)} \tag{5.10}$$

is called a log-normal random variable. The parameters are $\mu \in \mathbb{R}$ and $\sigma > 0$. We shall denote $X \sim LogN(\mu, \sigma^2)$ a distribution with these parameters.

Figure 5.13 presents the density and distribution for a $LogNormal(0, 1)$ random variable.

Proposition 5.55 *The function defined buy the formula (5.10) is a probability density function.*

Proof: Since $\sigma > 0$, we can see that $f(x) \geq 0$ for every x. Next, by using the change of variables

$$log\, x = y, \qquad x = e^y$$

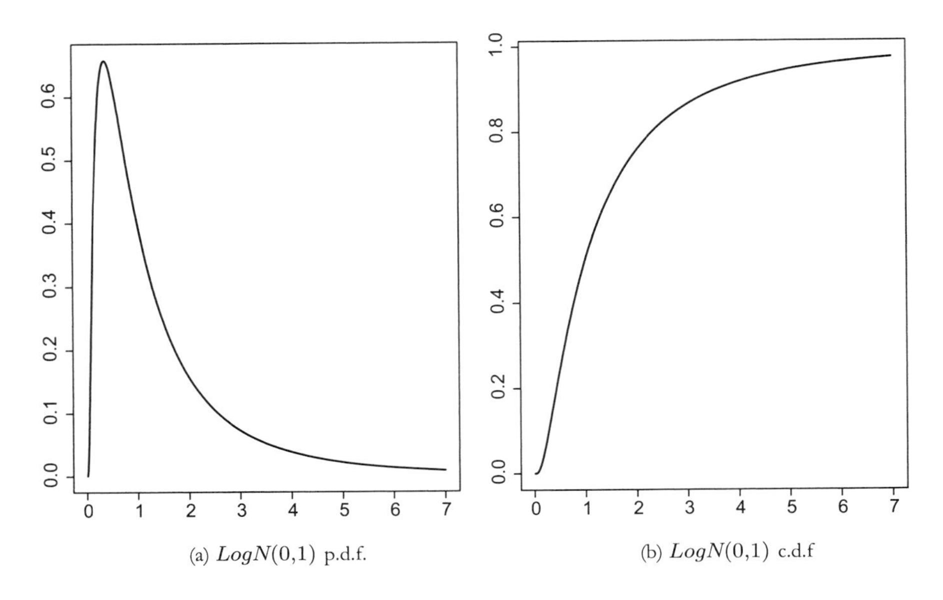

(a) $LogN(0,1)$ p.d.f. (b) $LogN(0,1)$ c.d.f

FIGURE 5.13 ***LogNormal* distribution.**

we obtain

$$\int_{\mathbb{R}} f(x)\,dx = \frac{1}{\sqrt{2\pi}\sigma}\int_0^{\infty} x^{-1}e^{-\frac{(\log x-\mu)^2}{2\sigma^2}}$$
$$= \frac{1}{\sqrt{2\pi}\sigma}\int_{-\infty}^{\infty} e^{-\frac{(y-\mu)^2}{2\sigma^2}} = 1$$

since this is the density of a normal random variable with mean μ and variance σ^2 (which we have proven is integrating to one). The calculation of the variance is left as an exercise. ■

The cumulative distribution function of the log-normal distribution can be written in terms of the cumulative function of the normal distribution. Specifically,

$$F_X(t) = \Phi\left(\frac{\log t - \mu}{\sigma}\right),$$

where

$$\Phi(z) = \mathbf{P}(Z \le z)$$

is the c.d.f. of a normal 0, 1 ($Z \sim N(0, 1)$). Therefore the probabilities involving the log-normal random variables can be obtained from the probability table of the standard normal distribution.

Proposition 5.56 *Let X be a random variable with density given by (5.10). Then*

$$\mathbf{E}X = e^{\mu+\frac{1}{2}\sigma^2}, \qquad VX = \left(e^{\sigma^2} - 1\right) e^{2\mu+\sigma^2}.$$

Proof: We compute the expectation using the same change of variable used in the previous proof ($\log x = y$):

$$\begin{aligned}\mathbf{E}X &= \frac{1}{\sqrt{2\pi}\sigma}\int_0^\infty x^{-1} e^{-\frac{(\log x-\mu)^2}{2\sigma^2}} \\ &= \frac{1}{\sqrt{2\pi}\sigma}\int_{-\infty}^\infty y e^{-\frac{(y-\mu)^2}{2\sigma^2}} = \mu,\end{aligned}$$

again using the formula for the expectation of a normal. ■

In the following result we show the relation between the log-normal distribution and the normal distribution.

Proposition 5.57 *Let Z be a r.v. with $N(\mu, \sigma^2)$ distribution. Then the random variable*

$$X = e^Z$$

follows a log normal distribution with parameters μ and σ^2.

Proof: The derivation follows easily by computing $\mathbf{E}F(Z)$ (for an F arbitrary bounded measurable function) using the density of Z and the change of variables $x = \log y$. ■

Remark 5.58 *The reciprocal is also true: if X has log normal law with parameter μ and σ^2 then the r.v. $Y = \log X \sim N(\mu, \sigma^2)$. Again the proof is simply retracing the argument in the proposition above.*

5.4.9 LAPLACE DISTRIBUTION

Definition 5.59 *We will say that a random variable X has Laplace distribution if its density is given by*

$$f(x) = \frac{\theta}{2} e^{-\theta|x|}, \tag{5.11}$$

with $\theta > 0$. We will denote $X \sim Laplace(\theta)$.

Figure 5.14 presents the density and distribution for a $Laplace(1)$ random variable.

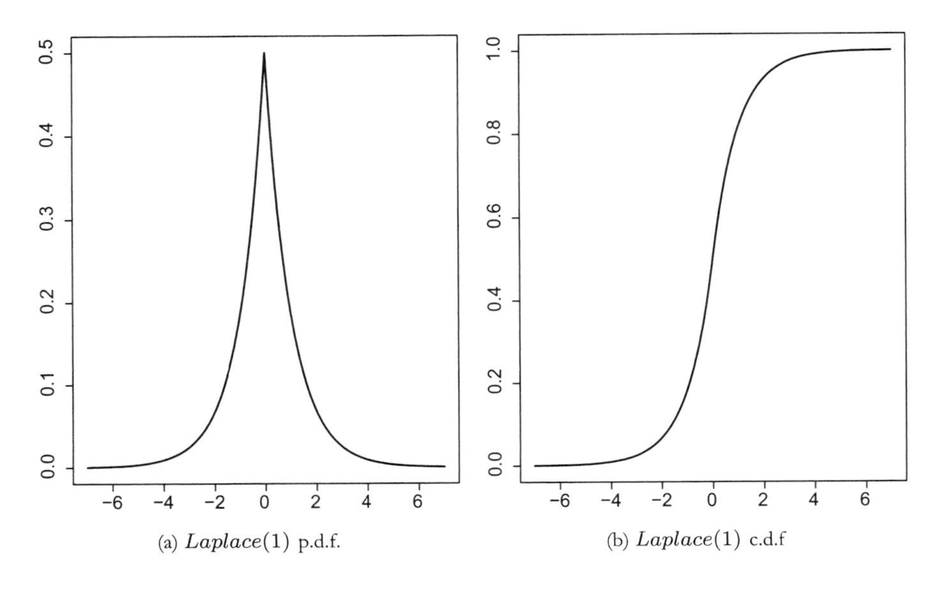

(a) *Laplace*(1) p.d.f. (b) *Laplace*(1) c.d.f

FIGURE 5.14 ***Laplace*** **distribution.**

Proposition 5.60 *The function in (5.11) is a probability density function.*

Proof: The function is positive and

$$\begin{aligned}\int_{\mathbb{R}} f(x)dx &= \frac{\theta}{2}\int_{\mathbb{R}} e^{-\theta|x|}\\ &= \frac{\theta}{2}\left(\int_0^{\infty} e^{-\theta x}dx + \int_{-\infty}^{0} e^{\theta x}dx\right)\\ &= \theta\int_0^{\infty} e^{-\theta x}dx = 1.\end{aligned}$$

■

Proposition 5.61 *Let X be an r.v. with Laplace(θ) distribution. Then*

$$\mathbf{E}X = 0 \quad \textit{and} \quad \mathbf{E}|X| = \frac{1}{\theta}.$$

Proof: First

$$\mathbf{E}X = \int_{\mathbb{R}} xe^{-\theta|x|}dx = 0$$

since the function $xe^{-\theta|x|}$ is odd and therefore its integral on a symmetric interval around the origin vanishes. Moreover,

$$\begin{aligned}\mathbf{E}|X| &= \int_{\mathbb{R}} |x|e^{-\theta|x|}dx \\ &= \theta \int_0^{\infty} xe^{-\theta x}dx = \frac{1}{\theta}.\end{aligned}$$

■

The Laplace distribution may be obtained as the product of a random variable with exponential distribution and an independent random variable with uniform distribution (discrete).

Proposition 5.62 *Let X be an r.v. with distribution exp(θ), $\theta > 0$ and let S be an r.v. with uniform law over $\{-1, +1\}$, independent of X (i.e., S takes values ± 1 with probability $1/2$). Set $Y = XS$. Show that Y has a Laplace distribution with parameter θ.*

Proof: Since the random variable S takes only values -1 and 1 and

$$\mathbf{P}(S = -1) = \mathbf{P}(S = 1) = \frac{1}{2},$$

let us compute the cumulative distribution function of Y by the law of total probability. For every $t \in \mathbb{R}$ we can write, using the independence of X and S,

$$\begin{aligned}\mathbf{P}(Y \leq t) &= \mathbf{P}(S = 1, X \leq t) + \mathbf{P}(S = -1, -X \leq t) \\ &= \mathbf{P}(S = 1)\mathbf{P}(X \leq t) + \mathbf{P}(S = -1)\mathbf{P}(X \geq -t) \\ &= \frac{1}{2}\left(\mathbf{P}(X \leq t) + 1 - \mathbf{P}(X \leq -t)\right)\end{aligned}$$

and by differentiating with respect to t we obtain that the density of Y is given by

$$f_Y(t) = \frac{1}{2}(f_X(t) + f_X(-t)) = \frac{\theta}{2}e^{-\theta|t|},$$

which indeed is the function in (5.11). ■

5.4.10 DOUBLE EXPONENTIAL DISTRIBUTION

The double exponential is a slight generalization of the Laplace distribution. We mention it here since it has its uses in mathematical finance where it is utilized to model the jump distribution.

Definition 5.63 *A random variable X with density,*

$$f(x) = \begin{cases} p\alpha_1 e^{-\alpha_1 x} & if\, x > 0 \\ (1-p)\alpha_2 e^{\alpha_2 x} & if\, x \leq 0 \end{cases}, \qquad \alpha_1, \alpha_2 > 0, \tag{5.12}$$

is said to have a double exponential density.

We can easily obtain that (5.12) defines a valid probability density by observing that the density is nothing more than a weighted sum of exponential densities. Furthermore, using the same observation, we easily obtain the mean respectively variance of this distribution as

$$\mathbf{E}X = p\frac{1}{\alpha_1} - (1-p)\frac{1}{\alpha_2}$$

and

$$V(X) = p\frac{1}{\alpha_1^2} + (1-p)\frac{1}{\alpha_2^2}.$$

EXERCISES

Problems with Solution

5.1 Consider the function $f : \mathbb{R} \to \mathbb{R}$ such that

$$f(x) = \frac{2}{\pi(1+x^2)} 1_{[0,\infty)}(x).$$

(a) Show that f is a probability density function.

(b) Take X a random variable with density f. Show that the expectation of X does not exist.

Solution: Point (a) follows since the antiderivative of $\frac{1}{x^2}$ is $\arctan(x)$ and $\arctan(\infty) = \frac{\pi}{2}$, $\arctan(0) = 0$. Concerning point (b), by definition we have

$$\mathbf{E}X = \frac{2}{\pi}\int_0^\infty \frac{x}{1+x^2}dx;$$

But the function $\frac{x}{1+x^2}$ is not integrable on $[0, \infty[$. Its primitive is $\ln 2(1+x^2)$ which explodes at infinity. ■

5.2 Let X be an r.v. with uniform distribution on $[0, 1]$. Show that $X^2 \sim$ *Beta*$(\frac{1}{2}, 1)$.

Solution: Let $F : \mathbb{R} \to \mathbb{R}$ be an arbitrary bounded measurable function. Then

$$\mathbf{E}F(X^2) = \int_0^1 F(x^2)\, dx = \int_0^1 F(y)\frac{1}{2\sqrt{y}}\, dy$$

where we made the change of variables $x = \sqrt{y}$ with $dx = \frac{1}{2\sqrt{y}}\, dy$. This implies that the density of X^2 is

$$\frac{1}{2}y^{-\frac{1}{2}} 1_{[0,1]}(y) = \frac{1}{\beta(\frac{1}{2}, 1)} y^{\frac{1}{2}-1}(1-y)^{1-1} 1_{[0,1]}(y),$$

which is exactly the density of the beta distribution $Beta(\frac{1}{2}, 1)$. ■

5.3 Let X be a continuous random variable with density f.
(a) Show that $Y = X^2$ is a continuous random variable with density

$$g_Y(y) = \frac{1}{2\sqrt{y}}\left(f(\sqrt{y}) + f(-\sqrt{y})\right) 1_{[0,\infty)}(y).$$

(b) Retrieve the results in exercise 5.2 and in Proposition 5.35.

Solution: We study the distribution function F_Y of the r.v. $Y = X^2$. If $x \le 0$ then obviously

$$F_Y(x) = \mathbf{P}(X^2 \le x) = 0.$$

For $x > 0$ we have

$$F_Y(x) = \mathbf{P}(-\sqrt{x} \le X \le \sqrt{x}) = F_X(\sqrt{x}) - F_X(-\sqrt{x}),$$

where F_X denotes the distribution function of X. The expression of the density g_Y is obtained when we differentiate with respect to x. ■

5.4 Suppose X is a Cauchy random variable with density $\frac{1}{\pi}\frac{1}{1+x^2}$.
(a) Is the r.v. X integrable?
(b) Compute the density function of the r.v. $Y = \frac{1}{X}$.
(c) Let V be an r.v. with uniform law on the interval $\left(-\frac{\pi}{2}, \frac{\pi}{2}\right)$. Find the law of $W = \tan V = \frac{\sin V}{\cos V}$.
(d) Find the value $a_k = \mathbf{P}\left(X \le \frac{k}{4}\right)$, for $k = 1, 2, 3$.

Solution: Let us compute the density of $Y = \frac{1}{X}$. Take F bounded measurable and

$$
\begin{aligned}
\mathbf{E}F(Y) = \mathbf{E}F(\frac{1}{X}) &= \int_{\mathbb{R}} F(\frac{1}{x})\frac{1}{\pi}\frac{1}{1+x^2}dx \\
&= \frac{1}{\pi}\int_0^{\infty} F(\frac{1}{x})\frac{1}{\pi}\frac{1}{1+x^2}dx \\
&+ \frac{1}{\pi}\int_{-\infty}^{0} F(\frac{1}{x})\frac{1}{\pi}\frac{1}{1+x^2}dx
\end{aligned}
$$

We separate the integral over $\mathbb{R}$ into two parts so that we can make the change of variable $\frac{1}{x} = y$ (note that the function $x \to \frac{1}{x}$ is not bijective on $\mathbb{R}$ but its restrictions to the positive and negative axis are). Thus

$$
\begin{aligned}
\mathbf{E}F(Y) &= \frac{1}{\pi}\int_0^{\infty} F(y)\frac{1}{\pi}\frac{1}{1+\frac{1}{y^2}}\frac{1}{y^2}\,dy \\
&+ \frac{1}{\pi}\int_{-\infty}^{0} F(y)\frac{1}{\pi}\frac{1}{1+\frac{1}{y^2}}\frac{1}{y^2}\,dy \\
&= \frac{1}{\pi}\int_{\mathbb{R}} F(y)\frac{1}{1+y^2}dy
\end{aligned}
$$

which implies that Y follows also a Cauchy distribution. Therefore, $1/Y$ the inverse of a Cauchy law has also a Cauchy distribution.

Let us look at part (c). Note that the density of V is

$$
g(x) = \frac{1}{\pi}1_{(-\frac{\pi}{2},\frac{\pi}{2})}(x)
$$

and the density of W is computed as follows. For any F bounded measurable, we have

$$
\begin{aligned}
\mathbf{E}F(W) &= \frac{1}{\pi}\int_{-\frac{\pi}{2}}^{\frac{\pi}{2}} F(\tan(v))\,dv \\
&= \frac{1}{\pi}\int_{\mathbb{R}} \frac{1}{1+x^2}F(x)\,dx
\end{aligned}
$$

with the change of variable

$$
\tan(v) = x, \qquad v = \arctan(x), \qquad dv = \frac{1}{1+x^2}\,dx.
$$

The meaning of point (c) is to show that the tangent of a uniformly distributed random variable on the interval $(-\pi/2, \pi/2)$ has a Cauchy distribution. ■

5.5 Let $X \sim \mathcal{N}(0,1)$.

(1) Calculate the distribution of

$$Y = X^2 \quad \text{and} \quad Z = e^X.$$

(2) For every $t \in \mathbb{R}$ calculate the expectation and the variance of $Y = e^{tX}$.

(3) For which values of $a > 0$ the r.v. $Z = e^{aX^2}$ is integrable? In this case, compute the integral.

(4) For which values of $a > 0$ the r.v. $Z = e^{aX^2}$ is square integrable? Compute its variance.

Solution: About part (2) we can write

$$\begin{aligned}\mathbf{E}Y = \mathbf{E}e^{tX} &= \frac{1}{\sqrt{2\pi}} \int_{\mathbb{R}} e^{tx} e^{-\frac{x^2}{2}} dx = \frac{1}{\sqrt{2\pi}} e^{\frac{t^2}{2}} \int_{\mathbb{R}} e^{-\frac{1}{2}(x-t)^2} dx \\ &= \frac{1}{\sqrt{2\pi}} e^{\frac{t^2}{2}} \int_{\mathbb{R}} e^{-\frac{y^2}{2}} dy = e^{\frac{t^2}{2}},\end{aligned}$$

where we used the change of variables $x - t = y$.

In part (3), since Z is a positive random variable, we obtain

$$\begin{aligned}\mathbf{E}\,|Z| = \mathbf{E}Z &= \frac{1}{\sqrt{2\pi}} \int_{\mathbb{R}} e^{ax^2} e^{-\frac{x^2}{2}} dx \\ &= \frac{1}{\sqrt{2\pi}} \int_{\mathbb{R}} e^{x^2(a-\frac{1}{2})} dx\end{aligned}$$

and this integral is finite if and only if

$$a < \frac{1}{2}.$$

Using the change of variables $x\sqrt{\frac{1}{2} - a} = \sqrt{2}y$, we obtain that

$$\mathbf{E}Z = \frac{1}{\sqrt{1-2a}}.$$

Part (4) follows from part 3 since

$$\mathbf{E}(e^{aX^2})^2 = \mathbf{E}e^{2aX^2} = \frac{1}{\sqrt{1-4a}}$$

which is finite if and only if

$$a < \frac{1}{4}.$$

■

5.6 Let $a > 0$. Define

$$f : \mathbb{R} \to \mathbb{R}$$

by

$$f(x) = ae^{-a(x-x_0)} 1_{(x_0,\infty)}(x).$$

(a) Prove that f is a density
(b) Compute $\mathbf{E}X$ and $V(X)$, where X is an r.v. with p.d.f f.

Solution: For $x_0 = 0$, we are in the case of the exponential law with parameter a. Follow the proofs of Propositions 5.18 and 5.19. ■

5.7 Let X be a random variable with probability distribution $\mathcal{U}[0, 1]$. Set

$$Y_1 = \max(X, 1 - X) \quad \text{and} \quad Y_2 = \min(X, 1 - X).$$

(a) Compute the cumulative distribution functions of Y_1 and Y_2.
(b) Calculate $\mathbf{E}(Y_1 - Y_2)$

Solution: We compute the cumulative distribution function of Y_1. For every $t \in [0, 1]$

$$F_{Y_1}(t) = \mathbf{P}(Y_1 \leq t) = \mathbf{P}(X \leq t, 1 - X \leq t)$$
$$= = \mathbf{P}(1 - t \leq X \leq t).$$

and this is zero if $1 - t > t$ that means $t < \frac{1}{2}$ while for $t \geq \frac{1}{2}$ we obtain

$$F_{Y_1}(t) = \int_{1-t}^{t} dx = 2t - 1.$$

For $t \geq 0$ we clearly have $F_{Y_1}(t) = 0$ while for $t \geq 1$, $F_{Y_1}(t) = 1$. By differentiating with respect to t we see that the density of Y_1 is

$$2 1_{[\frac{1}{2},1]}(x)$$

and thus Y_1 has uniform distribution on the interval $[\frac{1}{2}, 1]$. Similarly Y_2 will have a uniform distribution on $[0, \frac{1}{2}]$.

For the second question, use the identity

$$\max(a, b) - \min(a, b) = |a - b|$$

if $a, b \in \mathbb{R}$. ■

5.8 Let X be a random variable with density

$$f(x) = \frac{1}{2} \cos(x) 1_{(\frac{\pi}{2},\frac{\pi}{2})}(x).$$

Calculate the expectation and the variance of the random variable

$$Y = \sin X.$$

Solution: The expectation is

$$\begin{aligned}\mathbf{E}Y &= \int_{\mathbb{R}} x f(x)\, dx = \frac{1}{2}\int_{-\frac{\pi}{2}}^{\frac{\pi}{2}} x\cos x\, dx \\ &= 0\end{aligned}$$

because we integrate an odd function $x \cos x$ on an interval symmetric around the origin. Then we can write:

$$\begin{aligned}V(Y) = \mathbf{E}Y^2 &= \frac{1}{2}\int_{-\frac{\pi}{2}}^{\frac{\pi}{2}} x^2 \cos x\, dx \\ &= \left[x^2 \sin x\right]_{x=-\frac{\pi}{2}}^{x=\frac{\pi}{2}} - 2\int_{-\frac{\pi}{2}}^{\frac{\pi}{2}} x \sin x\, dx \\ &= \left(\frac{\pi}{2}\right)^2 - 1.\end{aligned}$$

■

5.9 Let U be a random variable with uniform distribution on the interval $[0, 1]$ and let $\lambda > 0$. Set

$$X = -\frac{1}{\lambda}\ln U.$$

Find the cumulative distribution function and the law of X.

Solution: Since $\mathbf{P}(U \in (0, 1)) = 1$ it follows easily that

$$\mathbf{P}(\ln U > 0) = 1.$$

Therefore, for any for any $t \geq 0$

$$1 - F_X(t) = \mathbf{P}\left(-\frac{1}{\lambda}\ln U \geq t\right) = 1.$$

For $t \geq 0$, we have

$$\begin{aligned}\mathbf{P}(X \geq t) = \mathbf{P}(\ln U \leq -\lambda t) &= \mathbf{P}(U \leq e^{-\lambda t}) \\ &= \int_0^{e^{-\lambda t}} dx = e^{-\lambda t}.\end{aligned}$$

Consequently

$$F_X(t) = (1 - e^{-\lambda t})1_{t \geq 0}.$$

Note that this is the c.d.f of an exponential therefore X follows an exponential law with parameter λ. ■

5.10 Show that the n-th moment of the log normal distribution with parameters μ and σ is given by

$$\mathbf{E}X^n = e^{n\mu + \frac{1}{2}n^2\sigma^2}.$$

Solution: Recall the density of the log normal distribution from the formula (5.10). Using this density we obtain:

$$\begin{aligned}\mathbf{E}X^n &= \frac{1}{\sqrt{2\pi}\sigma}\int_0^{\infty} x^{n-1} e^{-\frac{(\log x - \mu)^2}{2\sigma^2}} dx \\ &= \frac{1}{\sqrt{2\pi}\sigma}\int_{\mathbb{R}} e^{ny} e^{-\frac{(y-\mu)^2}{2\sigma^2}} dy.\end{aligned}$$

Notice that the above expression means that

$$\mathbf{E}X^n = \mathbf{E}e^{nY}$$

where $Y \sim N(\mu, \sigma^2)$. Completing the square under the integral we obtain

$$\begin{aligned}\mathbf{E}X^n &= e^{-\frac{1}{2}\frac{\mu}{\sigma^2}} \frac{1}{\sqrt{2\pi}\sigma}\int_{\mathbb{R}} e^{-\frac{1}{2\sigma^2}(y-(\mu+n\sigma))^2} dy \\ &= e^{-\frac{1}{2}\frac{\mu}{\sigma^2}} e^{-\frac{(n\sigma^2+\mu)^2}{2\sigma^2}} \\ &= e^{n\mu + \frac{1}{2}n^2\sigma^2}.\end{aligned}$$

■

5.11 Let X be with law $Exp(\lambda)$, $\lambda > 0$, that is, with density

$$f(x) = \begin{cases} \lambda e^{-\lambda x} & \text{if } x > 0, \\ 0 & \text{otherwise.} \end{cases}$$

(a) Prove that for every $s > 0$ and $t > 0$,

$$\mathbf{P}(X \geq s + t | X \geq s) = \mathbf{P}(X \geq t).$$

Remark 5.64 *Do you recall this property from somewhere? This property of the distribution is useful when modeling the time to break of a material where no external signs of aging are detected, or the waiting time (of a bus at a stop, etc.) when no other extra information is given besides the time of individual's arrival.*

(b) X arrives in a bank. There are n customer service counters which are all occupied. The service time for the counters are independent exponentially distributed random variables with parameter $\lambda > 0$. The moment a customer service window becomes available X will be served. What is the average waiting time before being served? Compare this time with the expected time in the situation when there is only one customer office and n people are before X (one at the office and $n - 1$ in the queue).

(c) A deposit is lighted by n light bulbs continuously on. The lifetime of each light bulb is exponentially distributed with parameter $\lambda > 0$. What is the average time after which the first bulb burns out? What is the average time after which the deposit is not lit at all if the maintenance department does not change the burnt out bulbs?

Solution: We only give the solution for the first part. It is immediate to calculate for every $t > 0$

$$\mathbf{P}(X > t) = \lambda \int_t^\infty e^{-\lambda x} dx = e^{-\lambda t}$$

and for every $t \leq 0$

$$\mathbf{P}(X > t) = 1.$$

Fix $s, t > 0$. We have

$$\begin{aligned}\mathbf{P}(X \geq s + t | X \geq s) = \mathbf{P}(X \geq t) &= \frac{\mathbf{P}(X > s + t, X > s)}{\mathbf{P}(X > s)} \\ &= \frac{\mathbf{P}(X > s + t)}{\mathbf{P}(X > s)} \\ &= \frac{e^{-\lambda(s+t)}}{e^{-\lambda s}} = e^{-\lambda t} \\ &= \mathbf{P}(X \geq t).\end{aligned}$$

■

5.12 If X is a random variable such that

$$\mathbf{P}(X > 0) = 1 \quad \text{and} \quad \mathbf{P}(X > t) > 0$$

for every $t > 0$ and

$$\mathbf{P}(X > t + s | X > t) = \mathbf{P}(X > s) \qquad \text{for every } s, t \geq 0,$$

then X has an exponential distribution.

Compare with exercise 5.11.

Solution: Let us define the function $h : \mathbb{R}_+ \to \mathbb{R}$,

$$h(t) = \mathbf{P}(X > t)$$

for all $t \geq 0$. Notice that h is decreasing, $h(0) = 1$ and $h(t) \geq 0$ for every $t \geq 0$. Using the definition of conditional probability and the hypothesis,

$$h(t+s) = h(t)h(s) \qquad \text{for all } s, t \geq 0.$$

This can be extended by induction to

$$h(t_1 + \cdots + t_n) = h(t_1)\dots h(t_n) \qquad \text{for all } t_1, \dots, t_n \geq 0 \text{ and } n \in \mathbb{N}^*. \tag{5.13}$$

Relation (5.13) implies that

$$h(1) = \left[f\left(\frac{1}{n}\right)\right]^n, \quad \forall n \in \mathbb{N}^*.$$

Therefore, for every positive rational numbers $\frac{p}{q}$ with $p, q \in \mathbb{N}$, $q \neq 0$

$$h\left(\frac{p}{q}\right) = \left[h\left(\frac{1}{q}\right)\right]^p = h^{\frac{p}{q}}(1).$$

We choose $\lambda > 0$ such that $h(1) = e^{-\lambda}$ and then

$$h(r) = e^{-\lambda r} \qquad \text{for every } r \in \mathbb{Q}_+.$$

For an arbitrary $t \in \mathbb{R}_+ \setminus \mathbb{Q}_+$ we choose $r, s \in \mathbb{Q}_+$ such that $r < t < s$ and since h is decreasing

$$e^{-\lambda r} = h(r) \geq h(t) \geq h(s) = e^{-\lambda s}.$$

By taking $s - r$ arbitrary small, and since h is continuous, we get

$$h(t) = e^{-\lambda t} \qquad \text{for all } t \geq 0.$$

For $t < 0$ we have by monotonicity and positivity of the c.d.f; that

$$0 \leq F(t) \leq F(0) = 1 - \mathbf{P}(X > 0) = 0.$$

then, $F(t) = 0$ for $t < 0$. As a consequence, X has exponential distribution with parameter

$$\lambda = -\ln \mathbf{P}(X > 1).$$

■

5.13 Consider $(Y_n, n \geq 1)$ a sequence of independent identically distributed random variables with common law $\mathcal{U}([a, b])$. For every $n \geq 1$ we define

$$I_n = \inf(Y_1, \dots, Y_n)$$

and

$$M_n = \sup(Y_1, \dots, Y_n).$$

(a) Calculate the c.d.f. of M_n.
(b) Derive the expression of the density of M_n.
(c) Derive the expression of the density of I_n.

Solution: We calculate the cumulative distribution function of M_n. For every t we have

$$(M_n \leq t) = \mathbf{P}(Y_1 \leq t, Y_2 \leq t, \dots, Y_n \leq t)$$

and

$$\begin{aligned}\mathbf{P}(M_n \leq t) &= \mathbf{P}(Y_1 \leq t, \dots, Y_n \leq t)\\ &= \mathbf{P}(Y_1 \leq t).....\mathbf{P}(Y_n \leq t)\end{aligned}$$

by the independence of Y_i's. From Proposition 5.16, for every i we have

$$\mathbf{P}(Y_i \leq t) = \begin{cases} 0 & \text{if } t < a, \\ \frac{t-a}{b-a} & \text{if } t \in [a, b] \\ 1 & \text{if } t > b. \end{cases}$$

and thus

$$\mathbf{P}(M_n \leq t) = \begin{cases} 0 & \text{if } t < a, \\ \left(\frac{t-a}{b-a}\right)^n & \text{if } t \in [a, b] \\ 1 & \text{if } t > b. \end{cases} \tag{5.14}$$

Now, differentiating with respect to t in (5.14), we obtain the following expression for the density of M_n:

$$f_{M_n}(t) = \begin{cases} 0 & \text{if } t < a \text{ or } t > b, \\ n\left(\frac{t-a}{b-a}\right)^{n-1} \frac{1}{b-a} & \text{if } t \in [a, b] \end{cases}$$

For the random variable I_n it is easier to compute $\mathbf{P}(I_n \geq t)$ rather than the c.d.f. We have

$$\begin{aligned}\mathbf{P}(I_n \geq t) &= \mathbf{P}(Y_1 \geq t, \dots, Y_n \geq t)\\ &= \mathbf{P}(Y_1 \geq t).....\mathbf{P}(Y_n \geq t)\end{aligned}$$

and since for every $i = 1, \dots, n$

$$\mathbf{P}(Y_i \geq t) = \begin{cases} 1 & \text{if } t < a, \\ 1 - \frac{t-a}{b-a} & \text{if } t \in [a, b] \\ 0 & \text{if } t > b. \end{cases}$$

we get

$$\mathbf{P}(I_n \geq t) = \begin{cases} 1 & \text{if } t < a, \\ \left(1 - \frac{t-a}{b-a}\right)^n & \text{if } t \in [a, b] \\ 0 & \text{if } t > b. \end{cases}$$

We differentiate with respect to t and we obtain

$$f_{I_n}(t) = \begin{cases} 0 & \text{if } t < a \text{ or } t > b, \\ n\left(1 - \frac{t-a}{b-a}\right)^{n-1} \frac{-1}{b-a} & \text{if } t \in [a, b]. \end{cases}$$

■

5.14 **The triangular distribution** Let

$$f(x) = c(1 - |x|) \text{ if } x \in [-1, 1]$$

and $f(x) = 0$ otherwise. Find the constant c that makes f a probability density function.

Solution: f is positive and

$$\begin{aligned} \int_{\mathbb{R}} f(x)dx &= c \int_{-1}^{0} (1 + x)\, dx + c \int_{0}^{1} (1 - x)\, dx \\ &= c\left(1 - \frac{1}{2}\right) + c\left(1 - \frac{1}{2}\right) = c, \end{aligned}$$

so $c = 1$. ■

5.15 Let $a \in \mathbb{R}$. For each of the next functions, determine if they can be considered as density functions. If yes, give the necessary and sufficient conditions on a, compute the c.d.f., the expectation and the variance.

(a)

$$f(x) = axe^{-x} \qquad \text{if } x > 0$$

and $f(x) = 0$ otherwise.

(b)

$$f(x) = x^3 \qquad \text{if } x \in [-a, a]$$

and $f(x) = 0$ otherwise.

(c)

$$f(x) = a\ln(x) \qquad \text{if } x \in [1, e]$$

and $f(x) = 0$ otherwise.

Solution: (**a**) Note that a should satisfy $a \geq 0$. Then, by integrating by parts, we obtain

$$\int_{\mathbb{R}} f(x)\, dx = \int_0^{\infty} axe^{-x}\, dx$$
$$= a\left[e^{-x}\right]_{x=0}^{x=\infty} = a,$$

thus

$$a = 1.$$

The c.d.f. is

$$F(b) = 0 \qquad \text{if } b \geq 0$$

and

$$F(b) = \int_0^b xe^{-x}\, dx = 1 - (b+1)e^{-b} \qquad \text{if } b > 0.$$

Integrating by parts we will find

$$\mathbf{E}X = 2 \quad \text{and} \quad V(X) = 2.$$

(**b**) We will show that f is not a probability density function. In fact, the condition $f(x) \geq 0$ for every $x \in \mathbb{R}$ implies in particular

$$f(a) \geq 0 \quad \text{and} \quad f(-a) \geq 0$$

so

$$a^3 \geq 0 \quad \text{and} \quad -a^3 \geq 0$$

and this gives $a = 0$. So $f(x) = 0$ for every $x \in \mathbb{R}$ and we cannot have

$$\int_{\mathbb{R}} f(x)\, dx = 1.$$

(**c**) we will show that f is a density if and only if $a = 1$. First, a should be bigger or equal than 0 and

$$\int_{\mathbb{R}} f(x)\, dx = a \int_1^e \ln(x)\, dx$$
$$= [x \ln x]_1^e - a \int_1^e \frac{x}{x}\, dx$$
$$= ae - a(e-1) = a$$

so $a = 1$. Then

$$F(b) = \mathbf{P}(X \leq b) = \int_{-\infty}^{b} f(x)\, dx$$

$$= \begin{cases} 0 & \text{if } b \leq 1 \\ 1 + b\ln(b) - b & \text{if } b \in [1, e] \\ 1 & \text{if } b > e \end{cases}$$

An integration by parts yields

$$\mathbf{E}X = \frac{e^2 + 1}{4}, \qquad \mathbf{E}X^2 = \frac{2e^3 + 1}{9}$$

and

$$V(X) = \frac{2}{9}e^3 + \frac{7}{144} - \frac{1}{16}e^4 - \frac{1}{8}e^2.$$

■

5.16 Let

$$f(x) = 12x^2 - 12x^3.$$

when x lies in the interval $(0, 1)$ and assume $f(x) = 0$ otherwise.

(a) Verify that f is a probability density function.

(b) Find the probability $\mathbf{P}(1/4 \leq X \leq 3/4)$.

Solution: Note that

$$f(x) = 12x^2(1 - x) \geq 0$$

for every $x \in (0, 1)$ and

$$\int_{-\infty}^{\infty} f(x)\, dx = \int_0^1 (12x^2 - 12x^3)\, dx$$

$$= \left[4x^3\right]_0^1 - \left[3x^4\right]_0^1$$

$$= 1.$$

Also

$$\mathbf{P}(1/4 \leq X \leq 3/4) = \int_{\frac{1}{4}}^{\frac{3}{4}} (12x^2 - 12x^3)\, dx$$

$$= \left[4x^3\right]_{1/4}^{3/4} - \left[3x^4\right]_{1/4}^{3/4}.$$

■

5.17 Let X be a continuous r.v. with density f given by

$$f(x) = \frac{3}{64}(x+2)^2 \text{ if } |x| \le k$$

and $f(x) = 0$ otherwise.

(**a**) Find k

(**b**) Give the density of the random variables

$$Y = \sqrt{X+2} \quad \text{and} \quad Z = |X|.$$

(**c**) Compute $\mathbf{E}Y$, $V(Y)$.

(**d**) Compute $\mathbf{E}Z$, $V(Z)$.

Solution: (**a**) We have

$$\begin{aligned}\int_{\mathbb{R}} f(x)\,dx &= \frac{3}{64}\int_{-k}^{k}(x+2)^2\,dx \\ &= \frac{3}{64}\left[\frac{1}{3}(x+2)^3\right]_{x=-k}^{x=k} \\ &= \frac{1}{32}(k^3+12k).\end{aligned}$$

We need to have

$$\frac{1}{32}(k^3+12k) = 1,$$

which is equivalent to

$$k^3 + 12k - 32 = 0.$$

The function $f(k) = k^3 + 12k - 32$ is increasing (its derivative is strictly positive) and satisfies $f(2) = 0$. So, $k = 2$ is the unique solution.

(**b**) First we obtain the c.d.f. of X

$$\begin{aligned}F(b) = \mathbf{P}(X \le b) &= \int_{-\infty}^{b} f(x)\,dx \\ &= \begin{cases} 0 & \text{if } b \in (-\infty, -2] \\ \frac{1}{64}(x+2)^3 & \text{if } b \in (-2, 2] \\ 1 & \text{if } b > 2 \end{cases}\end{aligned}$$

Then the c.d.f of Y will be

$$G(y) = \mathbf{P}(\sqrt{X+2} \leq y)$$
$$= \begin{cases} 0 & \text{if } y \in (-\infty, 0], \\ \frac{1}{64}y^6 & \text{if } y \in (0, 2], \\ 1 & \text{if } y > 2. \end{cases}$$

By differentiating G with respect to y we obtain the density of Y

$$g(y) = \begin{cases} 0 & \text{if } y \in (-\infty, 0], \\ \frac{1}{64}y^5 & \text{if } y \in (0, 2], \\ 0 & \text{if } y > 2. \end{cases}$$

The r.v. Z can be treated similarly. The expectation of Y is:

$$\mathbf{E}Y = \int_0^2 \frac{6}{64}y^6 dy = \frac{12}{7}.$$

■

5.18 Let X be a random variable with uniform law $U[0, 1]$ and set $Y = 2X$. Compute the density of the random variable $X + Y$.

Solution: Obviously

$$X + Y = 3X$$

and the random variable $Z = 3X$ follows an uniform distribution on the interval $[0, 3]$. this can be seen, for example, by using the c.d.f. Then

$$F_Z(x) = \frac{1}{3}1_{[0,3]}(x).$$

■

5.19 Could the following functions be the cumulative distribution functions of a continuous random variable?

(a)

$$F(x) = \begin{cases} 0 & \text{if } x < 0, \\ \frac{2}{x} & \text{if } x \in [1, 2], \\ 1 & \text{otherwise.} \end{cases}$$

(b)

$$F(x) = \begin{cases} 0 & \text{if } x < 0, \\ x & \text{if } x \geq 0. \end{cases}$$

(c)

$$F(x) = \begin{cases} 0 & \text{if } x < 0, \\ x & \text{if } x \in [0, 1), \\ 1 & \text{if } x \geq 1. \end{cases}$$

Solution: (**a**) F is not a c.d.f. because F is not continuous at $x = 1$ and F is not increasing on $[1, 2)$.

(**b**) F is not a c.d.f because

$$\lim_{b \to \infty} F(b) = \infty.$$

(**c**) F could be a c.d.f, it does satisfy the required conditions. ■

5.20 Let X be an r.v. with density

$$f(x) = xe^{-\frac{1}{2}x^2} \qquad \text{if } x \geq 0$$

and

$$f(x) = 0 \qquad \text{if } x < 0.$$

(**a**) Identify the law of X

(**b**) Give the law of $Y = X^2$

(**c**) Compute $\mathbf{E}Y$ and VY.

Solution: (**b**) We compute the c.d.f of Y.

$$\begin{aligned} G(y) &= \mathbf{P}(Y \leq y) \\ &= \mathbf{P}(X^2 \leq y) \\ &= \mathbf{P}(X \in (-\sqrt{y}, \sqrt{y})). \end{aligned}$$

Therefore,

$$G(y) = 0 \qquad \text{if } y < 0$$

and

$$G(y) = F(\sqrt{y}) - F(-\sqrt{y}),$$

where F denotes the cdf of X. By differentiating we find the density of Y

$$g(y) = 0 \qquad \text{if } y \leq 0$$

and

$$g(y) = \frac{1}{2} \exp\left(-(\sqrt{y})^2\right) \qquad \text{if } y > 0.$$

Consequently, Y follows an exponential law with parameter $\frac{1}{2}$ and thus

$$\mathbf{E}Y = 2 \quad \text{and} \quad V(Y) = 4.$$

■

5.21 Let $a \in \mathbb{R}$ and define

$$f(x) = -x^2 + (a+b)x - ab \qquad \text{if } x \in [a, b]$$

and

$$f(x) = 0 \qquad \text{otherwise.}$$

Prove that f is a p.d.f. if and only if

$$b = a + 6^{\frac{1}{3}}.$$

Solution: By factorizing the function $-x^2 + (a+b)x - ab = (x - a)(b - x)$ we will see that it is always positive. On the other hand,

$$\int_{\mathbb{R}} f(x)\,dx = -\frac{(a-b)^3}{6}$$

and this gives the conclusion. ■

5.22 Let

$$f(x) = -x^2 + bx$$

on $[0, 1]$ and $f(x) = 0$ otherwise.

(a) Find the parameter b.

(b) Find the probability that the random variable lies between 0 and 1.

(c) Find the probability that the random variable lies between 0 and $\frac{1}{2}$.

Solution: Since

$$\int_{\mathbb{R}} f(x)\,dx = \int_0^1 (-x^2 + bx)\,dx = -\frac{1}{3} + \frac{b}{2}$$

we obtain $b = \frac{8}{3}$.

Clearly

$$P(0 \le X \le 1) = 1$$

while

$$P\left(0 \le X \le \frac{1}{2}\right) = \int_0^{1/2} (-x^2 + bx)\,dx = -\frac{1}{24} + \frac{8}{24} = \frac{7}{24}.$$

■

5.23 Let X be an r.v. with law $Exp(1)$. Define

$$Y = 1 - e^{-X}.$$

Find the law of Y.

Solution: The c.d.f of Y can be written as

$$G(y) = 1 - \mathbf{P}(e^{-X} \leq 1 - y).$$

Thus

$$G(y) = 1 - 0 \text{ if } 1 - y \leq 0$$

and

$$G(y) = 1 - (1 - \mathbf{P}(X \leq -\ln(1 - y))) \qquad \text{if } y < 1.$$

Finally,

$$G(y) = 1 \text{ if } y \geq 1,\ G(y) = 0 \qquad \text{if } y < 0$$

and

$$G(y) = y \qquad \text{if } y \in (0, 1).$$

We recognize the cdf of the uniform law $U(0, 1)$. ∎

Problems without Solution

5.24 Let X be a random variable with uniform distribution $U[-1, 1]$. Show that the density of the random variable $Y = X^3$ is given by

$$f_Y(y) = \frac{1}{6y^{\frac{2}{3}}} 1_{[-1,1]}(y).$$

Hint. Study the cumulative distribution function of Y

$$F_Y(y) = \mathbf{P}(X^3 \leq y)$$

with $y \in \mathbb{R}$. Distinguish the cases $y \leq -1$, $-1 \leq y \leq 0$, $0 \leq y \leq 1$ and $y \geq 1$.

5.25 **Buffon's needle problem.** Suppose that a needle is tossed at random onto a plane ruled with parallel lines a distance L apart, where the length of the needle is $l \leq L$.

What is the probability that the needle intersects one of the parallel lines?

Hint: Consider the angle that is made by the needle with the parallel lines as a random variable α uniformly distributed in the interval $[0, 2\pi]$ and the position of the midpoint of the needle as another random variable

ξ also uniform on the interval $[0, L]$. Then express the condition "needle intersects the parallel lines" in terms of the position of the midpoint of the needle and the angle α.

5.26 A random variable X has distribution function

$$F(x) = a + b \arctan \frac{x}{2}, \qquad -\infty < x < \infty.$$

Find:

(a) The constants a and b.

(b) The probability density function of X.

5.27 What is the probability that two randomly chosen numbers between 0 and 1 will have a sum no greater than 1 and a product no greater than $\frac{15}{64}$?

5.28 Choose a point A at random in the interval $[0, 1]$. Let L_1 (respectively L_2) be the length of the bigger (respectively smaller) segment determined by A on $[0, 1]$. Calculate:

(a) $\mathbf{P}(L_1 \leq x)$ for $x \in \mathbb{R}$.

(b) $\mathbf{P}(L_2 \leq x)$ for $x \in \mathbb{R}$.

5.29 Two friends decide to meet at the Castle gate of Stevens Institute. They each arrive at that spot at some random time between a and $a + T$. They each wait for 15 minutes and then leave if the other did not appear. What is the probability that they meet?

5.30 Recall the Laplace density:

$$f(x) = \begin{cases} \frac{1}{2}\lambda e^{\lambda x}, & \text{if } x \leq 0, \\ \frac{1}{2}\lambda e^{-\lambda x}, & \text{if } x > 0, \end{cases}$$

(a) Find the distribution function $F(x)$ for a Laplace random variable.

Now, let X and Y be independent exponential random variables with parameter λ. Let I be independent of X and Y and equally likely to be 1 or -1.

(b) Show that $X - Y$ is a Laplace random variable.

(c) Show that IX is a Laplace random variable.

(d) Show that W is a Laplace random variable where

$$W = \begin{cases} X, & \text{if } I = 1, \\ -Y, & \text{if } I = -1. \end{cases}$$

5.31 For any random variable X,

(a) Show that if $X \geq 0$ and $\mathbf{E}X = 0$ then $X = 0$ almost surely.

(b) Show that if $V(X) = 0$ then $X = \mathbf{E}X$ almost surely.

5.32 Recall that Laplace's law with parameter $\theta > 0$ has the density

$$f(x) = \frac{\theta}{2} e^{-\theta|x|}.$$

(a) Compute the c.d.f. F of Laplace's law. Calculate its inverse F^{-1}.
(b) Let $p \in (0, 1)$. Check that the binary variable defined by

$$X = 1 \qquad \text{if } U < p$$

and

$$X = 0 \qquad \text{otherwise}$$

follows Bernoulli's law with parameter p.

5.33 Let X be a random variable with density

$$f(t) = Kt^2 1_{[-\alpha,\alpha]}(t).$$

(a) Find K in terms of α.
(b) Find the cumulative distribution function of X.
(c) Calculate $P(X > \frac{\alpha}{2})$.
(d) Calculate $\mathbf{E}X$.

5.34 Let X a uniformly distributed r.v. on $[0, 1]$.
(a) Find the law of X^2 using its cumulative distribution function.
(b) Calculate the density of this law.

5.35 Let $X : \Omega \to [0, +\infty)$ a positive continuous random variable and for every $t \in \mathbb{R}$ let

$$F(t) = \mathbf{P}(X \le t)$$

the c.d.f. of X. Show that

$$\mathbf{E}(X) = \int_0^{+\infty} [1 - F(t)]\, dt.$$

5.36 Let $X : \Omega \to \mathbb{R}$ be an integrable continuous r.v. Show that

$$\mathbf{E}(X) = \int_0^{+\infty} \mathbf{P}(X > t)\, dt - \int_{-\infty}^{0} \mathbf{P}(X < t)\, dt.$$

Hint: Use the previous problem

5.37 Let f be a density function defined:

$$f(x) = Axe^{-x} \qquad \text{for } x \ge 0$$

and zero otherwise.

(**a**) Find the parameter A.

(**b**) Find the probability that a random variable X with density f lies between 0 and 1.

Hint: One gets

$$A = 1 \quad \text{and} \quad \mathbf{P}(0 \leq X \leq 1) = 1 - \frac{2}{e}.$$

5.38 Let X be a random variable with Gamma law $\Gamma(a, \lambda)$, $a, \lambda > 0$. Prove that the r.v. cX follows the Gamma law $\Gamma(a, \frac{\lambda}{c})$ for every $c > 0$.

5.39 Let $X \sim t_n$ a t distributed r.v. with pdf given by Eq. (5.8). Calculate the first two moments of the density ($\mathbf{E}[X]$ and $\mathbf{E}[X^2]$) when $n \geq 2$. Show that the expectation does not exist when $n = 1$ and likewise show that the variance does not exist unless $n > 2$.

CHAPTER SIX

Generating Random Variables

6.1 Introduction/Purpose of the Chapter

In this chapter we talk about methods used for simulating random variables. In today's world where computers are part of any scientific activity, it is very important to know how to simulate a random experiment to find out what expectations one may have about the results of the phenomenon. Applying the Central Limit Theorem (Chapter 12) and simulation methods allows us to draw conclusions about the expectations even if the distributions involved are very complex. However, in order to apply simulation methods, we need to be capable of generating random variables with the needed distribution.

Throughout this chapter we assume as given a Uniform(0,1) random number generator. Any software is capable of producing such random numbers, and the typical name for a uniform random variable is RAND. For a more recent development and a very efficient way to generate exponential and normal random variables without going to uniform, we refer the reader to Rubin and Johnson (2006). The ziggurat method developed by Marsaglia and Tsang (2000) remains one of the most efficient ways to produce uniforms, and it is used in Matlab. The Mersene twister is another efficient way to create these random numbers (this is the default method in R).

Handbook of Probability, First Edition. Ionuţ Florescu and Ciprian Tudor.

6.2 Vignette/Historical Notes

Before modern computers, researchers requiring random numbers would either (a) generate them through rolling dice and shuffling cards or (b) use tables of random numbers generated previously.

The first attempt to provide researchers with a supply of random numbers was in 1927 when the Cambridge University Press published a table of 41,600 digits developed by Leonard H. C. Tippet. In 1947, the RAND Corporation generated numbers by the electronic simulation of a roulette wheel; the results were eventually published in 1955 as A Million Random Digits with 100,000 Normal Deviates (Corporation, 2001, reprinted). John von Neumann was a pioneer in computer-based random number generators. In 1949, Derrick Henry Lehmer invented the linear congruential generator, used in most pseudo-random number generators today. With the advances and the widespread use of computers, algorithmic pseudo-random number generators replaced random number tables. Today, "true" random number generators capable of generating numbers without using a cycle as most pseudo-random number generators are very rare and indeed not used on a large scale.

6.3 Theory and Applications

6.3.1 GENERATING ONE-DIMENSIONAL RANDOM VARIABLES BY INVERTING THE CUMULATIVE DISTRIBUTION FUNCTION (c.d.f.)

Let X be a one-dimensional random variable defined on any probability space $(\Omega, \mathcal{F}, \mathbf{P})$ with cumulative distribution function $F(x) = \mathbf{P}(X \leq x)$. Before we proceed with the main lemma which provides the most classical generation method, let us formalize the notion of a quantile of a distribution.

Definition 6.1 *Suppose F is a distribution function. Let $\alpha \in (0, 1)$ be a constant. Define*

$$x^+(\alpha) = \inf\{z \in \mathbb{R} : F(z) > \alpha\},$$
$$x^-(\alpha) = \inf\{z \in \mathbb{R} : F(z) \geq \alpha\},$$

where $x^-(\alpha)$ is defined as the α-percentile (or α-quantile) of the distribution F.

Note that since α is fixed, there is nothing random in the definition above, and $x^+(\alpha)$ and $x^-(\alpha)$ are just numbers.

Furthermore, if F is continuous at the point, then $x^+(\alpha) = x^-(\alpha) := x_\alpha$ and in this case the distribution admits a unique α-percentile (equal to x_α).

However, if the function F is discontinuous at the point which happens in one of the two cases described in Figure 6.1, then we need to have a general

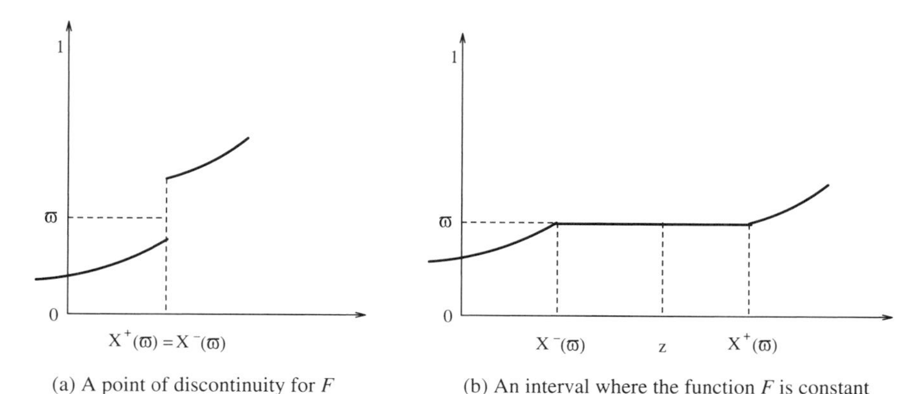

(a) A point of discontinuity for F

(b) An interval where the function F is constant

FIGURE 6.1 **Points where the two variables $X^{\pm}$ may have different outcomes.**

definition of α-percentile. This is why we chose $x^{-}(\alpha) = \inf\{z \in \mathbb{R} : F(z) \geq \alpha\}$ as the choice for the percentile.

In fact, look at Figure 6.1a. In this situation there is no problem since $x^{+}(\alpha) = x^{-}(\alpha)$. The issue appears when the α level is corresponding to the horizontal line in Figure 6.1b. In this case, any point between $x^{-}(\alpha)$ and $x^{+}(\alpha)$ is in fact an α-percentile. In this situation, $F^{-1}(\alpha)$ is the whole interval $[x^{-}(\alpha), x^{+}(\alpha)]$.

A percentile as defined above is a special case of a quartile. Both notions were born in statistics from observing practical applications. In statistics, one tries to infer the distribution of a random variable by observing repeated outcomes of the variable. Quantiles are outcomes corresponding to dividing the probabilities into equal parts. To be more specific, if we divide the probability into 2 equal parts, we obtain three points of which only the middle one is interesting (the others are minimum and maximum). This point is the median.

Similarly, if we divide the total probability into three equal parts, the two points that correspond to the 0.33 and 0.67 probability are called terciles. Quartiles are the best-known percentiles, with 0.25 and 0.75 probability corresponding to the first and third quartiles, respectively. The second quartile is again the median. To sum up, here are the most popular types of quantiles:

- The 2-quantile is called the median denoted M.
- The 3-quantiles (T_1, T_2) are called terciles.
- The 4-quantiles $(Q_1, M\ Q_3)$ are called quartiles.
- The 5-quantiles $(QU_1, \ldots, QU_4)$ are called quintiles.
- The 6-quantiles $(S_1, \ldots, S_5)$ are called sextiles.
- The 10-quantiles $(D_1, \ldots, D_9)$ are called deciles.
- The 12-quantiles $(Dd_1, \ldots, Dd_{11})$ are called duo-deciles.
- The 20-quantiles $(V_1, \ldots, V_{19})$ are called vigintiles.

- The 100-quantiles are called percentiles (denoted %).
- The 1000-quantiles are called permilles (denoted ‰).

Now that we understand what is a percentile, we come back to random numbers generation methodologies. All the distribution generation methods in this section are based on the following lemma.

Lemma 6.2 *The random variable $U = F(X)$ is distributed as a $U(0, 1)$ random variable. If we let $F^{-1}(u)$, denote the inverse function, that is,*

$$F^{-1}(u) = \{x \in \mathbb{R} \mid F(x) = u\},$$

then the variable $F(U)$ has the same distribution as X.

Note: The $F^{-1}(u)$ as defined in the lemma is a set. The set may contain a single number when the function is continuous. Please recall the discussion above and the issue when the distribution function is not continuous.

Proof: The proof is simple if F is a bijective function. Note that in this case we have

$$\mathbf{P}(U \leq u) = \mathbf{P}(F(X) \leq u).$$

But recall that F is a probability itself so the result above is zero if $u < 0$ and 1 if $u \geq 1$. If $0 < u < 1$ since F is an increasing function, we can write

$$\mathbf{P}(U \leq u) = \mathbf{P}(X \leq F^{-1}(u)) = F(F^{-1}(u)) = u,$$

and this is the distribution of a $U(0, 1)$ random variable.

If F is not bijective, the proof still works but we have to work with sets. The relevant case is once again $0 < u < 1$. Recall that F is increasing. Because of this and using the definition of F^{-1} above, we have

$$F^{-1}\left((-\infty, u]\right) = \{x \in \mathbb{R} \mid F(x) \leq u\}.$$

If the set $F^{-1}(u)$ has only one element (F is bijective at u), there is no problem; the set above is just $(-\infty, F^{-1}(u)]$ and the same proof works. If $F^{-1}(u) = \{x \in \mathbb{R} \mid F(x) = u\}$ has more than one element, then let $x_{\max}$ be the maximum element in the set (this exists and is in the set since $u < 1$) and we may write

$$F^{-1}\left((-\infty, u]\right) = (-\infty, x_{\max}].$$

Thus we have

$$\begin{aligned}
\mathbf{P}(U \leq u) &= \mathbf{P}(F(X) \leq u) = \mathbf{P}\left(X \in F^{-1}\left((-\infty, u]\right)\right) \\
&= \mathbf{P}\left(X \in (-\infty, x_{\max}]\right) = \mathbf{P} \circ X^{-1}((-\infty, x_{\max}]) \\
&= F(x_{\max})
\end{aligned}$$

by the definition of distribution function. Now recall that $x_{\max} \in F^{-1}(u)$, thus $F(x_{\max}) = u$. Once again we reach the distribution of a uniform random variable. ■

This lemma is very useful for generating random variables with prescribed distribution. We only have to calculate the function F^{-1} to be able to generate any distributions using only the uniform random number generator. This method works best when the distribution function F and its inverse F^{-1} have analytical formulas easy to compute. The best example of this situation is the exponential distribution.

EXAMPLE 6.1 Generating an exponential random variable

Suppose we want to generate an Exponential(λ) random variable—that is, a variable with density:

$$f(x) = \lambda e^{-\lambda x}\mathbf{1}_{\{x>0\}}.$$

Note that the expectation of this random variable is $1/\lambda$. This distribution can also be parameterized using $\lambda = 1/\theta$, in which case the expectation will be θ. The two formulations are equivalent.

We may calculate the distribution function in this case as

$$F(x) = (1 - e^{-\lambda x})\mathbf{1}_{\{x>0\}}.$$

We need to restrict this function to $F : (0, \infty) \to (0, 1)$ to have a bijection. In this case for any $y \in (0, 1)$ the inverse is calculated,

$$F(x) = 1 - e^{-\lambda x} = y \Rightarrow x = -\frac{1}{\lambda}\log(1 - y)$$
$$\Rightarrow F^{-1}(y) = -\frac{1}{\lambda}\log(1 - y).$$

So, to generate an Exponential(λ) random variable, first generate U (a Uniform(0,1) random variable) and simply calculate

$$-\frac{1}{\lambda}\log(1 - U);$$

this will have the desired distribution.

As a note, a further simplification may be made since $1 - U$ has the same distribution as U; we obtain the same exponential distribution by taking

$$-\frac{1}{\lambda}\log U.$$

We use one or the other formulation to generate exponentials and never both. Even though using U and $1 - U$ will produce two exponentially distributed numbers, they are going to be related—not independent.

For all discrete random variables the distribution function is a step function. In this case the c.d.f. F is not bijective, thus we need to restrict it somehow to obtain the desired distribution. The main issue is that the function is not surjective, so we need to know what to do when a uniform is generated.

EXAMPLE 6.2 Generating rolls of a six-sided fair die

Suppose we want to generate the rolls of a fair six sided die. The c.d.f. is easy to calculate as

$$F(x) = \begin{cases} 0 & \text{if } x < 1, \\ i/6 & \text{if } x \in [i, i+1) \text{ with } i = 1, \dots, 5, \\ 1 & \text{if } x \geq 6. \end{cases}$$

The inverse function is then

$$F^{-1}(0) = (-\infty, 1),$$

$$F^{-1}\left(\frac{1}{6}\right) = [1, 2),$$

$$\dots$$

$$F^{-1}\left(\frac{5}{6}\right) = [5, 6),$$

$$F^{-1}(1) = [6, \infty).$$

We can pick a point in the codomain, but that would not help since the inverse function will only be defined on the discrete set $\{0, 1/6, \dots, 5/6, 1\}$. Instead we extend the inverse function to $(0, 1)$ in the following way:

$$F^{-1}(y) = \begin{cases} 1 & \text{if } y \in \left(0, \frac{1}{6}\right) \\ i+1 & \text{if } y \in \left[\frac{i}{6}, \frac{i+1}{6}\right) \quad \text{with } i = 1, \dots, 5. \end{cases}$$

Thus, we first generate U a Uniform(0,1) random variable. Depending on its value, the roll of die is simulated as

$$Y = i + 1, \qquad \text{if } U \in \left[\frac{i}{6}, \frac{i+1}{6}\right) \quad \text{with } i = 0, \dots, 5.$$

EXAMPLE 6.3 Generating any discrete random variable with finite number of outcomes

The example above can be easily generalized to any discrete probability distribution. Suppose we need to generate the outcomes of the discrete random variable Y which takes n values $a_1, \dots, a_n$ each with probability $p_1, \dots, p_n$ respectively so that $\sum_{j=1}^{n} p_j = 1$.

To generate such outcomes, we first generate a random variable U as a Uniform(0,1) random variable. Then we find the index j such that

$$p_1 + \cdots + p_{j-1} \leq U < p_1 + \cdots + p_{j-1} + p_j.$$

The generated value of the Y variable is the outcome a_j.

Note that theoretically, the case when the generated uniform values of U are exactly equal to p_j do not matter since the distribution is continuous and the probability of this event happening is zero. However, in practice, such events do matter since the cycle used to generate the random variable is finite and thus the probability of the event is not zero—it is extremely small but not zero. This is dealt with by throwing away 1 if it is generated and keeping the algorithm as above.

Remark 6.3 *The previous example also covers the most commonly encountered need for generating random variables—the tossing of a coin or generating a Bernoulli(p) random variable. Specifically, first we generate U, a Uniform(0,1). If $U < p$ output* 1, *else output* 0.

6.3.2 GENERATING ONE-DIMENSIONAL NORMAL RANDOM VARIABLES

Generating normal (Gaussian) random variables is important because this distribution is the most widely encountered distribution in practice. In the Monte Carlo-type methods, one needs to generate millions of normally distributed random numbers; in practice, this means the precision with which these numbers are generated is crucial. Imagine if one in 1000 numbers are off, which means that in 1 million simulated numbers about 1000 are bad and thus the simulated path has 1000 places where the increment distribution is off.

Let us remark that if we know how to generate X, a normal variable with mean 0 and variance 1, that is, with density

$$f(x) = \frac{1}{\sqrt{2\pi}} e^{-\frac{x^2}{2}},$$

then we know how to generate any normal variable Y with mean μ and standard deviation σ by simply taking $Y = \mu + \sigma X$.

Thus, it is enough to learn how to generate $N(0, 1)$ random variables. The inversion methodology presented in the previous section cannot be applied directly since the normal CDF does not have an explicit functional form and therefore inverting it directly is impossible. However, one of the fastest and better methods (in fact the default in the R programming language) uses a variant of the inversion method.

Specifically, the algorithm developed by Wichura (1988) calculates quantiles corresponding to the generated probability values. The algorithm has two subroutines to deal with the hard-to-estimate quantiles from the tails of the gaussian distribution. First, the algorithm generates a probability p using a Uniform(0,1) distribution. Then it calculates the corresponding normally distributed value z_p by inverting the distribution function:

$$p = \int_{-\infty}^{z_p} \frac{1}{\sqrt{2\pi}} e^{-x^2/2} dx = \Phi(z_p),$$

$z_p = \Phi^{-1}(p)$. The respective subroutines PPND7 or PPND16 are chosen depending on the generated uniform value $|p - 0.5| \leq 0.425$ or $|p - 0.5| > 0.425$. These routines are polynomial approximations of the inverse function $\Phi^{-1}(\cdot)$. The algorithm has excellent precision (of the order 10^{-16}), and it runs relatively fast (only requires a logarithmic operation besides the polynomial operations).

Other methods of generating normally distributed numbers are presented below. All of them take advantage of transformations of random variables. Some are particular cases of more general methodology presented later in the chapter.

6.3.2.1 Taking Advantage of the Central Limit Theorem. Generate a number n of Uniform(0,1) random numbers. Then calculate their sum $Y = \sum_{i=1}^{n} U_i$. The exact distribution of Y is the so-called *Irwin–Hall* distribution, named after Joseph Oscar Irwin and Philip Hall, which has the probability density

$$f_Y(y) = \frac{1}{2(n-1)!} \sum_{k=0}^{n} (-1)^k \binom{n}{k} (y-k)^{n-1} \text{sign}(y-k),$$

where $\text{sign}(x) = |x|/x$ is the sign function (taking value -1 if x is negative and 1 if x is positive). This particular distribution has mean $n/2$ and variance $n/12$, as is very easy to see using the individual values for the uniforms. However, the Central Limit Theorem (Chapter 12) guarantees that as $n \to \infty$ the distribution of Y approaches the normal distribution. A simple algorithm designed around this result would generate, say, 12 uniforms and then it would calculate:

$$Y = \frac{\sum_{i=1}^{12} U_i - 12/2}{\sqrt{12/12}} = \sum_{i=1}^{12} U_i - 6$$

This new number is approximately distributed as an $N(0, 1)$ random variable. Since we use the 12 uniform numbers the range of the generated values is $[-6, 6]$, as opposed to the real normally distributed numbers which are taking values in $\mathbb{R}$.

Of course, taking a larger n will produce normals with better precision; however, observe that we need n uniform numbers to create one normal so the algorithm slows down considerably as n gets larger.

6.3.2.2 The Box–Muller Method. This method of generating normally distributed random numbers is named after George Edward Pelham Box and Mervin Edgar Muller, who developed the algorithm in 1958. The algorithm uses two independent Uniform(0,1) random numbers U and V. Then two random variables X and Y are calculated using

$$X = \sqrt{-2\ln U}\cos 2\pi V,$$
$$Y = \sqrt{-2\ln U}\sin 2\pi V.$$

Then the two random numbers X and Y have the standard normal distribution and are independent.

The independence and normality is easy to derive since for a bivariate normal random vector (XY), the variable $X^2 + Y^2$ is distributed as a chi-square random variable with two degrees of freedom. This is the same as an exponential random variable, and note that the quantity $-2\ln U$ has this exact distribution. Furthermore, the projection on the axes is determined by the angle between $[0, \pi]$, and this angle is chosen by the random variable V. Please note that the method produces two independent random numbers that can both be used later, but the method requires calculating the logarithm and the square root and applying trigonometric functions; thus the practical implementation may be quite slow.

6.3.2.3 The Polar Rejection Method. This method is due to Marsaglia and is a modification of the Box–Muller algorithm, which does not require computation of the trigonometric functions sin and cos. In this method, two random numbers U and V are drawn from the Uniform $(-1, 1)$ distribution, and then $S = U^2 + V^2$ is calculated. If S is greater or equal to one, then the method starts over by regenerating two uniforms; otherwise two new numbers

$$X = U\sqrt{-2\frac{\ln S}{S}},$$
$$Y = V\sqrt{-2\frac{\ln S}{S}}$$

are calculated. These X and Y are independent, standard normal random numbers. The Marsaglia polar rejection method is both faster and more accurate (because it does not require the approximation of three complicated functions) than the Box–Muller method. The drawback, though, is that unlike the Muller method, it may reject many generated uniforms until it reaches values that are acceptable.

In fact, a simple calculation shows that the probability of the generated pair representing coordinates of points inside the unit circle is $\pi/4$, or about 79% of generated pairs lie inside the circle.

6.3.3 GENERATING RANDOM VARIABLES. REJECTION SAMPLING METHOD

The polar method is in fact a particular case of the more general rejection sampling method presented next. In rejection sampling (also named the accept–reject method), the objective is to generate a random variable X having the *known* density function $f(x)$. The idea of this method is to use a different but easy-to-generate-from distribution $g(x)$. The method is very simple and originally was presented by John von Neumann. The idea of this algorithm lies with the Buffon needle problem (throwing the needle and accepting or rejecting, depending whether or not the needle touches the lines).

First, determine a constant M such that

$$\frac{f(x)}{g(x)} < M, \qquad \forall x$$

Once such M is determined, the algorithm is

Step 1. Generate a random variable Y from the distribution $g(x)$

Step 2. Accept $X = Y$ with probability $f(Y)/Mg(Y)$. If reject go back to step 1.

The accept–reject step can be easily accomplished using a Bernoulli random variable. Specifically, step 2 is as follows:

Step 2. Generate $U \sim$ Uniform(0, 1) and accept the generated Y if

$$U < \frac{f(Y)}{Mg(Y)};$$

go back to step 1 if reject.

Proposition 6.4 *The random variable X created by the rejection sampling algorithm above has the desired density $f(x)$.*

Proof: Let N be the number of necessary iterations to obtain the final number X. Let us calculate the distribution of X. Since each trial is independent, we have

$$\mathbf{P}\{X \leq x\} = \mathbf{P}\left\{Y \leq x \,\middle|\, U \leq \frac{f(Y)}{Mg(Y)}\right\} = \frac{\mathbf{P}\left(\{Y \leq x\} \bigcap \left\{U \leq \frac{f(Y)}{Mg(Y)}\right\}\right)}{\mathbf{P}\left\{U \leq \frac{f(Y)}{Mg(Y)}\right\}}.$$

Now the numerator is

$$\begin{aligned}
&\mathbf{P}\left(\left\{U \leq \frac{f(Y)}{Mg(Y)}\right\} \mid \{Y \leq x\}\right) \mathbf{P}(\{Y \leq x\}) \\
&= \int_{-\infty}^{x} \mathbf{P}\left(\left\{U \leq \frac{f(y)}{Mg(y)}\right\} \mid \{Y = y\}\right) g(y)\, dy \\
&= \int_{-\infty}^{x} \frac{f(y)}{Mg(y)} g(y)\, dy = \frac{1}{M} \int_{-\infty}^{x} f(y) dy.
\end{aligned}$$

Similarly the denominator is

$$\begin{aligned}
&\mathbf{P}\left\{U \leq \frac{f(Y)}{Mg(Y)}\right\} = \int_{-\infty}^{\infty} \mathbf{P}\left\{U \leq \frac{f(y)}{Mg(y)} \mid \{Y = y\}\right\} g(y)\, dy \\
&= \int_{-\infty}^{\infty} \frac{f(y)}{Mg(y)} g(y)\, dy = \frac{1}{M} \int_{-\infty}^{\infty} f(y)\, dy = \frac{1}{M}.
\end{aligned}$$

Taking the ratio of the two probabilities shows that X has the desired distribution. ■

Note that calculating the denominator in the proof above shows that the probability of accepting the generated number is always $1/M$. So if the constant M is close to 1, then the method works very efficiently. However, this is dependent on the shape of the densities f and g. If the density g is close in shape with f, then the method works very well. Otherwise, a large number of generated variates are needed to obtain one random number with density f.

Corollary 6.5 (Slight generalization) *Suppose that we need to generate from density $f(x) = C f_1(x)$ where we know that the functional form f_1 and C is a normalizing constant, potentially unknown. Then suppose we can find a density $g(x)$ easy to generate from and a constant M such that*

$$\frac{f(x)}{g(x)} < M, \qquad \forall x.$$

Then the rejection sampling procedure described above will create random numbers with density f.

The corollary is proven in exactly the same way as the main proposition. Sometimes the constant is hard to calculate, and this is why the corollary is useful in practice.

EXAMPLE 6.4 Generating from densities where only the functional form is known

Let us exemplify the practical value of the corollary. Suppose I want to generate from the density:

$$f(x) = Cx^2(\sin x)^{\cos x}|\log x|, \qquad x \in \left(\frac{\pi}{6}, \frac{\pi}{2}\right).$$

The constant C is chosen to make the density f integrate to 1. Note that actually calculating C is impossible. A plot of this density may be observed in Figure 6.2.

We wish to apply the rejection sampling to generate from the distribution f. To this end, we will use the uniform distribution on the interval $\left(\frac{\pi}{6}, \frac{\pi}{2}\right)$, and we shall calculate the constant M so that the resulting function is majoring the distribution. To do so, we calculate the maximum of the function, namely

$$m = \max_{x\in\left(\frac{\pi}{6}, \frac{\pi}{2}\right)} x^2(\sin x)^{\cos x}|\log x| = 1.113645,$$

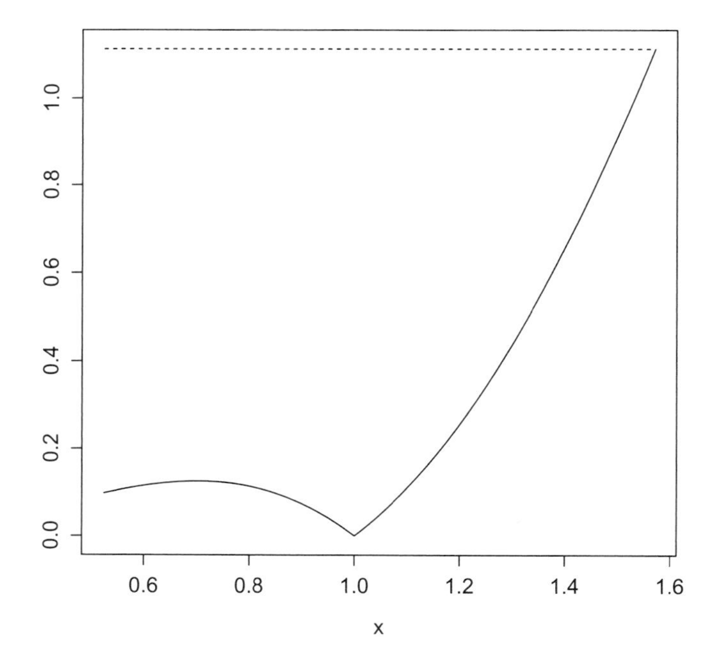

FIGURE 6.2 The function defining the density $f(\cdot)$ (continuous line) and the uniform distribution $M * g(\cdot)$ (dashed line) without the scaling constant C.

and we take

$$M = Cm\left(\frac{\pi}{2} - \frac{\pi}{6}\right).$$

With this constant M we are guaranteed that $f(x) < Mg(x)$ for every $x \in \left(\frac{\pi}{6}, \frac{\pi}{2}\right)$. To see this, recall that the density of the uniform on the desired interval $x \in \left(\frac{\pi}{6}, \frac{\pi}{2}\right)$ is constant $g(x) = \left(\frac{\pi}{2} - \frac{\pi}{6}\right)^{-1}$. Furthermore, the ratio that needs to be calculated is

$$\frac{f(x)}{Mg(x)} = \frac{x^2(\sin x)^{\cos x}|\log x|}{m}.$$

Obviously, this ratio is very good (approaches 1) when x is close to $\pi/2$, and it is close to 0 (as it should) when x is close to 1.

The following code is written in R and implements the rejection sampling for the example. The output of this code may be seen in Figure 6.3. Any line which starts with the # character is a comment line.

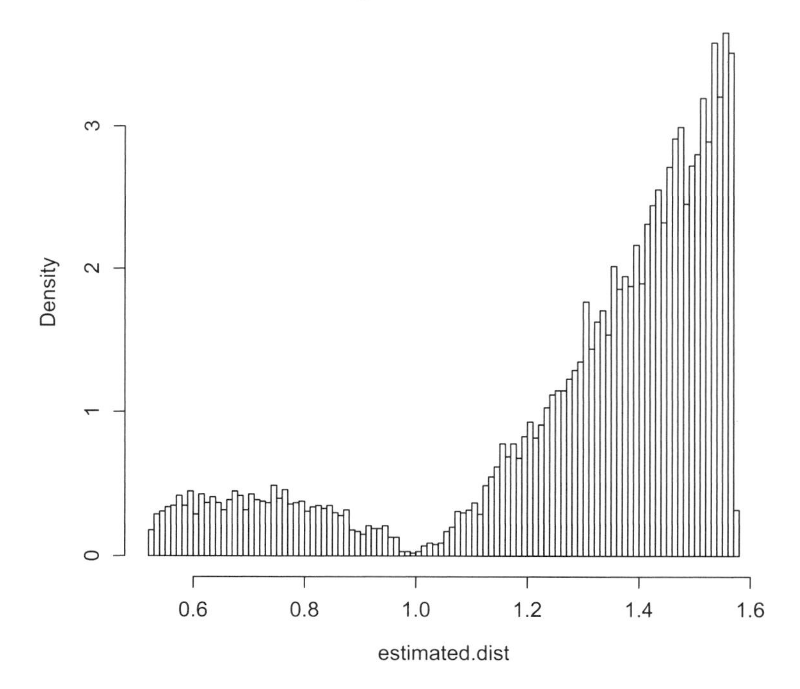

FIGURE 6.3 The resulting histogram of the generated values. This should be close in shape to the real function if the simulation is working properly. Note that this is a proper distribution and contains the scaling constant C.

```
##Example 6.4: Rejection Sampling R code
#We calculate the constant m used later

m=max(x^2*(sin(x)^cos(x))*abs(log(x)))

# Next defines the function used to calculate the ratio f(x)/M*g(x)

ratio.calc=function(x)
{return(x^2*(sin(x)^cos(x))*abs(log(x))/m)}

#The next function returns n generated values from the distribution
f(x)

random.f=function(n)
{GeneratedRN=NULL;
     for(i in 1:n)
          {OK=0;
     while(OK!=1){
          Candidate=runif(1,pi/6,pi/2);
          U=runif(1);
     if(U<ratio.calc(Candidate)){OK=1;GeneratedRN=c(GeneratedRN,
     Candidate)}}
          }
return(GeneratedRN)}

#Now we call the function we just created to generate 10,000
numbers

estimated.dist=random.f(10000)

#Finally, to check here is the histogram of these numbers

hist(estimated.dist,nclass=75)
```

Next we present a seemingly more complex example which has a simpler solution than the rejection sampling method.

EXAMPLE 6.5 An example generating from a beta mixture

As another exemplification let us generate from the mixture of beta distributions:

$$f(x) = 0.5\,\beta(x; 10, 3) + 0.25\,\beta(x; 3, 15) + 0.25\,\beta(x; 7, 10), \qquad x \in (0, 1),$$

where the $\beta(x, a, b)$ denotes the Beta distribution p.d.f. with shape parameters a and b:

$$\beta(x, a, b) = \frac{\Gamma(a+b)}{\Gamma(a)\Gamma(b)} x^{a-1}(1-x)^{b-1},$$

and $\Gamma(\cdot)$ is the gamma function:

$$\Gamma(x) = \int_0^\infty t^{x-1} e^{-t}\, dt.$$

A plot of the resulting distribution may be observed in Figure 6.4.

The mixture of distributions is always a distribution since the individual pdf's integrate to 1. Please note that the beta distribution is always distributed on (0, 1); and since we can use the uniform distribution on (0.1) to generate candidate values, the constant M can be chosen as the maximum value of the mixture gamma density function f. The code presented next uses the rejection sampling to generate random numbers from the desired distribution.

```
#We implement the mixture density
f=function(x)
{return(0.5*dbeta(x,10,3)+0.25*dbeta(x,3,15)+0.25*dbeta(x,7,10))}

#We calculate the constant M
M=max(f(x))
```

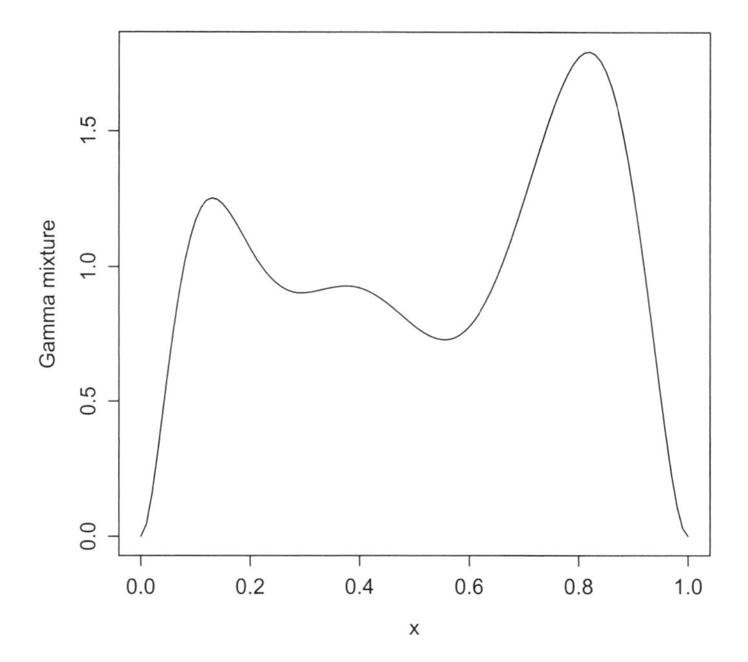

FIGURE 6.4 The mixture gamma density function (continuous line).

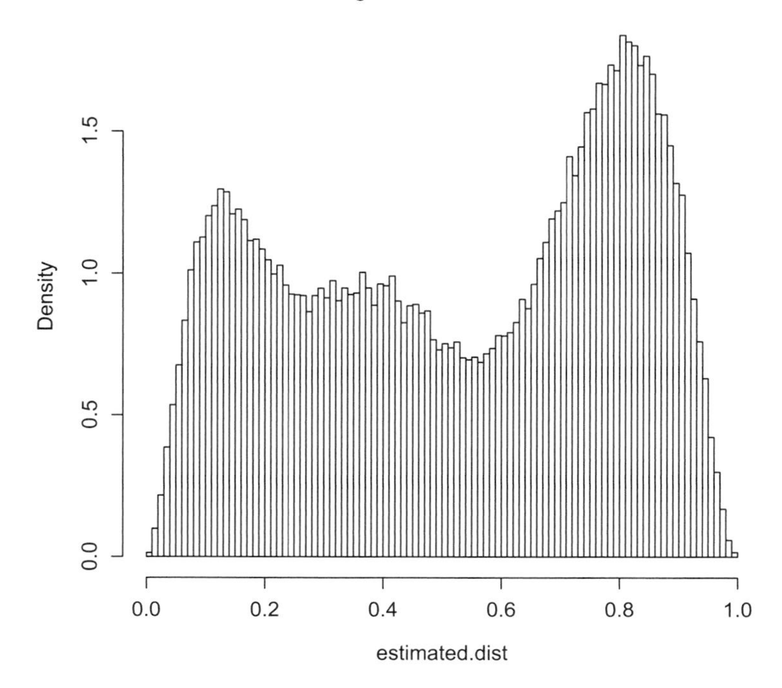

FIGURE 6.5 The resulting histogram of the generated values from the gamma mixture density. We had to use 100,000 generated values to see the middle hump.

```
#The next routine generates n numbers with desired distribution
random.mixturebeta=function(n)
{GeneratedRN=NULL;
for(i in 1:n)
{ OK=0;
while(OK!=1){
Candidate=runif(1);
U=runif(1);
if(U<f(Candidate)/M){OK=1;GeneratedRN=c(GeneratedRN,Candidate)}}
}
return(GeneratedRN)}

#Finally we verify by plotting the histogram of generated values
estimated.dist=random.mixturebeta(100000)
hist(estimated.dist,nclass=100,freq=F)
```

The resulting histogram may be observed in Figure 6.5. Once again the simulated values look good. However, we want to remark that the generation of simulated values may be made faster by taking advantage of the specific form of the distribution.

We note that the distribution in Example 6.5 is a type of distribution obtained by mixing three classical distributions. Such random variables are much easier to generate (and much faster) as the next section details.

6.3.4 GENERATING FROM A MIXTURE OF DISTRIBUTIONS

Suppose that the density we need to generate from is a mixture of simple, easy to generate from distributions. Specifically,

$$f(x) = \sum_{i=1}^{n} w_i f_i(x|\theta_i),$$

where the weights w_i sum to 1, and the densities f_i may all be different and dependent on the vectors of parameters θ_i. It is much easier to generate from such distributions, provided that we have implemented already generators for each of the distributions f_i. The idea is that the weights determine which distribution generates the respective random number.

Specifically, we first generate a random variable U as a Uniform(0,1) random variable. Then we find the weight index j such that

$$w_1 + \cdots + w_{j-1} \leq U < w_1 + \cdots + w_{j-1} + w_j.$$

Next, the desired random number is generated from the distribution f_j. Let us exemplify this generating strategy by continuing Example 6.5.

EXAMPLE 6.6 (Continuation of Example 6.5)

The objective is once again to generate random numbers from the mixture beta distribution:

$$f(x) = 0.5\,\beta(x; 10, 3) + 0.25\,\beta(x; 3, 15) + 0.25\,\beta(x; 7, 10), \qquad x \in (0, 1)$$

We will implement the code in R once again and we will take this opportunity to display some advanced R programming features.

The following code implements the generation of random numbers one by one, just as one would do it in C or in some other low-level language.

```
## Method of generation for mixture distributions.

random.mixturebeta.v2=function(n)
{GeneratedRN=NULL;
for(i in 1:n)
{U=runif(1);
RandomVal=ifelse(U<0.5,rbeta(1,10,3),ifelse(U<0.75,rbeta(1,3,15),
rbeta(1,7,10)))
```

```
GeneratedRN=c(GeneratedRN,RandomVal)
}
return(GeneratedRN)}

#And calling the function
random.mixturebeta.v2(100000)
```

The function *ifelse*(*CONDITION, VALUEIFYES, VALUEIFNO*) is R-specific, but the code above does not take advantage of the amazing strength of R which is working with vectors and large objects. The next function accomplishes the same thing, but it is much faster as we shall see.

```
## Method of generation for mixture distributions (optimized code).
random.mixturebeta.v2.optimal=function(n)
{U=runif(n); GeneratedRN=rep(0,n)
beta1=(U<0.5);beta2=(U>=0.5)&(U<0.75);beta3=(U>=0.75);
n1=GeneratedRN[beta1]; n2=GeneratedRN[beta2];n3=GeneratedRN[beta3];
GeneratedRN[beta1]=rbeta(n1,10,3);
GeneratedRN[beta2]=rbeta(n2,3,15)
GeneratedRN[beta3]=rbeta(n3,7,10)
return(GeneratedRN)}
```

In the code above, the *beta*1, *beta*2, and *beta*3 are vectors containing values TRUE and FALSE, depending on whether the respective condition is satisfied. When such vectors (containing TRUE and FALSE values) are applied as indices as in *GeneratedRN*[*beta*1], they select from the vector *GeneratedRN* only those values which correspond to the TRUE indices. This allows us to operate inside vectors very fast without going through vectors one by one. Furthermore, the code takes advantage of the internal R method of generating beta distributions, one of the best and fastest available in any statistical software.

We did not plot the resulting histograms of the generated values for the later two functions since they are very similar with Figure 6.5. However, the next table provides running times for the two methods as well as the optimized algorithm above.

Table 6.1 presents some interesting conclusions when looking at the two method and at the special optimized R implementation. Each of the three methods was run 30 times and a garbage collection was performed before each run so that interferences from the other processes run by the operating system were minimized. The first two columns directly compare the two methods.

TABLE 6.1 Average Running Time (in Seconds) for 30 Runs of the Three Methods[a]

	Rejection Sampling	Mixture Gen	Mixture Gen (Optimal)
Average time (sec):	20.122	20.282	0.034
Standard deviation:	0.409	0.457	0.008

[a]Each run generates a vector of 100,000 random numbers.

For each number having the desired distribution, the rejection sampling procedure generates minimum two uniform random numbers; and since it may reject the numbers produced, the number of uniforms generated may actually be larger. On the other hand, the mixture-generating algorithm always generates one uniform and one beta distributed number. Both algorithms produce numbers one after another until the entire set of 100,000 values is produced. The fact that the times are so close to each other tells us that generating uniforms only (even more than two) may be comparable in speed with generating from a more complex distribution.

When comparing the numbers in the second column with the numbers in the third column and we recall that it actually is the same algorithm, the only difference being that the optimized version works with vectors, we can see the true power of R on display. The 100,000 numbers are generated in about one-third of a second; this basically means that simulating a path takes nothing at all if done in this way and thus long simulations may be made significantly quicker by rethinking the code.

6.3.5 GENERATING RANDOM VARIABLES. IMPORTANCE SAMPLING

Rejection sampling works even if one only knows the approximate shape of the target density. As we have seen from the examples, any candidate distribution $g(\cdot)$ may be used, but in all examples we used the uniform distribution for it. This happens to always be the case in practice. There are two reasons for this. One is the fact that generating random uniforms is the most common random number generator as well as the fastest. Two, the constant M must be chosen so that $f(x) < Mg(x)$ for all x in the support of f. This is generally hard to assess unless one uses the uniform distribution and the constant M becomes related to the maximum value of $f(\cdot)$ over its support as we have already seen. However, this translates into rejecting a number of generated values which is *proportional* with the difference in the area under the constant function M and the area under $f(\cdot)$ (look at the ratio between the difference in areas and the area under the dashed line in Figure 6.2).

As we have seen, this is not so bad for one-dimensional random variables. However, it gets really bad really quickly as the dimension increases. Importance sampling tries to deal with this by sampling more from certain parts of the $f(\cdot)$ density.

It is important to realize that, unlike the methods presented thus far, the importance sampling method does not generate random numbers with specific density $f(\cdot)$. Instead the purpose of the importance sampling method is to estimate expectations. Specifically, suppose that X is a random variable (or vector) with known density function $f(x)$ and h is some other known function on the domain of the random variable X, then the importance sampling method will help us estimate:

$$\mathbf{E}[h(X)].$$

If we recall that the probability of some set A may be expressed as an expectation, namely

$$\mathbf{P}(X \in A) = \mathbf{E}[\mathbf{1}_A(X)],$$

then we see that the importance sampling method may be used to calculate any probabilities related to the random variable X as well. For example, probability of the tails of the random variable X is

$$\mathbf{P}(|X| > M) = \mathbf{E}[\mathbf{1}_{(-\infty,-M)}(X)] + \mathbf{E}[\mathbf{1}_{(M,\infty)}(X)],$$

for some suitable M, and both may be estimated using importance sampling.

6.3.5.1 The Idea and Estimating Expectations Using Samples. The laws of large numbers (either weak or strong) say that if $X_1, \ldots, X_n$ are i.i.d. random variables drawn from the distribution $f(\cdot)$ with finite mean $\mathbf{E}[X_i]$, then the sample mean converges to the theoretical mean $\mathbf{E}[X]$ (in probability or almost surely). Either way, the theory says that if we have a way to draw samples from the distribution $f(\cdot)$, be these $x_1, \ldots, x_n$, then for each function $h(\cdot)$ defined on the codomain of X we must have

$$\frac{1}{n}\sum_{i=1}^{n} h(x_i) \to \int h(x)f(x)\,dx = \mathbf{E}[h(X)].$$

Therefore the idea of estimating expectations is to use generated numbers from the distribution $f(\cdot)$. However, in many cases we may not draw from the density f and instead we use some easy to sample from density g. The method is modified using the following observation:

$$\mathbf{E}_f[h(X)] = \int h(x)f(x)\,dx = \int h(x)\frac{f(x)}{g(x)}g(x)\,dx = \mathbf{E}_g\left[h(X)\frac{f(X)}{g(X)}\right],$$

where we used the notations $\mathbf{E}_f$ and $\mathbf{E}_g$ to denote expectations with respect to density f respectively g. The expression above is only correct if the support of g includes the support of f; otherwise we can have points where $f(x) \neq 0$ and $g(x) = 0$ and thus the ratio $f(x)/g(x)$ becomes undefined.

6.3.5.2 The Algorithm Description. Combining the approximating idea with the expression above, it is now easy to describe the importance sampling algorithm to estimate $\mathbf{E}_f[h(X)]$.

1. Find a distribution g which is easy to sample from, and its support includes the support of f (i.e., if $f(x) = 0$ for some x will necessarily imply $g(x) = 0$).
2. Draw n sampled numbers from the distribution g: $x_1, \ldots, x_n$.

3. Calculate and output the estimate:

$$\frac{1}{n}\sum_{i=1}^{n} h(x_i)\frac{f(x_i)}{g(x_i)} = \sum_{i=1}^{n} h(x_i)\frac{f(x_i)}{ng(x_i)}.$$

The reason this method is called importance sampling is the so-called importance weight $\frac{f(x_i)}{ng(x_i)}$ given to x_i. The ratio $\frac{f(x_i)}{g(x_i)}$ may be interpreted as the number modifying the original weight $1/n$ given to each observation x_i. Specifically, if the two densities are close to each other at x_i then the ratio $\frac{f(x_i)}{g(x_i)}$ is close to 1 and the overall weight given to x_i is close to the weight $1/n$ (the weight of x_i if we would be able to draw directly from f). Suppose that x_i is in a region of f, which is very unlikely (small values of f). Then the ratio $\frac{f(x_i)}{g(x_i)}$ is going to be close to 0, and thus the weight given to this observation is very low. On the other hand, if x_i is from a region where f is very likely, then the ratio $\frac{f(x_i)}{g(x_i)}$ is going to be large and thus the weight $1/n$ is much increased.

6.3.5.3 Observations.

First note that the weights $\frac{f(x_i)}{ng(x_i)}$ may not sum to 1. However, their expected value is 1:

$$\mathbf{E}_g\left[\frac{f(X)}{g(X)}\right] = \int \frac{f(x)}{g(x)}g(x)\,dx = \int f(x)\,dx = 1,$$

thus the sum $\sum_{i=1}^{n} \frac{f(x_i)}{ng(x_i)}$ tends to be close to 1.

Second, the estimator

$$\hat{\mu} = \sum_{i=1}^{n} h(X_i)\frac{f(X_i)}{ng(X_i)}.$$

is unbiased and we can calculate its variance. That is,

$$\begin{aligned} \mathbf{E}[\hat{\mu}] &= \mathbf{E}_f[h(X)], \\ Var(\hat{\mu}) &= \frac{1}{n}Var_g\left(h(X)\frac{f(X)}{g(X)}\right). \end{aligned} \tag{6.1}$$

Third, the variance of the estimator obviously depends on the choice of the distribution g. However, we may actually determine the best choice for this distribution. Minimizing the variance of the estimator with respect to the distribution g means minimizing:

$$Var_g\left(h(X)\frac{f(X)}{g(X)}\right) = \mathbf{E}_g\left[h^2(X)\left(\frac{f(X)}{g(X)}\right)^2\right] - \mathbf{E}_f^2[h(X)].$$

The second term does not depend on g, while using the Jensen inequality in the first term provides

$$\mathbf{E}_g\left[\left(h(X)\frac{f(X)}{g(X)}\right)^2\right] \geq \left(\mathbf{E}_g\left[|h(X)|\frac{f(X)}{g(X)}\right]\right)^2 = \left(\int |h(x)|f(x)\,dx\right)^2.$$

However, the right side is not a distribution, but it does provide the *optimal importance sampling distribution*:

$$g^*(x) = \frac{|h(x)|f(x)}{\int |h(x)|f(x)\,dx}.$$

This is not really useful from a practical perspective since typically sampling from $f(x)h(x)$ is harder than sampling from $f(x)$. However, it does tell us that the best results are obtained when we sample from $f(x)$ in regions where $|h(x)|f(x)$ is relatively large. As a consequence of this, using the importance sampling is better at calculating $\mathbf{E}[h(X)]$ than using a straight Monte Carlo approximation (i.e., sampling directly from f and taking a simple average of the $h(x_i)$ values).

6.3.5.4 Practical Considerations. In practice it important that the estimator has finite variance (otherwise it never improves with n). To see this, please observe the formula for variance in (6.1). Here are sufficient conditions for the finite variance of the estimator $\hat{\mu}$:

- There exists some M such that $f(x) < Mg(x)$ for all x and $Var_f(h(X)) < \infty$, or
- The support of f is compact, f is bounded above, and g is bounded below on the support of f.

Remark 6.6 *Choosing the distribution g is crucial. For example, if f has support on $\mathbb{R}$ and has heavier tails than g, the weights $w(X_i) = f(X_i)/g(X_i)$ will have infinite variance and the estimator will fail.*

6.3.6 APPLYING IMPORTANCE SAMPLING

EXAMPLE 6.7

For this example we will showcase the importance of the choice of distribution g. The example (distributional form) is due to Nick Whiteley in his lecture notes on machine learning.

Suppose we want to estimate $\mathbf{E}[|X|]$ where X is distributed as a Student-t random variable with 3 degrees of freedom. In the notation used above,

$h(x) = |x|$ and $f(x)$ is the t-density function

$$\frac{\Gamma\left(\frac{\nu+1}{2}\right)}{\sqrt{\nu\pi}\Gamma\left(\frac{\nu}{2}\right)}\left(1+\frac{x^2}{\nu}\right)^{-\frac{\nu+1}{2}},$$

with degrees of freedom $\nu = 3$, and $\Gamma(x)$ is a notation for the gamma function used earlier in this chapter.

Please note that the target density does not have compact support; thus the use of a uniform density for g is not possible (see the Remark 6.6). To exemplify the practical aspects of the importance sampling algorithm, we shall use two candidate densities:

1. $g_1(x)$, the density of a t distribution with 1 degree of freedom.
2. $g_2(x)$, the standard normal density (N(0,1)).

   To compare we will also use the following:

3. A straight Monte Carlo where we generate directly from the distribution f.

The plot of these densities may be observed in Figure 6.6.

We know that the optimal choice is $|h(x)|f(x)$ (plotted in Figure 6.6); however, generating from this density is very complex. We may also observe that while the t density with 1 degree of freedom dominates the tails of the target distribution f, the candidate normal density is bellow the tails of the density so we expect the estimator produced using the normal density to perform badly (the weights $f(X_i)/g(X_i)$ have infinite variance).

Next, we present the R-code used for the importance sampling example.

```
#Straight Monte Carlo:
n=10:1500
nsim=100

straightMC=NULL;
for(i in n)
{mu=NULL;
for(j in 1:nsim)
{a=rt(i,3);mu=c(mu,mean(abs(a)))}
straightMC=cbind(straightMC,c(i,mean(mu),sd(mu)))
}

#Importance Sampling using first candidate:

usingt1=NULL;
for(i in n)
{mu=NULL;
for(j in 1:nsim)
{a=rt(i,1);mu=c(mu,mean(abs(a)*dt(a,3)/dt(a,1)))}
usingt1=cbind(usingt1,c(i,mean(mu),sd(mu)))
}
```

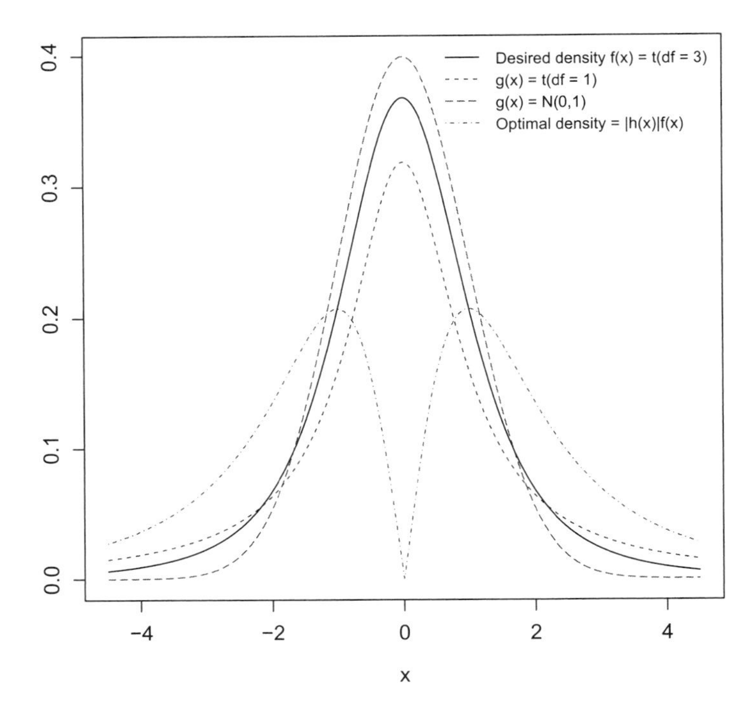

FIGURE 6.6 Candidate densities for the importance sampling procedure as well as the target density.

```
#Importance Sampling using second candidate:

usingnorm=NULL;
for(i in n)
{mu=NULL;
for(j in 1:nsim)
{a=rnorm(i);mu=c(mu,mean(abs(a)*dt(a,3)/dnorm(a)))}
usingnorm=cbind(usingnorm,c(i,mean(mu),sd(mu)))
}
```

We plot the results of this code in Figure 6.7. To see the evolution of the estimator, we use n (the number of generated samples) between 10 and 1500. The images present the behavior of this estimator as n increases. The estimator should converge to the right number, and the best estimator would have the tightest confidence bounds. To calculate the confidence intervals, we repeat the process 100 times for each value of n, which produces an estimate for the standard deviation of the estimator.

To have meaningful comparisons, we also generate from the f distribution and we simply use the average of h calculated at the generated values (this is the straight Monte Carlo technique).

Looking at the first two plots (Figures 6.7a and 6.7b), we see that in fact the importance sampling estimator is much better (superoptimal) than the straight

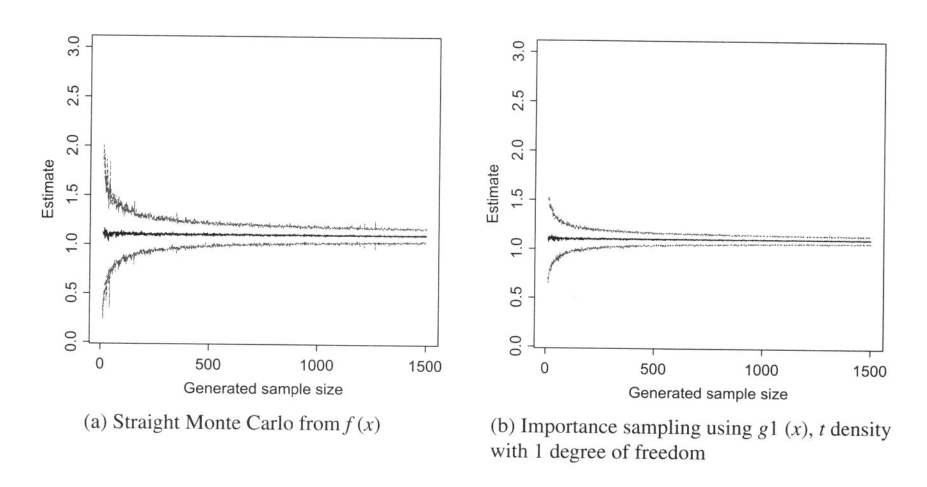

(a) Straight Monte Carlo from $f(x)$

(b) Importance sampling using $g1\ (x)$, t density with 1 degree of freedom

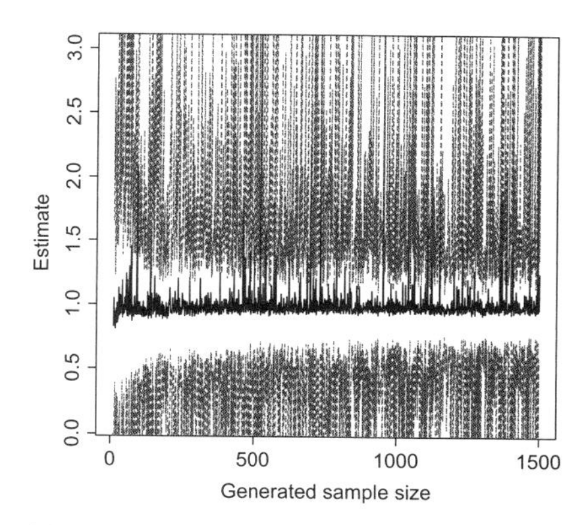

(c) Importance sampling using $g2\ (x)$, the standard normal density

FIGURE 6.7 The evolution of different importance sampling estimators of E[$|X|$]. The black line is the estimator while the gray lines give the estimated 95% confidence interval for the estimator.

Monte Carlo estimator. The reason is easy to see by looking at the plot of densities (Figure 6.6) and observing that the density g_1 is in fact closer to the optimal density $|h(x)|f(x)$ than f. Furthermore, Figure 6.7c displays the suspected poor performance of the importance sampling estimator obtained when generating observations from the normal density. The confidence intervals never decrease in width (since the variance never converges), and the convergence of the estimator itself to the right value is very slow.

6.3.7 PRACTICAL CONSIDERATION: NORMALIZING DISTRIBUTIONS

Oftentimes, in practical application, only the functional form of the distribution is known. For instance, the importance sampling methodology may be used to estimate an integral of the form

$$\int_A h(x)f(x)\,dx,$$

for some domain A. However, what if the function f does not integrate to 1 on the domain A? Or what if it integrates to a finite value but is not a density (it does integrate to 1 on the whole domain)? It turns out we can still use the methodology in this case as well. However, we need to use the so called *self-normalizing weights*.

Specifically, suppose that we need to estimate the expectation, namely:

$$\mathbf{E}[h(X)] = \int h(x)f(x)\,dx,$$

but the density f is only known up to a constant $f(x) = C_f\varphi(x)$, where the function $\varphi(\cdot)$ is known but the constant C_f is unknown.

The idea is to estimate the constant C_f as well. Note that since f is a density, we must have

$$C_f = \frac{1}{\int \varphi(x)\,dx} = \frac{1}{\int \frac{\varphi(x)}{g(x)}g(x)\,dx}.$$

Thus, C_f is approximately estimated by taking samples x_i from the distribution g and constructing:

$$\frac{1}{\sum_{i=1}^{n} \frac{\varphi(x_i)}{g(x_i)}}.$$

So proceeding exactly as in the straight importance sampling case, an estimator is

$$\tilde{\mu} = \frac{\sum_{i=1}^{n} h(X_i)\frac{\varphi(X_i)}{g(X_i)}}{\sum_{i=1}^{n} \frac{\varphi(X_i)}{g(X_i)}}.$$

Note that the estimator does not depend on the unknown constant C_f. Further note that the weights associated with each variable X_i drawn from the distribution g are normalized (the weights sum to 1).

Thus, the algorithm to estimate $\mathbf{E}_f[h(X)]$ is:

1. Find a distribution g which is easy to sample from and with its support including the support of f.
2. Draw n sampled numbers from the distribution g: $x_1, \ldots, x_n$.
3. Calculate and output the estimate:

$$\frac{\sum_{i=1}^{n} h(x_i)\frac{\varphi(x_i)}{g(x_i)}}{\sum_{i=1}^{n} \frac{\varphi(x_i)}{g(x_i)}}.$$

The estimator obtained is strongly consistent (it converges fast to the right estimate); however, the estimator is biased (the expected value of the estimator is under or over the target it estimates).

Lemma 6.7 *Suppose we are given a random variable X with density $f(x) = C_f\varphi(x)$ and a function h defined on the support of the function f. Let g be another density function such that its support includes the support of the function f, and define $w(x) = \varphi(x)/g(x)$. Suppose that $X_1, \ldots, X_n$ are i.i.d. random variables with density g. Then the estimator*

$$\tilde{\mu} = \frac{\sum_{i=1}^{n} h(X_i)w(X_i)}{\sum_{i=1}^{n} w(X_i)}$$

is strongly consistent, that is,

$$\tilde{\mu} \xrightarrow{a.s.} \mathbf{E}_f[h(X)] = \mu;$$

furthermore,

$$\mathbf{E}_g[\tilde{\mu}] = \mu + \frac{\mu\, Var_g(w(X_1)) - Cov_g(w(X_1), h(X_1)w(X_1))}{n} + O(n^{-2}),$$

$$Var_g(\tilde{\mu}) = \frac{Var_g(h(X_1)w(X_1) - \mu w(X_1))}{n} + O(n^{-2}).$$

The consistence result is immediate using the strong law for each of the two integrals in the estimator's formula. However, the proof of the biasedness of the estimator is quite computational, and we refer the reader to recent work (Whiteley and Johansen, 2011) for more details and a proof.

You may also wonder how it is possible for the estimator to be biased but still converge to the right place. Just look at the expressions in Lemma 6.7 and observe that the bias goes to 0 with n.

6.3.8 SAMPLING IMPORTANCE RESAMPLING

As we mentioned, the importance sampling technique is primarily used to calculate expectations. However, it is possible to adapt the technique to obtain samples from the distribution f. To do this, recall that

$$\mathbf{P}(X \in A) = \mathbf{E}_f[\mathbf{1}_A(X)].$$

Thus, following the procedure already detailed, one approximates the probability with

$$\mathbf{P}(X \in A) = \frac{1}{N}\sum_{x_i} \mathbf{1}_A(x_i)\frac{f(x_i)}{g(x_i)}.$$

Thus only the generated random numbers that fall into the set A are actually counted.

One may take this argument further and obtain an approximate discrete distribution for f by

$$\hat{p}(x) = \sum_{i=1}^{N} \frac{f(x_i)}{g(x_i)} \mathbf{1}_{\{x_i\}}(x) = \sum_{i=1}^{N} w(x_i) \mathbf{1}_{\{x_i\}}(x),$$

where the x_i values are generated from the density g and obviously the corresponding weights $w(x_i) = \frac{f(x_i)}{g(x_i)}$ may not sum to 1. To generate M independent random variables from the distribution f, one normalizes this discrete distribution $\hat{p}$ and just generates from it M values where M is much larger than N. Once the M new values are obtained—denoted, say, $\{y_1, \ldots, y_M\}$—the new estimated f distribution is

$$\tilde{p}(x) = \frac{1}{M} \sum_{i=1}^{M} \mathbf{1}_{\{y_i\}}(x).$$

Please note that since the values y_i are generated from the x_i values, the new sample $\{y_1, y_2, \ldots, y_M\}$ contains repeated observations. This method is known as "*sampling importance resampling*" and is due to Rubin (1998). In fact, the technique is a simple bootstraping technique and it is not clear to us whether the technique is better than simply using the $\hat{p}$ distribution directly to generate random variables. However, the paper is cited extensively in computer vision and machine learning literature, so we decided to include the technique here.

6.3.9 ADAPTIVE IMPORTANCE SAMPLING

As we have shown in Example 6.7, the proper choice of g can lead to superefficient estimation algorithms. Specifically, using a g distribution which is close to $|h(x)|f(x)$ is much more efficient than using a straight Monte Carlo method (i.e., with $g(x) = f(x)$). However, as the dimension of x increases (x becomes a random vector with many dimensions), it becomes more complicated to obtain a suitable $g(x)$ from which to draw the samples. A strategy to deal with this problem is the *adaptive importance sampling* technique, which seems to have originated in the structural safety literature (Bucher, 1988).

The method considers a parameterized distribution $g(x|\theta)$, where θ is a parameter vector which is adaptable depending on the sampling results. The idea of the method is to try and minimize the variance of the estimator $\hat{\mu}$. Specifically, consider a parameterized distribution $g(x|\theta)$. We want to minimize

$$\mathbf{E}_g[f^2(X)w^2(X)] - \mathbf{E}_f^2[h(X)],$$

with respect to g where $w(x|\theta) = \frac{f(x)}{g(x|\theta)}$. Note that the second term does not depend on g at all and that minimizing involves calculating the derivative of the first term with respect to θ. Since derivative and expectation commute on a probability space if the expectation exists, the problem reduces to finding the roots of the derivative:

$$D(\theta) = 2\mathbf{E}_g\left[f^2(x)w(x|\theta)\frac{\partial w}{\partial \theta}(x|\theta)\right].$$

A Newton–Raphson iteration procedure would find the minimum of the original expression by using

$$\theta_{n+1} = \theta_n - (\nabla D(\theta_n))^{-1} D(\theta_n).$$

However, the expectation $D(\theta_n)$ is hard to compute exactly. Furthermore, the inverse of the gradient of D (or the hessian) of the original expression to be minimized is even harder to calculate. Instead, the algorithm simply replaces the expression $(\nabla D(\theta_n))^{-1}$ with a learning constant and the $D(\theta_n)$ with its sample value.

To describe the algorithm, we give a pseudo-code below. The technique starts with a general distribution $g(x|\theta)$ capable of many shapes as the θ parameter varies in some parameter space Θ. Then the method *adapts* the parameter θ to fit the problem at hand. To this end:

- Start with an initial parameter value θ_0 (which produces some indifferent shape of the distribution g), and let $n = 0$.
- Do the following until the difference $|\theta_{n+1} - \theta_n| < \varepsilon$, where ε is a prespecified tolerance level.
  - Generate N values $x_1, \ldots, x_N$ from the distribution $g(x|\theta_n)$.
  - Update the θ value using

$$\theta_{n+1} = \theta_n - \alpha\frac{1}{N}\sum_{i=1}^{N} f^2(x_i)w(x|\theta_n)\frac{\partial w}{\partial \theta_n}(x_i|\theta_n).$$

- Check the condition $|\theta_{n+1} - \theta_n| < \varepsilon$. If not satisfied, let $n = n + 1$ and repeat the loop.

6.4 Generating Multivariate Distributions with Prescribed Covariance Structure

This section talks about a multivariate distribution problem. Suppose that we have a random number generator (from any distribution) and thus we can generate a vector with all components independent. But what if we want to generate a vector with a given correlation structure?

We look at a related problem involving normals.

6.4.1.1 Question. How do we generate a normal vector with known and given mean vector μ and covariance matrix Σ?

Suppose that we can generate $X = (X_1, \dots, X_n)$ a vector of n independent standard normal random variables ($X_i \sim N(0, 1)$). Clearly the correlation (and covariance matrix) is I_n, the identity matrix with 1 on the diagonal and 0 everywhere else. The mean vector is all 0.

Suppose that we can find a n-dimensional square matrix R such that $R^T R = \Sigma$. R^T denotes the transpose of the matrix R.

Then the vector $R^T X + \mu$ is an n-dimensional vector of normals having the required structure.

This claim is very easy to prove by just using the properties of the mean and covariance of vectors. First, note that $R^T X + \mu$ is a vector with n dimensions and that each component is a linear combination of independent normals. We know that this is a normal. Therefore the Gaussian vector is determined completely by its mean and covariance matrix.

Now by simply calculating the mean and covariance of the vector $R^T X + \mu$, we see that we obtain the desired distribution.

The practical application of the theory then involves finding the matrix R. The decomposition presented next is an implementation of exactly this.

Cholesky Decomposition. Given a symmetric positive definite matrix Σ, there exists U, an upper triangular matrix, and D, a diagonal matrix, such that

$$\Sigma = U^T D U.$$

If we can calculate the U and D matrices from the Cholesky decomposition, then we can write

$$\Sigma = U^T D U = (U^T \sqrt{D})(\sqrt{D} U) = (\sqrt{D} U)^T (\sqrt{D} U),$$

where $\sqrt{D}$ is the diagonal matrix having the elements on the diagonal equal to the squared root of the respective diagonal elements in D. Therefore the matrix R is just simply

$$R = \sqrt{D} U.$$

Cholesky Decomposition in Practice. If one wants to code the Cholesky decomposition in C, one has to look at the method or to use libraries that readily compute the U and D matrices. A source is the numerical recipes book Press et al. (2007). The 2nd edition from 1992 is still free and available online. Section 2.9 in the older edition presents an algorithm written in C which performs the Cholesky decomposition.

You may also use MATLAB or R. In both programs the command is "chol()"—the argument of the function is the Σ matrix for which the decomposition is required. Also note that the R function provides the upper triangular matrix $R = \sqrt{D} U$ directly, skipping the D and $\sqrt{D}$ calculation. The matrix R is normally all that is needed in practice.

As a remark, please note that the input matrix should be symmetric (easy to check) and positive definite (harder a bit). Any symmetric positive definite matrix has positive eigenvalues. So to check if the matrix is positive definite, run the function "eigen(matrixname)" in R and inspect the resulting eigenvalues. If they are all positive, then the matrix is good to go.

6.4.1.2 Generating a Vector of Correlated Normals—Practice. If you followed the theoretical presentation above, it should be easy to generate the vectors required. The researcher would generate a sample of n standard normals and put them in an n-dimensional vector X_1; then $R^T X_1 + \mu$ is the first vector with the desired correlation structure. The process is repeated M times to create a sample of M observations.

6.4.1.3 Extension to Generating Correlated Brownian motions. Since a popular problem is to generate Brownian motions, we shall give a pseudo-code that will accomplish this. When generating Brownian motions, one needs to take advantage of the fact that that the increments of a one-dimensional Brownian motion over any interval of length Δt are independent and distributed as $N(0, \Delta t)$.

Using this, the goal is to generate

$$\mathbf{B}(t) = \begin{pmatrix} B_1(t) \\ B_2(t) \\ \vdots \\ B_n(t) \end{pmatrix}$$

such that the correlation matrix of the Brownian motion is ρt for some given matrix ρ.

The algorithm generates the increments and then it sums them.

- First calculate the Cholesky decomposition of the correlation matrix $\overline{\rho} = R^T R$. Please note that we can calculate very easily the Cholesky decomposition of the matrix $\rho \Delta t$ by simply multiplying R with $\sqrt{\Delta t}$.
- Start with the initial values and decide on a time increment Δt.
- For every interval $[t, t + \Delta t]$, do the following:

1. Generate n independent Normal(0,1) random variables and construct a vector $\Delta \mathbf{x}$ with these components.
2. Calculate the vector of increments on the interval as

$$\Delta \mathbf{B}(t) = R^T \Delta \mathbf{x} \sqrt{\Delta t},$$

where $R^T \Delta x$ represents regular multiplication of two matrices and the last operation is multiplication with a scalar (all elements of the matrix get multiplied with Δt).

3. The process value at some time $T = k\Delta t$ is

$$\mathbf{B}(T) = \sum_{i=1}^{k} \Delta\mathbf{B}(i\Delta t)$$

Please note that $R = \sqrt{D}U$ is upper triangular and we are multiplying with R^T which should be lower triangular. If at step 2 the matrix multiplying the normals is anything else, you have a mistake in the code. Of course you could also generate independent Normal$(0, \sqrt{t})$ in step 1 directly and only multiply with R^T to obtain the right increments.

EXERCISES

6.1 Look at the Box–Muller and the two resulting variables X and Y. Calculate the joint and marginal distributions of these variables and show that they are independent.

6.2 Look at the polar rejection method. Show that the two variables given by this algorithm are independent.

6.3 Consider the following normal mixture density:

$$f(x) = 0.7\frac{1}{\sqrt{2\pi 9}}e^{-\frac{(x-2)^2}{18}} + 0.3\frac{1}{\sqrt{2\pi 4}}e^{-\frac{(x+1)^2}{8}}.$$

(a) Calculate the expected value of a random variable with this distribution.

(b) Write code and implement to generate random variables with this distribution. Use your mind or whatever methods you learned in this chapter.

(c) Use the previous part to generate 1000 independent random numbers with this density. Calculate the sample mean (average of generated numbers). Compare with answer in part a.

(d) Repeat the previous part, but this time generate 10,000 numbers. Comment.

6.4 Consider the random variable X with the density:

$$f(x) = C\cos(x)\sin(x), \qquad x \in [0, \pi/2].$$

(a) Calculate the constant C which makes above a probability density.

(b) Sketch this density.

(c) Implement the importance sampling method to generate random numbers from the density. Create a histogram by generating 1000 such numbers and compare with the previous part.

(d) Generate 1000 such numbers and use them to estimate the probability:

$$\mathbf{P}(X > \pi/12).$$

(e) Calculate the same probability by integrating the density. Are the two numbers close?

(f) Generate 10, 000 and use them to estimate

$$\mathbf{E}\left[e^{\sin(x)}\right].$$

6.5 Repeat the previous exercise for the density

$$f(x) = Ce^{-\tan(x)}, \qquad x \in [0, \pi/2].$$

Skip part (e) in the previous problem.

CHAPTER SEVEN

Random Vectors in $\mathbb{R}^n$

7.1 Introduction/Purpose of the Chapter

Often, we have more than one random variable describing the same object. For example, height and weight of a person are two different random variables. Each of the variables may be considered separately; but usually they have probabilistic ties, which means they have to be studied jointly. There is only one case where considering variables separately or jointly is identical: the case of independent variables. In probability theory, a random vector is any finite collection of real-valued random variables. More precisely, a random vector is a measurable function $X : \Omega \to \mathbb{R}^N$. It is usually denoted by

$$X = (X_1, X_2, \ldots, X_N),$$

where each component is one-dimensional. A random vector is sometimes called a multidimensional random variable. The X_i, $i = 1, \ldots, N$, are the components of the vector X. Each component is a random variable from Ω to $\mathbb{R}$.

7.2 Vignette/Historical Notes

The study of random vectors is very important today. In the big data studies we deal with many characteristics measures for each individual. In statistics the study of these vectors is called the multivariate analysis. Anderson (1958), in this book entitled *An Introduction to Multivariate Analysis* educated a generation of

---

Handbook of Probability, First Edition. Ionuţ Florescu and Ciprian Tudor.

theorists and applied statisticians on the statistical principles of this area. Analysis of variance (ANOVA), multivariate regression, and principal components analysis (PCA) are all techniques born from the need of describing the distribution of complex large-dimensional distributions. Today, the evolution of these distributions in time (multivariate time series) is the subject of intense research.

Interestingly enough, the marginal distributions take their name from the initial bivariate joint tables (which were discrete) and therefore the distribution of the individual components was written in the margins of the table—thus marginal distribution.

7.3 Theory and Applications

7.3.1 THE BASICS

As is the case for random variables, the random vectors are characterized by their distribution function.

Definition 7.1 *Let $X = (X_1, \dots, X_N)$ be a random vector. Its (cumulative) distribution function $F_X : \mathbb{R}^N \to [0, 1]$ is given by*

$$F_X(x_1, \dots, x_N) = \mathbf{P}\left(\{X_1 \le x_1\} \bigcap \{X_2 \le x_2\} \bigcap \dots \bigcap \{X_N \le x_N\}\right)$$

for every $x_1, \dots, x_N \in \mathbb{R}$. The right-hand term is usually denoted by

$$\mathbf{P}(X_1 \le x_1, \dots, X_N \le x_N).$$

The cumulative distributions function of a random vector has the following properties.

Proposition 7.2 *If $(X_1, \dots, X_n)$ is a random vector with c.d.f. F, then:*

1. *The inclusion–exclusion formula holds:*

$$\begin{aligned}&\mathbf{P}(a_1 \le X_1 < b_1, \dots, a_n \le X_n \le b_n)\\&= F(b_1, \dots, b_n) - \sum_{i=1}^{n} F(b_1, \dots, b_{i-1}, a_i, b_{i+1}, \dots, b_n\\&+ \sum_{i,j=1;\, i<j}^{n} F(b_1, \dots, a_i, \dots, a_j, \dots, b_n) - \cdots\\&+(-1)^n F(a_1, \dots, a_n)\end{aligned}$$

   whenever $a_i, b_i \in \mathbb{R}$ with $a_i < b_i$ for every $i = 1, \dots, n$.
2. *The function F is increasing in each of its arguments.*
3. *The function F is left continuous in each argument.*

4. We have

$$\lim_{x_1,\ldots,x_n\to\infty} F(x_1,\ldots,x_n) = 1,$$

and for each $k \in \{1,\ldots,n\}$ we obtain

$$\lim_{x_k\to-\infty} F(x_1,\ldots,x_n) = 0.$$

Proof: The proof is left as an exercise. The reader is referred to the proof of Proposition 3.9 as a reference. ■

The random vectors are characterized by the probability distribution which is defined next. This notion generalizes the distribution of real-valued random variables to vectors.

Definition 7.3 (Distribution of random vectors) *Let $(\Omega, \mathcal{F}, \mathbf{P})$ be a probability space and let $X : \Omega \to \mathbb{R}^N$ be a random vector.*

The law (or the distribution) of X, denoted by $\mathbf{P}_X$, is a function defined on the Borel sets of $\mathbb{R}^N$ (denoted $\mathcal{B}(\mathbb{R}^N)$), with values in $[0, 1]$ by

$$\mathbf{P}_X(B) = \mathbf{P}(\{\omega : X(\omega) \in B\}) = \mathbf{P}\left(X^{-1}(B)\right) = \mathbf{P} \circ X^{-1}(B)$$

for every $B \in \mathcal{B}(\mathbb{R}^N)$.

Proposition 7.4 *The application $\mathbf{P}_X$ is a probability on the measurable space $\left(\mathbb{R}^N, \mathcal{B}(\mathbb{R}^N)\right)$. Consequently, $\left(\mathbb{R}^N, \mathcal{B}(\mathbb{R}^N), \mathbf{P}_X\right)$ is a probability space.*

Proof: The proof is analogous to the proof of Proposition 3.6. ■

7.3.2 MARGINAL DISTRIBUTIONS

A random vector is made of several random variables. These random variables are called the components of the vector. Each subset of components has its own probability distribution, and all these distributions are called the marginal distributions of the vector.

Definition 7.5 (Marginal distribution) *Let $X = (X_1,\ldots,X_N)$ be a random vector. For any subset $\{i_1,\ldots,i_k\}$, the distribution of the vector $(X_{i_1},\ldots,X_{i_k})$ is called a marginal distribution and it may be calculated from the overal distribution. In particular, the law of the random variable X_i, for $i \in \{1,\ldots,N\}$ is called the i-th marginal distribution of the vector X.*

EXAMPLE 7.1 A simple example of a vector with independent components

> Roll independently two dice. Denote by X and Y the outcomes of the two dice. Define the random vector
>
> $$Z = (X, Y).$$
>
> Describe the vector Z.

Recall that any random variable or vector is entirely characterized by its distribution. To describe the distribution of Z, we first note that $Z(\Omega) = X(\Omega) \times Y(\Omega) = \{1, \ldots, 6\} \times \{1, \ldots, 6\}$, which is a finite set (it has 36 elements), and

$$\mathbf{P}(Z = (i, j)) = \mathbf{P}(X = i, Y = j) = \mathbf{P}(X = i)\mathbf{P}(Y = j) = \frac{1}{6} \times \frac{1}{6} = \frac{1}{36}$$

for every $i, j \in \{1, \ldots, 6\}$ because each die is fair and the outcome of one die is independent of the outcome of the other die. The laws of X and Y (which constitute the two marginal distributions of the vector Z) are given by

$$X(\Omega) = Y(\Omega) = \{1, 2, 3, 4, 5, 6\}$$

and

$$\mathbf{P}(X = i) = \frac{1}{6} = \mathbf{P}(Y = j)$$

for every $i, j \in \{1, \ldots, 6\}$.

The distribution of the vector (X, Y) can be represented by the following table.

Y/X	1	2	3	4	5	6
1	1/36	1/36	1/36	1/36	1/36	1/36
2	1/36	1/36	1/36	1/36	1/36	1/36
3	1/36	1/36	1/36	1/36	1/36	1/36
4	1/36	1/36	1/36	1/36	1/36	1/36
5	1/36	1/36	1/36	1/36	1/36	1/36
6	1/36	1/36	1/36	1/36	1/36	1/36

Remark 7.6 *Given the law of a random vector $X = (X_1, \ldots, X_N)$, the marginal distributions can be easily obtained. Indeed, concerning the law of X_1, we can write*

$$\mathbf{P}_{X_1}(A) = \mathbf{P}(X_1 \in A) = \mathbf{P}(X \in A \times \mathbb{R} \times \cdots \times \mathbb{R}) = \mathbf{P}_X(A \times \mathbb{R} \times \cdots \times \mathbb{R})$$

for every $A \in \mathcal{B}(\mathbb{R})$. Moreover, concerning the c.d.f. of the components, the following rules applies: If (X, Y) is a random vector, then

$$F_X(x) = \lim_{y \to \infty} F_{(X,Y)}(x, y)$$

and

$$F_Y(y) = \lim_{x \to \infty} F_{(X,Y)}(x, y)$$

for every $x, y \in \mathbb{R}$. But the converse direction is not true: That is, if we know each marginal distribution, we cannot obtain in general the law of the vector. The law of the vector contains more complex information in the sense that it says how the components of the vector are connected one to each other.

Remark 7.7 *If the components of the vector X are independent, then the law of the vector can be obtained from the marginal laws. Indeed,*

$$\begin{aligned}&\mathbf{P}(X \in (A_1 \times \cdots \times A_N) \\ &= \mathbf{P}(X_1 \in A_1, \ldots, X_N \in A_N) = \mathbf{P}(X_1 \in A_1) \cdots \mathbf{P}(X_N \in A_N).\end{aligned}$$

We will treat separately the discrete and continuous random vectors.

7.3.3 DISCRETE RANDOM VECTORS

Definition 7.8 *We will say that a random vector is discrete if $X(\Omega)$ is a finite or countable subset of $\mathbb{R}^n$.*

In the single-variable case, the probability function of a discrete random variable X assigns nonzero probabilities to a countable number of distinct values of X such that the sum of the probabilities is equal to 1. Similarly, in the bivariate case (two-dimensional random vector) the joint probability function $p(x, y)$ assigns nonzero probabilities to only a countable number of pairs of values (x, y). Further, the nonzero probabilities must sum to 1.

Remark 7.9 *In this case the law of $X = (X_1, X_2, \ldots, X_N)$ is completely determined by the set $X(\Omega)$ and the probabilities*

$$\mathbf{P}_X(\{x\}) = \mathbf{P}(X = x) = \mathbf{P}(X_1 = x_1, X_2 = x_2, \ldots, X_N = x_N)$$

for every $x = (x_1, x_2, \ldots, x_N) \in X(\Omega)$. Further, we have

$$\sum_{x \in X(\Omega)} \mathbf{P}(X = x) = 1.$$

Proposition 7.10 *Let $X_1, \ldots, X_n$ be discrete random variables and put*

$$X := (X_1, \ldots, X_n).$$

Then the c.d.f. of the vector X can be computed as follows:

$$F_X(x_1, \ldots, x_n) = \mathbf{P}(X_1 \leq x_1, \ldots, X_n \leq x_n)$$
$$= \sum_{u_1 \in X_1(\Omega); u_1 \leq x_1} \cdots \sum_{u_n \in X_n(\Omega); u_n \leq x_n} \mathbf{P}(X_1 = u_1, \ldots, X_n = u_n)$$

for every $x_1, \ldots, x_n \in \mathbb{R}$.

Proof: This follows since

$$(X_1 \leq x_1, \ldots, X_n \leq x_n) = \bigcup_{u_1 \in X_1(\Omega); u_1 \leq x_1} \cdots \bigcup_{u_n \in X_n(\Omega); u_n \leq x_n} (X_1 = u_1, \ldots, X_n = u_n).$$

■

EXAMPLE 7.2 Another die example

Consider the experiment of tossing a red and green die where X_1 is the number of the red die and X_2 is the number on the green die. Let $X = (X_1, X_2)$.

Now let us find the distribution of the vector: $F_X(2, 3)$.

$$\begin{aligned}
F_X(2,3) &= \mathbf{P}(X_1 \leq 2, X_2 \leq 3) \\
&= \sum_{u_1 \leq 2, u_2 \leq 3} \mathbf{P}(X_1 = u_1, X_2 = u_2) \\
&= \mathbf{P}(X_1 = 1, X_2 = 1) + \mathbf{P}(X_1 = 1, X_2 = 2) + \mathbf{P}(X_1 = 1, X_2 = 3) \\
&\quad + \mathbf{P}(X_1 = 2, X_2 = 1) + \mathbf{P}(X_1 = 2, X_2 = 2) + \mathbf{P}(X_1 = 2, X_2 = 3) \\
&= \frac{1}{36} + \frac{1}{36} + \frac{1}{36} + \frac{1}{36} + \frac{1}{36} + \frac{1}{36} \\
&= \frac{6}{36} = \frac{1}{6}.
\end{aligned}$$

Once the joint probability function has been determined for discrete random variables $X_1, X_2, \ldots, X_n$, calculating joint probabilities involving the vector $X_1, X_2, \ldots, X_n$ is straightforward.

The following proposition show how the marginal laws can be obtained from the law of the vector.

Proposition 7.11 *Let* $Z = (X, Y)$ *be a random vector with values in* $\mathbb{R}^2$. *Then*

$$X(\Omega) = \{x \in \mathbb{R}, \exists y \in \mathbb{R} \text{ such that } (x, y) \in Z(\Omega)\}$$

and

$$Y(\Omega) = \{y \in \mathbb{R}, \exists x \in \mathbb{R} \text{ such that } (x, y) \in Z(\Omega)\}.$$

For every $x \in X(\Omega)$ we have

$$\mathbf{P}(X = x) = \sum_{y \in Y(\Omega)} \mathbf{P}(X = x, Y = y)$$

and for every $y \in Y(\Omega)$ we obtain

$$\mathbf{P}(Y = y) = \sum_{x \in X(\Omega)} \mathbf{P}(X = x, Y = y)$$

Proof: The conclusion follows by writing, for every $x \in X(\Omega)$

$$\begin{aligned}\mathbf{P}(X = x) &= \mathbf{P}(X = x, Y \in \mathbb{R}) \\ &= \mathbf{P}\left(\bigcup_{y \in Y(\Omega)} (X = x, Y = y)\right) \\ &= \sum_{y \in Y(\Omega)} \mathbf{P}(X = x, Y = y).\end{aligned}$$

■

Clearly, the result extends to the n-dimensional case. Specifically, if $X = (X_1, \ldots, X_n)$ is a discrete vector in $\mathbb{R}^n$, then

$$\mathbf{P}(X_1 = x_1) = \sum_{x_2} \sum_{x_3} .. \sum_{x_n} \mathbf{P}(X = x_1, X_2 = x_2, \ldots, X_n = x_n)$$

for every $x_1 \in X_1(\Omega)$. Similarly for any subset of random variables; for instance,

$$\mathbf{P}(X_1 = x_1, X_2 = x_2, X_{n-1} = x_{n-1}) = \sum_{x_n} \mathbf{P}(X = x_1, X_2 = x_2, \ldots, X_n = x_n).$$

EXAMPLE 7.3 Obtaining the marginal distributions

Consider a random vector (X, Y) whose distribution is described by the following table.

Y/X	2	4	6
1	0.1	0.05	0.1
3	0.2	0.15	0.15
5	0.1	0.1	0.05

Let's write the marginal distributions. We have

$$X(\Omega) = \{2, 4, 6\}$$

and

$$Y(\Omega) = \{1, 3, 5\}.$$

We next calculate the law of X (the marginal of X). To calculate $\mathbf{P}(X = 2)$, we add the probabilities on the column below the outcome 2.

$$\begin{aligned}\mathbf{P}(X = 2) &= \mathbf{P}(X = 2, Y = 1) + \mathbf{P}(X = 2, Y = 3) + \mathbf{P}(X = 2, Y = 5)\\ &= 0.1 + 0.2 + 0.1 = 0.4.\end{aligned}$$

In the same way, we obtain

$$\mathbf{P}(X = 4) = 0.05 + 0.15 + 0.1 = 0.3$$

and

$$\mathbf{P}(X = 6) = 0.1 + 0.15 + 0.05 = 0.3.$$

Proposition 7.12 *Let X, Y be two discrete random variables. Then X and Y are independent if and only if*

$$\mathbf{P}(X = x, Y = y) = \mathbf{P}(X = x)\mathbf{P}(Y = y) \tag{7.1}$$

for every $x \in X(\Omega)$ and $y \in Y(\Omega)$.

Proof: Suppose that X and Y are independent. Then we apply Definition 3.16 to the sets $A = \{x\}$ and $Y = \{y\}$ and the direct implication follows. For the reciprocal, assume (7.1) is true for all x and y. Take $A, B \subset X(\Omega)$. Then

$$(X \in A) \cap (Y \in B) = \{(X, Y) \in A \times B\}$$

and

$$A \times B = \bigcup_{x \in A, y \in B} \{(x, y)\}.$$

We can thus write

$$\begin{aligned}\mathbf{P}(X \in A, Y \in B) &= \sum_{x \in X(\Omega),\, y \in Y(\Omega)} \mathbf{P}(X = x, Y = x)\\ &= \sum_{x \in X(\Omega)} \sum_{y \in Y(\Omega)} \mathbf{P}(X = x)\mathbf{P}(Y = y)\\ &= \sum_{x \in X(\Omega)} \mathbf{P}(X = x) \sum_{y \in Y(\Omega)} \mathbf{P}(Y = y)\\ &= \mathbf{P}(X \in A)\mathbf{P}(Y \in B),\end{aligned}$$

which implies the independence of X and Y. ■

EXAMPLE 7.4

Consider a random vector (X, Y) with distribution

$$X(\Omega) = Y(\Omega) = \{0, 1\}$$

and

$$\mathbf{P}(X = 0, Y = 0) = (1 - p)^2, \qquad \mathbf{P}(X = 1, Y = 1) = p^2$$

and

$$\mathbf{P}(X = 0, Y = 1) = \mathbf{P}(X = 1, Y = 0) = p(1 - p).$$

Here $p \in (0, 1)$. Show that X, Y are independent.

Proof: We easily get

$$\mathbf{P}(X = 0) = 1 - p, \quad \mathbf{P}(X = 1) = p$$

and

$$\mathbf{P}(Y = 0) = 1 - p, \quad \mathbf{P}(y = 1) = p.$$

Then we can check that

$$\mathbf{P}(X = x, Y = y) = \mathbf{P}(X = x)\mathbf{P}(Y = y)$$

for every $x, y \in \{0, 1\}$. There are only 4 pairs of outcome, but all need to be verified. ■

EXAMPLE 7.5 Checking independence

Consider a random vector (X, Y) whose distribution is described by the following table.

Y/X	1	2	3
4	0.1	0.2	0.3
5	0.1	0.2	0.1

Check if the random variables X and Y are independent.

Solution: We have

$$X(\Omega) = \{1, 2, 3\} \quad \text{and} \quad Y(\Omega) = \{4, 5\}$$

with

$$\mathbf{P}(X = 1) = 0.2, \mathbf{P}(X = 2) = 0.4 \quad \text{and} \quad \mathbf{P}(X = 3) = 0.4$$

while

$$\mathbf{P}(Y = 4) = 0.6 \quad \text{and} \quad \mathbf{P}(Y = 5) = 0.4.$$

We can see, for example, that

$$\mathbf{P}(X = 1, Y = 4) = 0.1 \neq \mathbf{P}(X = 1)\mathbf{P}(Y = 4) = 0.2 \times 0.6.$$

Thus X and Y are not independent. ■

An important example of a discrete multidimensional law is the so-called multinomial law.

7.3.4 MULTINOMIAL DISTRIBUTION

The multinomial distribution is a generalization of the binomial distribution. Specifically, assume that n independent trials may each result in one of the k outcomes generically labeled $S = \{1, 2, \ldots, k\}$. The probability of outcome $1, 2, \ldots, k$ is respectively $p_1, \ldots, p_k$. Now define a vector $\mathbf{X} = (X_1, \ldots, X_k)$ where each of the X_i counts the number of outcomes $1, \ldots, k$ in the resulting sample of size n. We clearly must have $X_1 + \cdots + X_k = n$.

Furthermore, the joint distribution of the vector $\mathbf{X}$ is

$$f(x_1, \ldots, x_k) = \frac{n!}{x_1! \ldots x_k!} p_1^{x_1} \cdots p_k^{x_k} \mathbf{1}_{\{x_1 + \cdots + x_k = n\}}.$$

To see that the probability formula is correct, note that the product of probabilities is the probability that the sequence of n trials will contain exactly x_1 outcomes $1, \ldots, x_k$ outcomes k. Thus one only needs to count how many such outcomes exist; and the factorial term is precisely the number of ways in which we can divide the set $\{1, \ldots, n\}$ into k subsets, each subset having the respective number of terms.

In the same way that the binomial probabilities appear as coefficients in the binomial expansion of $(p + (1 - p))^n$, the multinomial probabilities are the coefficients in the multinomial expansion: $(p_1 + \cdots + p_k)^n$ so they will sum to 1, thus creating a proper probability mass function. This expansion in fact gives the name of the distribution.

Remark 7.13 *It is very easy to see that if we label the outcome i as a success and everything else a failure, then X_i simply counts successes in n independent trials and thus $X_i \sim Binom(n, p_i)$.*

We refer to exercises 7.17 and 7.18 for the proof.

As a consequence of this remark, the expected value of the random vector and the diagonal elements in the covariance matrix are easy to calculate as np_i and

$np_i(1 - p_i)$ respectively. The off-diagonal elements (covariances) are not complicated to calculate either. If we do the calculation, the multinomial's random vector first two moments are

$$\mathbf{E}[\mathbf{X}] = \begin{pmatrix} np_1 \\ np_2 \\ \vdots \\ np_k \end{pmatrix}$$

and

$$Cov(\mathbf{X}) = \begin{pmatrix} np_1(1-p_1) & -np_1p_2 & \dots & -np_1p_k \\ -np_1p_2 & np_2(1-p_2) & \dots & -np_2p_k \\ \vdots & \vdots & \ddots & \vdots \\ -np_kp_1 & -np_kp_2 & \dots & np_k(1-p_k) \end{pmatrix}$$

Important. As mentioned in the remark, the one-dimensional marginal distributions are binomial. However, in the joint distribution of $(X_1, \dots, X_r)$ the first r components are not multinomial.

However, if we group the rest of categories into one and we let $Y = X_{r+1} + \cdots + X_k$, we do obtain the multinomial again. It is easy to see that since the categories are linked ($X_1 + \cdots + X_k = n$) we also have that $Y = n - X_1 - \cdots - X_r$. Therefore, it is easy to see that the vector $(X_1, \dots, X_r, Y)$ or equivalently $((X_1, \dots, X_r, n - X_1 - \cdots - X_r)$ will have a multinomial distribution with associate probabilities $(p_1, \dots, p_r, p_Y) = (p_1, \dots, p_r, p_{r+1} + \cdots + p_k)$.

Let us present one last result about this distribution. Consider the conditional distribution of the first r components given the last $k - r$ components—that is, the distribution of

$$(X_1, \dots, X_r) \mid X_{r+1} = n_{r+1}, \dots, X_k = n_k$$

This conditional distribution is also multinomial with n the sample size replaced by $n - n_{r+1} - \cdots - n_k$ and probabilities

$$(p'_1, \dots, p'_r), \qquad \text{with } p'_i = \frac{p_i}{p_1 + \cdots + p_r}.$$

7.3.5 TESTING WHETHER COUNTS ARE COMING FROM A SPECIFIC MULTINOMIAL DISTRIBUTION

We conclude the section dedicated to this distribution with an important problem. Suppose we have observed and obtained a sample of n observations $(n_1, \dots, n_k)$ which we suspect comes from the multinomial distribution with probabilities $(\pi_1, \dots, \pi_k)$. Assuming that the right probability is the multinomial, suppose we want to test if all outcomes are equally likely.

To test the hypotheses

$$H_0 : (p_1, \ldots, p_k) = (\pi_1, \ldots, \pi_k),$$
$$H_a : \text{ Not all probabilities are equal}$$

we use the statistic

$$\sum_{i=1}^{k} \frac{(n_i - n\pi_i)^2}{n\pi_i},$$

which is also called the Pearson's Chi-squared statistic. This statistic is asymptotically distributed as a (univariate) Chi-squared random variable with $k - 1$ degrees of freedom when the sample size n goes to infinity.

Such a test would be very useful, for instance, when testing whether or not a six-sided die is fair—that is, whether all outcomes have the same chance.

EXAMPLE 7.6 Conditions for the multinomial distribution

Consider the following statistical experiment. Toss two dice three times, and record the outcome on each toss. This is a multinomial experiment because:

- The experiment consists of repeated trials. We toss the dice three times.
- Each trial can result in a discrete number of outcomes, 2 through 12.
- The probability of any of the outcomes is constant; it does not change from one toss to the next.
- The trials are independent; that is, getting a particular outcome on one trial does not affect the outcome on other trials.

7.3.6 INDEPENDENCE

The following result is very useful. It shows that the independence of two random variables implies the independence of every functions applied to these random variables.

Proposition 7.14 *Let X, Y be two discrete random variables on $(\Omega, \mathcal{F}, \mathbf{P})$. Then X and Y are independent if and only if for every functions $f, g : \mathbb{R} \to \mathbb{R}$, the random variables $f(X)$ and $g(Y)$ are independent.*

Proof: If $f(X)$ and $g(Y)$ are independent for every function f, g, let us to choose $f(x) = x$ and $g(y) = y$. This immediately implies X, Y are independent.

Conversely, suppose X, Y are independent and let $x' \in f(X)(\Omega)$ and $y' \in g(Y)(\Omega)$ outcomes of the random variables $f(X)$ and $g(Y)$. Then

$$(f(X) = x') = (X \in f^{-1}(x'))$$

and

$$(g(Y) = y') = (Y \in g^{-1}(y)).$$

We have

$$\begin{aligned}\mathbf{P}(f(X) = x', g(Y) = y') &= \mathbf{P}(X \in f^{-1}(x'), Y \in g^{-1}(y)) \\ &= \mathbf{P}(X \in f^{-1}(x'))\mathbf{P}(Y \in g^{-1}(y)) \\ &= \mathbf{P}(f(X) = x')\mathbf{P}(g(Y) = y')\end{aligned}$$

and thus $f(X)$ and $g(Y)$ are independent. ■

It is also possible to state a variant of Proposition 7.14 in terms of expectations.

Proposition 7.15 *Let X, Y be two discrete random variables on* $(\Omega, \mathcal{F}, \mathbf{P})$. *Then X and Y are independent if and only if for every function $f, g : \mathbb{R} \to \mathbb{R}$ such that $f(X)$ and $g(Y)$ are integrable, we obtain*

$$\mathbf{E}[f(X)g(Y)] = \mathbf{E}[f(X)]\mathbf{E}[g(Y)]. \tag{7.2}$$

Proof: If X, Y are independent, then relation (7.2) can be obtained from Proposition 7.14. Let us show the converse direction. Let $x \in X(\Omega)$ and $Y \in Y(\Omega)$ be fixed and define

$$f(s) = 1_x(s), \qquad g(s) = 1_y(s),$$

the indicator function of a point. Then

$$f(X) = 1_x(X), \qquad g(Y) = 1_y(Y).$$

Therefore, by definition we have

$$\mathbf{E}f(X) = \mathbf{P}(X = x), \qquad \mathbf{E}g(Y) = \mathbf{P}(Y = y). \tag{7.3}$$

On the other hand,

$$\mathbf{E}f(X)g(Y) = \mathbf{E}1_{(X=x)}1_{(Y=y)} = \mathbf{E}1_{(X=x, Y=y)} = \mathbf{P}(X = x, Y = y). \tag{7.4}$$

Using (7.3) and (7.4), we obtain

$$\mathbf{P}(X = x, Y = y) = \mathbf{P}(X = x)\mathbf{P}(Y = y)$$

and thus X is independent by Y. ■

Remark 7.16 *As a particular case of Proposition 7.15, we obtain that if X and Y are two independent integrable random variables, then XY is integrable and*

$$\mathbf{E}(XY) = \mathbf{E}X\mathbf{E}Y.$$

Consequently,

$$Cov(X, Y) = 0.$$

Also it can be seen that, if X, Y are square integrable, then

$$V(X + Y) = V(X) + V(Y).$$

This follows from the fact that

$$V(X + Y) = V(X) + V'Y) + 2Cov(X, Y).$$

Remark 7.17 *The fact that $Cov(X, Y) = 0$ does not imply the independence of X and Y.*

7.3.7 CONTINUOUS RANDOM VECTORS

The following definition is an extension of Definition 5.1.

Definition 7.18 *A function $f : \mathbb{R}^n \to \mathbb{R}$ is called a probability density on $\mathbb{R}^n$ if it is Lebesque integrable and positive and we have*

$$\int_{\mathbb{R}^n} f(x)dx = \int_{\mathbb{R}} \cdots \int_{\mathbb{R}} f(x_1, \ldots, x_n)\, dx_1 \cdots dx_n = 1.$$

Definition 7.19 *A random vector*

$$X = (X_1, \ldots, X_n)$$

has a density $f : \mathbb{R}^n \to \mathbb{R}$ if for any Borel set D in $\mathbb{R}^n$, we have

$$\mathbf{P}\left((X_1, \ldots, X_n) \in D\right) = \int_D f(x_1, \ldots, x_n)\, dx_1 \cdots dx_n.$$

It follows that the distribution function of the vector X is given by

$$\begin{aligned} F_X(x) &= F_{X_1,\ldots,X_n}(x_1, \ldots, x_n) = \mathbf{P}\left(X_1 \le x_1, \ldots, X_n \le x_n\right) \\ &= \int_{-\infty}^{x_1} \cdots \int_{-\infty}^{x_n} f(x_1, \ldots, x_n)\, dx_1 \cdots dx_n. \end{aligned}$$

As in the case of discrete random vectors, we define the marginal distributions of the components of the random vector $X = (X_1, \ldots, X_n)$.

Proposition 7.20 *Let $(X_1, \ldots, X_n) = X$ be a random vector with density f. Then the density of the random variable X_i $(1 \leq 1 \leq n)$ is given by*

$$f_{X_i}(x_i) = \int_{\mathbb{R}^{n-1}} f(x_1, \ldots, x_n) dx_1 \cdots x_{i-1} dx_{i+1} \cdots dx_n \tag{7.5}$$

for every $1 \leq i \leq n$.

Proof: Indeed, for every $a \in \mathbb{R}$, we obtain

$$\begin{aligned}
\mathbf{P}(X_i \leq a) &= \mathbf{P}\left(X_i \leq a, X_1, \ldots, X_{i-1}, X_{i+1}, \ldots, X_n \in \mathbb{R}\right) \\
&= \mathbf{P}\left((X_1, \ldots, X_n) \int (\mathbb{R} \times \mathbb{R} \cdots \times (-\infty, a) \times \mathbb{R} \cdots \times \mathbb{R})\right) \\
&= \int_{\mathbb{R}} dx_1 \cdots \int_{\mathbb{R}} dx_{i-1} \int_{-\infty}^{a} dx_i \int_{\mathbb{R}} dx_{i+1} \cdots \int_{\mathbb{R}} dx_n f(x_1, \ldots, x_n) \\
&= \int_{-\infty}^{a} dx_i \left(\int_{\mathbb{R}^{n-1}} dx_1 \cdots x_{i-1} dx_{i+1} \cdots dx_n f(x_1, \ldots, x_n) \right)
\end{aligned}$$

and this shows that the density of X_i is by (7.5). ■

The above result can be generalized as follows.

Proposition 7.21 *Let $(X_1, \ldots, X_n)$ be a continuous random vector with c.d.f. F and density f. If $i_1, \ldots, i_k \in \{1, \ldots, n\}$ with $i_1 < i_2 < \cdots < i_k$, then:*

1. *The function*

$$F_{i_1, \ldots, i_k} : \mathbb{R}^k \to \mathbb{R}$$

given by

$$F_{i_1, \ldots, i_k}(x_{i_1}, \ldots, x_{i_k}) = \lim_{x_j \to \infty;\, j \neq i_1, \ldots, i_k} F(x_1, \ldots, x_n)$$

is the joint c.d.f. of the vector

$$(X_{i_1}, \ldots, X_{i_k}).$$

2. *The joint density of the vector*

$$(X_{i_1}, \ldots, X_{i_k})$$

is given by

$$f_{i_1, \ldots, i_k}(x_{i_1}, \ldots, x_{i_k}) = \int_{\mathbb{R}^{n-k}} f(x_1, \ldots, x_n)\, dx_{j_1} \cdots dx_{j_{n-k}}$$

where $j_1, \ldots, j_{n-k} \in \{1, \ldots, n\} \setminus \{i_1, \ldots, i_k\}$ and $j_1 < \cdots < j_{n-k}$.

EXAMPLE 7.7 Uniformly distributed vector

Let $f : \mathbb{R}^2 \to \mathbb{R}$ be given by

$$f(x, y) = \frac{1}{(b-a)(d-c)} 1_{[a,b]}(x) 1_{[c,d]}(y),$$

where $a < b$ and $c < d$. Then f is a density function on $\mathbb{R}^2$.

Indeed, $f(x, y) \geq 0$ for every $x, y \in \mathbb{R}$ and

$$\int_{\mathbb{R}^2} f(x, y) = \frac{1}{(b-a)(d-c)} \int_a^b dx \int_c^d dy = 1;$$

therefore f is a proper density.

In fact the above density is corresponding to the coordinates of a point uniformly distributed over the rectangle $[a, b] \times [c, d]$ in $\mathbb{R}^2$.

More general, a random vector $(X_1, \ldots, X_n)$ has an n-dimensional uniform distribution on the rectangle:

$$I = [a_1, b_1] \times \cdots \times [a_n, b_n]$$

in $\mathbb{R}^n$, where $a_i < b_i$ for every $i = 1, \ldots, n$ if the joint density of said vector is given by

$$f(x_1, \ldots, x_n) = \frac{1}{(b_1 - a_1) \cdots (b_n - a_n)} \mathbf{1}_I,$$

where we use the indicator notation.

EXAMPLE 7.8 Standard normal vector

Let $f : \mathbb{R}^2 \to \mathbb{R}$ be given by

$$f(x, y) = \frac{1}{2\pi} e^{-\frac{x^2+y^2}{2}}.$$

Then f is a density function on $\mathbb{R}^2$.

A random vector with such a density is called a two-dimensional standard normal vector or a bivariate normal vector. A detailed study of Gaussian random vectors is performed in one of the next chapters.

The two examples presented above are of densities coming from one-dimensional probability density functions. Recall that if the individual random variables are independent, the joint distribution is just the product of marginals.

Specifically, if f, g are two densities on $\mathbb{R}$, then

$$h(x, y) = f(x)g(y)$$

is a density on $\mathbb{R}^2$. And of course this can be generalized to $\mathbb{R}^n$.

Proposition 7.22 *Let X, Y be two continuous random variables and assume that the couple (X, Y) admits a density $f_{(X,Y)}$. Let us denote by f_X and f_Y the probability density functions of X and Y respectively. Then X and Y are independent if and only if*

$$f_{(X,Y)}(x, y) = f_X(x)f_Y(y) \tag{7.6}$$

for every $x, y \in \mathbb{R}$.

Proof: Suppose that relation (7.6) is true. Then for every $a, b \in \mathbb{R}$, we have

$$\begin{aligned}
\mathbf{P}(X \leq a, Y \leq b) &= \int_{-\infty}^{a} \int_{\infty}^{b} f_{X,Y}(x, y)dydx \\
&= \int_{-\infty}^{a} \int_{\infty}^{b} f_X(x)f_Y(y)dydx = \int_{-\infty}^{a} dx f_X(x) \int_{\infty}^{b} dy f_Y(y) \\
&= \mathbf{P}(X \leq a)\mathbf{P}(Y \leq b)
\end{aligned}$$

and then X is independent by Y. Suppose that X, Y are independent and let us show (7.6).

$$\begin{aligned}
\mathbf{P}(X \leq a, Y \leq b) = \mathbf{P}(X \leq a)\mathbf{P}(Y \leq b) &= \int_{-\infty}^{a} dx f_X(x) \int_{\infty}^{b} dy f_Y(y) \\
&= \int_{-\infty}^{a} \int_{\infty}^{b} f_X(x)f_Y(y)\, dydx,
\end{aligned}$$

and this implies that $f_X(x)f_Y(y)$ is the density of the vector (X, Y) by Definition 7.19 and using the fact that the indicator functions $1_{(-\infty,a)\times(\infty,b)}$, $a, b \in \mathbb{R}$ generates the Borel sets of $\mathbb{R}^2$. ■

The above result extends to a n-dimensional random vector.

Corollary 7.23 *Let X_i, $i = 1, \ldots, n$ be independent random variables with densities f_{X_i} respectively. Then the density of the vector $X = (X_1, \ldots, X_n)$ is*

$$f_X(x_1, \ldots, x_n) = \prod_{i=1}^{n} f_{X_i}(x_i)$$

for every $x_1, \ldots, x_n \in \mathbb{R}$.

Proof: The proof can be easily done by induction. ■

The following examples contain situations where the joint is not the product of the marginals. That is, the components are not independent.

EXAMPLE 7.9

Let $f : \mathbb{R}^2 \to \mathbb{R}$,

$$f(x, y) = \frac{1}{6}(x + 4y)\mathbf{1}_{(0,2)}(x)\mathbf{1}_{(0,1)}(y).$$

Check that f is a probability density function on $\mathbb{R}^2$. If (X, Y) is a random vector with density f, compute the marginal densities of X and Y.

One can easily check that $\int_{\mathbb{R}^2} f(x, y)\, dxdy = 1$. We can calculate the marginal densities by integrating out the other variable:

$$\begin{aligned} f_X(x) &= 1_{(0,2)}(x) \int_0^1 \frac{1}{6}(x + 4y)dy \\ &= 1_{(0,2)}(x) \left[xy + 2y^2\right]_{y=0}^{y=1} \\ &= \frac{1}{6}(x + 2)1_{(0,2)}(x) \end{aligned}$$

while

$$\begin{aligned} f_Y(y) &= 1_{(0,1)}(y) \int_0^2 dx \frac{1}{6}(x + 4y) \\ &= \frac{1}{6}(2 + 8y)1_{(0,1)}(y). \end{aligned}$$

Obviously $f(x, y) \neq f_X(x)f_Y(y)$, and thus the two components of the vector are not independent.

EXAMPLE 7.10

Define $f : \mathbb{R}^2 \to \mathbb{R}$,

$$f(x, y) = \lambda^2 e^{-\lambda x} 1_{[0,x]}(y),$$

where $\lambda > 0$. This function f is a density on $\mathbb{R}^2$. Check that the function defines a proper density and calculate the marginal distributions.

Indeed, f is clearly positive and

$$\begin{aligned}\int_{\mathbb{R}^2} f(x, y)\, dxdy &= \lambda^2 \int_0^{\infty} dx \int_0^x dy e^{-\lambda x} \\ &= \lambda^2 \int_0^{\infty} x e^{-\lambda x} dx = 1,\end{aligned}$$

where we used integration by parts to obtain the last value.

Let us compute the marginal densities. The density of the random variable X is

$$\begin{aligned}f_X(x) &= \int_{\mathbb{R}} f(x, y)\, dy \\ &= \lambda^2 e^{-\lambda x} 1_{(x>0)} \int_0^x dy = \lambda^2 x e^{-\lambda x} 1_{(x>0)}.\end{aligned}$$

We note that this is the density of the law $\Gamma(2, \lambda)$.

The density of Y is

$$\begin{aligned}f_Y(y) &= \int_{\mathbb{R}} f(x, y)\, dx \\ &= \lambda^2 1_{(y>0)} \int_y^{\infty} e^{-\lambda x} dx \\ &= \lambda e^{-\lambda y} 1_{(y>0)},\end{aligned}$$

which is an exponential law with parameter λ.

Please note that the two random variables X and Y are not independent because the joint is not equal to the product of the marginals. This is typically true when the function $f(x, y)$ cannot be separated as a product of functions of x and y only.

EXAMPLE 7.11 Uniformly distributed point on the disk

Let

$$f(x, y) = \frac{1}{\pi} \mathbf{1}_{\Delta(0,1)}(x, y),$$

where $\Delta(0, 1)$ denotes the unit disk centered in $(0, 0)$ and radius 1. Mathematically,

$$\Delta(0, 1) = \{(x, y) \in \mathbb{R}^2 \mid x^2 + y^2 \leq 1\}.$$

This is the density of the random vectors (X, Y), which represent the coordinates of a point uniformly chosen from the unit disk $\Delta(0, 1)$.

Using polar coordinates $x = r\cos\theta, y = r\sin\theta$, we have

$$\int_{\mathbb{R}^2} f(x,y)\,dxdy = \frac{1}{\pi}\int_0^1 dr \int_0^{2\pi} d\theta r = 1,$$

which shows that this is a proper density. The reader is left with the calculation of marginals and the verification whether or not the random variables representing the coordinates X and Y are independent of each other (they should not be).

Let $F : \mathbb{R}^n \to \mathbb{R}^m$ be such that all its partial derivatives exist. Let

$$F_1, F_2, \ldots, F_m : \mathbb{R}^n \to \mathbb{R}$$

be the real-valued component function of F. Then the Jacobian matrix associated with F is denoted by J_F and is a matrix with m lines and n rows whose coefficients are given by

$$(J_F)_{i,j} = \frac{\partial F_i}{\partial x_j}$$

for every $i = 1, \ldots, m$ and $j = 1, \ldots, n$.

7.3.8 CHANGE OF VARIABLES. OBTAINING DENSITIES OF FUNCTIONS OF RANDOM VECTORS

The following result is very useful. It gives a formula to obtain the density of a random variable Y which can be written as a function of X if only the density of the random vector X is known.

Theorem 7.24 *[Change of variables] Let X be a random vector with density f_X with support an open set $\mathcal{O} \subset \mathbb{R}^d$. Let $\mathcal{O}'$ be another open set in $\mathbb{R}^d$ and let*

$$\varphi : \mathcal{O} \to \mathcal{O}'$$

be a diffeomorphism (i.e., φ is bijective and both φ and its inverse φ^{-1} are differentiable). Then the density of the random vector $Y := \varphi(X)$ is given by

$$f_Y(y) = f_X(\varphi^{-1}(y))]\left|J(\varphi^{-1}(y))\right|,$$

where J denotes the Jacobian[1].

Proof: In the first place, let us recall the definition of the Jacobian. If $\varphi : \mathbb{R}^n \to \mathbb{R}^n$ and

$$\varphi(x_1, \ldots, x_n) = (\varphi_1(x_1, \ldots, x_n), \ldots, \varphi_n(x_1, \ldots, x_n)),$$

[1] The differentiability condition is needed to make sure that the Jacobian exists

then its Jacobian is defined as the determinant of the matrix of first-order derivatives or

$$J_\varphi = \begin{vmatrix} \frac{\partial \varphi_1}{\partial x_1} & \frac{\partial \varphi_1}{\partial x_2} & \cdots & \frac{\partial \varphi_1}{\partial x_n} \\ \frac{\partial \varphi_2}{\partial x_1} & \frac{\partial \varphi_2}{\partial x_2} & \cdots & \frac{\partial \varphi_2}{\partial x_n} \\ \vdots & \vdots & \cdots & \vdots \\ \frac{\partial \varphi_n}{\partial x_1} & \frac{\partial \varphi_n}{\partial x_2} & \cdots & \frac{\partial \varphi_n}{\partial x_n} \end{vmatrix}.$$

Now, let $X = (X_1, \dots, X_d)$. For any $F : \mathbb{R} \to \mathbb{R}$ arbitrary bounded and measurable function, we have

$$\begin{aligned} \mathbf{E}F(Y) &= \mathbf{E}F(\varphi(x) \\ &= \int_{\mathcal{O}} F(\varphi(x_1, \dots, x_d)) f_X(x_1, \dots, x_d) dx_1 ... dx_d \\ &= \int_{\mathcal{O}'} F(y_1, \dots, y_d) \left| J(\varphi^{-1}(y_1, \dots, y_d)) \right| f_X(\varphi^{-1}(y_1, \dots, y_d)) \end{aligned}$$

using the change of variables $\varphi(x_1, \dots, x_d) = (y_1, \dots, y_d)$.

Note that it is fine if you do not remember which order to take derivatives since in the end we use the absolute value of the determinant (if the order is changed that may only change the sign of the determinant not the magnitude). ■

EXAMPLE 7.12

Let X, Y be two independent random variables with the same density

$$f(x) = \frac{1}{x^2} 1_{[1,\infty)}(x). \tag{7.7}$$

Define

$$U = XY \quad \text{and} \quad V = \frac{X}{Y}.$$

Using Theorem 7.24 find the joint density of the random vector (U, V).

First let us note that the density of the couple (X, Y) is

$$f_{(X,Y)}(x, y) = \frac{1}{x^2} \frac{1}{y^2} 1_{[1,\infty)}(x) 1_{[1,\infty)}(y)$$

from the independence of X and Y. We will consider the transformation:

$$\varphi(x, y) = \left(xy, \frac{x}{y} \right)$$

which is a bijective function from $[1, \infty) \times [1, \infty)$ into the domain D where

$$D = \left\{(u, v) \in [1, \infty) \times [1, \infty) \mid \frac{u}{v} \geq 1\right\}.$$

The inverse is defined on D with values in $[1, \infty) \times [1, \infty)$ and it is calculated as

$$\varphi^{-1}(u, v) = \left(\sqrt{uv}, \sqrt{\frac{u}{v}}\right).$$

The Jacobian matrix is

$$J\varphi^{-1}(u, v) = \begin{pmatrix} \frac{\sqrt{v}}{2\sqrt{u}} & \frac{1}{2\sqrt{uv}} \\ \frac{sqrtu}{2\sqrt{v}} & \frac{-2\sqrt{u}}{v^{\frac{3}{2}}} \end{pmatrix}$$

and its determinant is

$$det\, J\varphi^{-1}(u, v) = \frac{5}{4v}.$$

Applying the Theorem 7.24, we obtain the density of the random vector (U, V) as

$$\begin{aligned} f_{(U,V)}(u, v) &= f_{(X,Y)}(\varphi^{-1}(u, v)) det J\varphi^{-1}(u, v) 1_D(u, v) \\ &= \frac{1}{u^2} \frac{5}{4v} 1_D(u, v). \end{aligned}$$

7.3.9 DISTRIBUTION OF SUMS OF RANDOM VARIABLES. CONVOLUTIONS

Any sum has at least two components. Let us start by analyzing the distribution of a sum containing only two variables. Let X, Y be two random variables. Let F, G be the distribution functions of X and Y, where $F(x) = \mathbf{P} \circ X^{-1}(-\infty, x]$, $G(x) = \mathbf{P} \circ Y^{-1}(-\infty, x]$. We are interested in the distribution of $X + Y$.

Definition 7.25 (Convolution) *The convolution is the distribution function of the random variable $X + Y$. We denote this distribution function with*

$$F * G = (F, G) \circ s^{-1} = \mathbf{P} \circ (X + Y)^{-1},$$

where:

$$\Omega \xrightarrow{(X(\omega), Y(\omega))} \mathbb{R}^2 \xrightarrow{s(x,y)=x+y} \mathbb{R}.$$

The next proposition presents properties of general convolution operators.

Proposition 7.26 (Properties of convolution) *Let F, G be two distribution functions. Then:*

(i) $F * G = G * F$.

(ii) $F * (G * H) = (F * G) * H$.

(iii) $\delta_a * \delta_b = \delta_{a+b}$ *with* $a, b \in \mathbb{R}$, *and* δ_a, δ_b *are delta distribution functions.*

(iv) *If* F, G *are discrete, then* $F * G$ *is discrete.*

This proposition is easy to prove from the definition, and the proof is left as an exercise (exercise 7.14).

In the special case when the random variables involved are independent and have densities, we can obtain the density of a sum of two independent random variables as a much simpler formula.

We next introduce the definition of a convolution of two functions.

Definition 7.27 (Convolution of two functions) *Suppose that* f, g *are two functions from* $\mathbb{R}$ *to* $\mathbb{R}$. *The convolution of* f *with* g *is defined by the function*

$$(f * g)(x) = \int_{\mathbb{R}} f(y)g(x-y)\,dy = \int_{\mathbb{R}} g(z)f(x-z)\,dz.$$

We note that the last equality is obtained by performing a change of variables $z = x - y$. Clearly this last inequality implies that the convolution product is commutative (i.e., $(f * g)(x) = (g * f)(x)$.

Returning to the random variables, in the special case when X and Y have joint density f, applying Theorem 7.24, then the sum of the random variables $X + Y$ also has a density and

$$f_{X+Y}(z) = \int f(z-y, y)\,dy.$$

If, in addition, X and Y are independent and have densities f_X and f_Y, then we have the density of $X + Y$:

$$f_{X+Y}(z) = f_X * f_Y(z) = \int f_X(z-y)f_Y(y)\,dy;$$

in other words, the density of $X + Y$ is obtained as the convolution of the density functions f_X and f_Y (as in Definition 7.27). Let us prove this result.

Proposition 7.28 *Let* X, Y *be two independent random variables with densities denoted by* f_X *and* f_Y *respectively. Then the density of the r.v.* $X + Y$ *is given by*

$$f_{X+Y}(x) = \int_{\mathbb{R}} f_X(y)f_Y(x-y)dy = (f_X * f_Y)(x),$$

where $*$ *denotes the convolution product.*

Proof: Since the random variables are independent, the density of the couple (X, Y) is

$$f_{(X,Y)}(x, y) = f_X(x) f_Y(y)$$

for every $x, y \in \mathbb{R}$.

Consider the function $\varphi : \mathbb{R}^2 \to \mathbb{R}^2$ defined as

$$\varphi(x, y) = (x + y, y).$$

Then φ satisfied the assumptions in the Theorem 7.24 and

$$\varphi^{-1}(u, v) = (u - v, v)$$

with

$$J\varphi^{-1}(u, v) = \begin{pmatrix} 1 & 0 \\ -1 & 1 \end{pmatrix}.$$

We then conclude using Theorem 7.24 that the density of $\varphi(X, Y) = (X + Y, Y)$ is

$$f_{(X+Y,Y)}(u, v) = f_{(X,Y)}(u - v, v) = f_X(u - v) f_Y(v)$$

and therefore the marginal distribution of the first component in $\varphi(X, Y)$ (which is the density of the random variable $X + Y$) is

$$\begin{aligned} f_{X+Y}(u) &= \int_{\mathbb{R}} f_{(X+Y,Y)}(u, v)\, dv \\ &= \int_{\mathbb{R}} f_X(u - v) f_Y(v)\, dv. \end{aligned}$$

which concludes the proof. ■

As an application, we next show that the sum of independent normal random variables is still a normal random variables with parameters equals to the sum of parameters.

Proposition 7.29 *Let $X \sim N(\mu_1, \sigma_1^2)$ and $Y \sim N(\mu_2, \sigma_2^2)$. Assume X and Y are independent. Then*

$$X + Y \sim N(\mu_1 + \mu_2, \sigma_1^2 + \sigma_2^2).$$

Proof: We will use Theorem 7.28, which gives the density of the sum of two independent random variables. Suppose first that

$$\mu_1 = \mu_2 = 0 \quad \text{and} \quad \sigma_1 = \sigma_2 = 1.$$

By Theorem 7.28 we can write that the density of $X + Y$ is

$$f_{X+Y}(x) = \frac{1}{2\pi} \int_{\mathbb{R}} e^{-\frac{y^2}{2}} e^{-\frac{(x-y)^2}{2}} dy.$$

We can further calculate f_{X+Y} as follows:

$$\begin{aligned} f_{X+Y}(x) &= \frac{1}{2\pi} e^{-\frac{x^2}{2}} \int_{\mathbb{R}} e^{-(y^2-xy)} dy \\ &= \frac{1}{2\pi} e^{-\frac{x^2}{2}} \int_{\mathbb{R}} e^{-(y-\frac{1}{2}x)^2} dy e^{\frac{x^2}{4}} \\ &= \frac{1}{2\sqrt{\pi}} e^{-\frac{x^2}{4}}, \end{aligned}$$

where we used the change of variables $y - \frac{1}{2}x = \frac{z}{\sqrt{2}}$. Therefore, $X + Y \sim N(0, 2)$. In other words, we have just shown that if $X \sim N(0, 1)$ and $Y \sim N(0, 1)$, then $X + Y \sim N(0, 2)$.

In the same way when $X, Y \sim N(0, 1)$ but using the function $\varphi(x, y) = (ax + by, y)$ we obtain the distribution:

$$aX + bY \sim N(0, a^2 + b^2).$$

Suppose now that $X \sim N(\mu_1, \sigma_1^2)$ and $Y \sim N(\mu_2, \sigma_2^2)$. Then we may write

$$X = \mu_1 + \sigma_1 Z_1 \quad \text{and} \quad Y = \mu_2 + \sigma_2 Z_2$$

where Z_1, Z_2 are two standard normal random variables. The independence of X and Y implies the independence of Z_1 and Z_2. Therefore. using Remark 5.12, we obtain

$$Z_1, Z_2 \sim N(0, 1)$$

and thus by the first part of the proof we have $Z_1 + Z_2 \sim N(0, 2)$ and $\sigma Z_1 + \sigma_2 Z_2 \sim N(0, \sigma_1^2 + \sigma^2)$. We then obviously obtain $X + Y \sim N(\mu_1 + \mu_2, \sigma_1^2 + \sigma_2^2)$. ■

Remark 7.30 *From the proof of Proposition 7.28 we can deduce more general results.*

(a) *If (X, Y) is a random vector with density $f_{(X,Y)}$, then the density of $X + Y$ is*

$$f_{X+Y}(z) = \int_{\mathbb{R}} f_{(X,Y)}(y, z - u)\, du$$

for almost every $z \in \mathbb{R}$.

(b) *More generally, if $X_1, X_2, \ldots, X_n$ are independent, then the density of $X_1 + \cdots + X_n$ is given by*

$$f_{X_1+\cdots X_n} = f_{X_1} * \cdots * f_{X_n}.$$

Part (b) in the remark can be easily obtained by induction.

We can also express the density of a product and of a ratio of two random variables.

Proposition 7.31 *If the random vector (X, Y) has joint density $f_{(X,Y)}$, then the density of the product XY is*

$$f_{XY}(z) = \int_{\mathbb{R}} \frac{1}{|u|} f_{(X,Y)}\left(u, \frac{z}{u}\right) du$$

for almost everywhere $z \in \mathbb{R}$ and the density of $\frac{X}{Y}$ is

$$f_{\frac{X}{Y}}(z) = \int_{\mathbb{R}} |v| f_{(X,Y)}(vz, v)\, dv$$

for a.e. $z \in \mathbb{R}$.

Proof: The proof follows the lines of the Example 7.12. ■

Remark 7.32 *As a consequence, if X, Y are independent random variables with density functions f_X and f_Y respectively, then*

$$f_{XY}(z) = \int_{\mathbb{R}} \frac{1}{|u|} f_X(y) f_Y\left(\frac{z}{u}\right) du$$

for a.e. $z \in \mathbb{R}$ and

$$f_{\frac{X}{Y}}(z) = \int_{\mathbb{R}} |v| f_X(vz) f_Y(v)\, dv$$

for a.e. $z \in \mathbb{R}$.

EXERCISES

Problems with Solution

7.1 Let X, Y be two random variables such that

$$\mathbf{P}\ (X = 0, Y = 1) = \frac{1}{5}$$

$$\mathbf{P}\ (X = 0, Y = 2) = \frac{1}{5}$$

$$\mathbf{P}\ (X = 1, Y = 0) = \frac{1}{5};$$

$$\mathbf{P}\ (X = 1, Y = 1) = \frac{1}{5}$$

$$\mathbf{P}\ (X = 1, Y = 2) = \frac{1}{5}.$$

(a) Find the law and the c.d.f. of X.
(b) Find the law and the c.d.f. of Y.
(c) Are the r.v. X and Y independent?
(d) Find the law and the c.d.f. of the r.v.

$$Z = X - Y.$$

Solution: The sum of the five probabilities in the statement of the problem is 1 and therefore the space is

$$(X, Y)(\Omega) = \{(0, 1), (0, 2), (1, 0), (1, 1), (1, 2)\}.$$

Consequently,

$$X(\Omega) = \{0, 1\} \quad \text{and} \quad Y(\Omega) = \{0, 1, 2\}.$$

To calculate the marginals:

$$\mathbf{P}(X = 0) = \mathbf{P}(X = 0, Y = 1) + \mathbf{P}(X = 0, Y = 2) = \frac{2}{5}$$

and

$$\mathbf{P}(X = 1) = \frac{3}{5}.$$

For Y we get

$$\mathbf{P}(Y = 0) = \frac{1}{5} \quad \text{and} \quad \mathbf{P}(Y = 1) = \mathbf{P}(Y = 2) = \frac{2}{5}.$$

The random variables X and Y are not independent, because for example

$$\frac{1}{5} = \mathbf{P}(X = 0, Y = 1) \neq \mathbf{P}(X = 0)\mathbf{P}(Y = 1) = \frac{2}{5}\frac{2}{5}.$$

The r.v. $Z = X - Y$ has the sample space:

$$Z(\Omega) = \{-2, -1, 0, 1\}$$

with

$$\mathbf{P}(Z = -2) = \mathbf{P}(X = 0, Y = 2) = \frac{1}{5},$$

$$\mathbf{P}(Z = -1) = \mathbf{P}(X = 0, Y = 1) + \mathbf{P}(X = 1, Y = 2) = \frac{2}{5}$$

$$\mathbf{P}(Z = 0) = \mathbf{P}(X = 1, Y = 1) = \frac{1}{5},$$

$$\mathbf{P}(Z = 1) = \mathbf{P}(X = 1, Y = 0) = \frac{1}{5}.$$

■

7.2 *Let* (X, Y) be a random vector with joint density

$$f(x, y) = \frac{1}{6}1_{[0,2]\times[0,3]}.$$

(a) Check that f is a proper density.
(b) Calculate the probability:

$$\mathbf{P}(1 \leq X \leq 3, -1 \leq Y \leq 2).$$

Solution: We can easily see that f is positive and $\int_{\mathbb{R}^2} f(x, y)\, dxdy = 1$. Next,

$$\begin{aligned}\mathbf{P}(1 \leq X \leq 3, -1 \leq Y \leq 2) &= \int_1^3 \int_{-1}^1 f(x, y)\, dydx \\ &= \int_1^2 \int_0^1 \frac{1}{6}\, dydx \\ &= \frac{1}{6}.\end{aligned}$$

■

7.3 Let $\lambda > 0$, and let (X, Y) denote a random vector with joint density f on $\mathbb{R}^2$ given by

$$f(x, y) = C\, \mathbf{1}_{\{x\geq 0,\, |y|\leq x\}}\, e^{-\lambda x}.$$

(a) For which values of C is the function f a proper density?

(b) Find the marginal distributions of X and Y. Calculate the mean, variance, and covariance of the variables X and Y.

Solution: First note that, since f should be positive, we must have $C > 0$. Then

$$\begin{aligned}
&\int_{\mathbb{R}}\int_{\mathbb{R}} f(x,y)\,dxdy \\
&= C\int_0^{\infty} dx \int_{-x}^{x} dye^{-\lambda x} = C\int_0^{\infty} dxe^{-\lambda x}2x \\
&= C\frac{-2}{\lambda}\left[xe^{-\lambda x}\right]_{x=0}^{x=\infty} + C\frac{2}{\lambda}\int_0^{\infty} e^{-\lambda x}dx = C\frac{2}{\lambda}\int_0^{\infty} e^{-\lambda x}dx \\
&= \frac{2C}{\lambda^2}.
\end{aligned}$$

This gives the value $C = \frac{\lambda^2}{2}$.

For the marginal distribution of X, we write

$$\begin{aligned}
f_X(x) &= \int_{\mathbb{R}} f(x,y)\,dy = \frac{\lambda^2}{2}1_{(0,\infty)}(x)e^{-\lambda x}\int_{-x}^{x} dy \\
&= \lambda^2 1_{(0,\infty)}(x)xe^{-\lambda x}.
\end{aligned}$$

The marginal density of Y is

$$\begin{aligned}
f_Y(y) &= \int_{\mathbb{R}} f(x,y)\,dx = \frac{\lambda^2}{2}\int_{|y|}^{\infty} dxe^{-\lambda x}dx \\
&= \frac{\lambda}{2}e^{-\lambda|y|}.
\end{aligned}$$

We leave the rest of the exercise for the reader; we note, however, that X has a gamma distribution with parameters $a = 2$ and $\lambda = 1$ and Y has a Laplace distribution. ■

7.4 Let X_1, X_2 be two independent random variables both with uniform distribution on $\{-1, +1\}$, meaning that

$$\mathbf{P}(X_1 = 1) = \mathbf{P}(X_1 = -1) = \mathbf{P}(X_2 = 1) = \mathbf{P}(X_2 = -1),$$

and let

$$X_3 = X_1X_2.$$

(a) Are the three random variables pairwise independent?

(b) Are the three random variables mutually independent?

Solution: First note that

$$\begin{aligned}\mathbf{P}(X_3 = 1) &= \mathbf{P}(X_1 = 1, X_2 = 1) + \mathbf{P}(X_1 = -1, X_2 = -1)\\ &= \mathbf{P}(X_1 = 1)\mathbf{P}(X_2 = 1) + \mathbf{P}(X_1 = -1)\mathbf{P}(X_2 = -1)\\ &= \frac{1}{2}.\end{aligned}$$

Let us prove that the three random variables are two by two (pairwise) independent. Clearly X_1 and X_2 are independent by hypothesis. We show that X_1, X_3 are independent.

$$\begin{aligned}\mathbf{P}(X_1 = 1, X_3 = 1) &= \mathbf{P}(X_1 = 1, X_1X_2 = 1)\\ &= \mathbf{P}(X_1 = 1, X_2 = 1)\\ &= \mathbf{P}(X_1 = 1)\mathbf{P}(X_2 = 1)\\ &= \frac{1}{4}\end{aligned}$$

and this is clearly equal to the product:

$$\mathbf{P}(X_3 = 1)\mathbf{P}(X_1 = 1).$$

In a similar way we can show that

$$\mathbf{P}(X_1 = i, X_3 = j) = \mathbf{P}(X_1 = i)\mathbf{P}(X_3 = j)$$

for every $i, j \in \{-1, 1\}$, which means that X_1 is independent by X_3.

We have

$$\mathbf{P}(X_i = 1) = \mathbf{P}(X_i = -1) = \frac{1}{2}$$

for $i = 1, 2, 3$ and

$$\begin{aligned}&\mathbf{P}(X_3 = 1, X_1 = 1, X_2 = 1)\\ &= \mathbf{P}(X_1X_2 = 1, X_1 = 1, X_2 = 1) = \mathbf{P}(X_1 = 1, X_2 = 1)\\ &= \mathbf{P}(X_1 = 1)\mathbf{P}(X_2 = 1) = \frac{1}{2}\frac{1}{2} = \frac{1}{4}.\end{aligned}$$

On the other hand the product of all three probabilities is

$$\begin{aligned}&\mathbf{P}(X_3 = 1)\mathbf{P}(X_2 = 1)\mathbf{P}(X_1 = 1) = \frac{1}{2}\frac{1}{2}\frac{1}{2} = \frac{1}{8}\\ &\neq \mathbf{P}(X_3 = 1, X_1 = 1, X_2 = 1).\end{aligned}$$

We conclude that the random variables X_1, X_2, X_3 are not mutually independent. ∎

7.5 Let (X, Y) be a random vector with probability density function $f : \mathbb{R}^2 \to \mathbb{R}$ given by

$$f(x, y) = 2e^{-x}e^{-y}(1 + e^{-x} + e^{-y})^{-3},$$

where $x, y \in \mathbb{R}$.

(a) Find the marginal distributions of X and Y. Are X and Y independent?

(b) Define

$$Z = X - Y.$$

Give the law of the random vector $(Z, -Y)$?

(c) Calculate the density of Z.

Solution: The density f is symmetric with respect to its two variables x and y. It is therefore clear that the densities of X and Y are the same (integrating out one variable with give the same result). We compute the density function for X.

$$\begin{aligned} f_X(x) &= 2e^{-x} \int_{\mathbb{R}} e^{-y}(1 + e^{-x} + e^{-y})^{-3} dy \\ &= 2e^{-x} \left[\frac{1}{2}(1 + e^{-x} + e^{-y})^{-2} \right]_{y=-\infty}^{y=\infty} \\ &= e^{-x} \frac{1}{(1 + e^{-x})^2} \end{aligned}$$

since

$$\lim_{y\to\infty} (1 + e^{-x} + e^{-y})^{-2} = (1 + e^{-x})^2$$

and

$$\lim_{y\to-\infty} (1 + e^{-x} + e^{-y})^{-2} = 0.$$

By the previous observation we also have

$$f_y(y) = e^{-y} \frac{1}{(1 + e^{-y})^2}$$

for every $y \in \mathbb{R}$.

The random variables X and Y are not independent because the product of the marginals is obviously not the joint density.

Let us find the density of the couple $(Z, -Y)$. Define the function $\varphi : \mathbb{R}^2 \to \mathbb{R}^2$ by

$$\varphi(x, y) = (x - y, -y).$$

This function φ satisfies the hypotheses of Theorem 7.24 and

$$\varphi^{-1}(u, v) = (u - v, -v)$$

with

$$J\varphi^{-1}(u, v) = \begin{pmatrix} 1 & 0 \\ -1 & -1 \end{pmatrix}$$

Applying Theorem 7.24 we find the density of $(Z, -Y)$ as

$$\begin{aligned} f_{(Z,-Y)}(u, v) &= f(\varphi^{-1}(u, v) \ \det J\varphi^{-1}(u, v) \\ &= 2e^{v-u}e^{v}(1 + e^{v-u} + e^{v})^{-3} \\ &= 2e^{2v}e^{-u}(1 + e^{v-u} + e^{v})^{-3} \end{aligned}$$

for every $u, v \in \mathbb{R}$.

To calculate the density of Z, we write

$$\begin{aligned} f_Z(u) &= \int_{\mathbb{R}} f_{(Z,-Y)}(u, v)\, dv \\ &= \int_{\mathbb{R}} 2e^{-u}e^{-v}\left(e^{-v} + e^{-u} + 1\right)^{-3} dv \\ &= e^{-u}\left[\left(e^{-v} + e^{-u} + 1\right)^{-3}\right]_{v=-\infty}^{v=\infty} \\ &= e^{-u}(1 + e^{-u})^{-3}. \end{aligned}$$

As can be easily observed, the random variable $Z = X - Y$ has the same distribution as either X or Y! ■

7.6 Let X_1 and X_2 be two independent random variables with the same density f given by

$$f(x) = 2x\, 1_{[0,1]}(x).$$

(**a**) Find the law of $Y = \frac{X_1}{X_2}$.

(**b**) Are the variables Y and X_2 independent?

Solution: Note first that the vector (X, Y) has the joint probability density function given by

$$f_{(X,Y)}(x, y) = 4xy 1_{[0,1]}(x)\mathbf{1}_{[0,1]}(y)$$

(Theorem 7.6). Consider the function $\varphi : \mathbb{R}^2 \to \mathbb{R}^2$ given by

$$\varphi(x, y) = \left(\frac{x}{y}, y\right).$$

This mapping is bijective from $[0, 1] \times [0, 1]$ into the domain

$$D = \{(u, v) \in [0, \infty) \times [0, 1] \mid uv \in [0, 1]\}$$

with

$$\varphi^{-1} : D \to [0, 1] \times [0, 1]$$

given by

$$\varphi^{-1}(u, v) = (uv, u).$$

The Jacobian matrix of φ^{-1} is

$$J\varphi^{-1}(u, v) = \begin{pmatrix} v & u \\ 0 & 1 \end{pmatrix}$$

and $det\, J\varphi^{-1}(u, v) = v$. By Theorem 7.24, the density of the random vector $\left(\frac{X_1}{X_2}, X_2\right)$ is

$$h(u, v) = 4uv^3 \mathbf{1}_{[0,1]}(v)\mathbf{1}_{[0,1]}(uv).$$

We can now obtain the density of $\frac{X_1}{X_2}$.

$$\begin{aligned} f_{\frac{X_1}{X_2}}(u) &= \int_{\mathbb{R}} h(u, v)\, dv \\ &= 4u1_{(x\geq 0)} \int_0^{1\wedge\frac{1}{u}} v^3\, dv \\ &= u1_{(x\geq 0)} \left[v^4\right]_{v=0}^{v=1\wedge\frac{1}{u}} \\ &= u\left(1 \wedge \frac{1}{u}\right)^4, \end{aligned}$$

where $x \wedge y$ denotes the smaller of the two numbers x and y. This gives the final expression

$$f_{\frac{X_1}{X_2}}(u) = u^{-3}\mathbf{1}_{(1,\infty)}(u) + u\mathbf{1}_{[0,1]}(u).$$

■

7.7 Assume that the vector (X_1, X_2) has joint density

$$f(x_1, x_2) = 2x_2 e^{-x_1} \mathbf{1}_{(x_1\geq 0,\ 0\leq x_2\leq 1)}.$$

(a) Check that f defines a proper probability density.

(b) Find the marginal laws of X_1 and X_2.

Solution: For the marginal distribution of X_1,

$$\begin{aligned} f_{X_1}(x_1) &= 1_{(x_1 \geq 0)} \int_0^1 2x_2 e^{-x_1} dx_2 \\ &= 1_{(x_1 \geq 0)} e^{-x_1} \left[x_2^2 \right]_{x_2=0}^{x_2=1} \\ &= 1_{(x_1 \geq 0)} e^{-x_1}. \end{aligned}$$

In the same way, we find

$$f_{X_2}(x_2) = 2x_2 1_{[0,1]}(x_2).$$

■

7.8 Let X be a standard normal random variable $N(0, 1)$ and let ε be a random variable independent of X such that

$$\mathbf{P}(\varepsilon = +1) = \mathbf{P}(\varepsilon = -1) = \frac{1}{2}.$$

Set

$$Y = \varepsilon X.$$

(a) What is the law of Y?
(b) Give the law of the random vector (X, Y)?
(c) Calculate $Cov(X, Y)$.
(d) Are the random variables X and Y independent?

Solution: We compute the cumulative distribution function of Y. For every $x \in \mathbb{R}$,

$$\begin{aligned} \mathbf{P}(Y < x) &= \mathbf{P}(Y < x, \varepsilon = 1) + \mathbf{P}(Y < x, \varepsilon = -1) \\ &= \mathbf{P}(X < x, \varepsilon = 1) + \mathbf{P}(-X < x, \varepsilon = 1) \\ &= \mathbf{P}(X < -x)\mathbf{P}(\varepsilon = 1) + \mathbf{P}(-X < x)\mathbf{P}(\varepsilon = -1) \\ &= \frac{1}{2}\mathbf{P}(X < x) + \frac{1}{2}\mathbf{P}(-X < x), \end{aligned}$$

where we used the independence of X and ε. Since X and $-X$ have the same law, namely $N(0, 1)$ (see Remark 5.26), we get

$$\mathbf{P}(Y < x) = \mathbf{P}(X < x)$$

for every $x \in \mathbb{R}$; therefore Y is also a standard normal $N(0, 1)$. Next

$$\begin{aligned} Cov(X, Y) &= \mathbf{E}XY - \mathbf{E}X\mathbf{E}Y = \mathbf{E}XY \\ &= \mathbf{E}X\varepsilon X = \mathbf{E}X^2\varepsilon = \mathbf{E}X^2\mathbf{E}\varepsilon = 1 \times 0 = 0. \end{aligned}$$

We see that the random variables X and Y are uncorrelated.

About the last question, we note that

$$\mathbf{E}X^2Y^2 = \mathbf{E}X^4\varepsilon^2 = \mathbf{E}X^4\mathbf{E}\varepsilon^2 = 3 \times 1 = 3$$

since

$$\mathbf{E}X^4 = 3 \quad \text{and} \quad \mathbf{P}(\varepsilon^2 = 1) = 1.$$

On the other hand,

$$\mathbf{E}X^2\mathbf{E}Y^2 = 1 \times 1 = 1.$$

Since $\mathbf{E}X^2Y^2 \neq \mathbf{E}X^2\mathbf{E}Y^2$, we easily see that the random variables X and Y cannot be independent.

This exercise provides an example of two uncorrelated standard normal random variables which are not independent. We will refer to this example later in the book. ∎

7.9 Let (X, Y) be a continuous random vector with density

$$f(x) = \lambda^2 e^{-\lambda(x+y)} 1_{(x,y \geq 0)}$$

with $\lambda > 0$. Compute the density of

$$Z = X + Y.$$

Solution: Prove first that X, Y are independent. Use Proposition 7.28 and check that

$$f_Z(z) = \lambda^2 z e^{-\lambda z}.$$

∎

7.10 Consider the function

$$f(x, y) = Ky\mathbf{1}_{[0,1]}(x)1_{[0,1]}(y).$$

(a) Find K so that f defines a proper density on $\mathbb{R}^2$.

(b) Assume that the random vector (X, Y) has the density f with the K computed in the previous part. Compute the marginal densities of X and Y. Are the random variables X and Y independent?

(c) Let $Z = X + Y$ with X, Y as above. Compute the density of Z and its cumulative distribution function (c.d.f.).

Solution: We have

$$\int_{\mathbb{R}} \int_{\mathbb{R}} dxdy\, f(x, y) = K \int_0^1 dx \int_0^1 dyy = K \left[\frac{1}{2}y^2\right]_{y=0}^{y=1} = \frac{K}{2}$$

and then $K = 2$. The density of X is

$$f_X(x) = 2\mathbf{1}_{[0,1]}(x)\int_0^1 dy = 2\mathbf{1}_{[0,1]}(x)\frac{1}{2} = \mathbf{1}_{[0,1]}(x).$$

So, X follows an uniform law over the interval $[0, 1]$. The density of Y is

$$\begin{aligned} f_Y(y) &= 2y\mathbf{1}_{[0,1]}(y)\int_0^1 dx \\ &= 2y\mathbf{1}_{[0,1]}(y). \end{aligned}$$

It is immediate to see that

$$f_{(X,Y)}(x, y) = f_X(x)f_Y(y)$$

for every $x, y \in \mathbb{R}$ so X and Y are independent.

The density of Y is

$$f_Y(y) = 2y\mathbf{1}_{[0,1]}(y).$$

To find the density of $X + Y$ we apply Proposition 7.28 and we obtain

$$\begin{aligned} f_{X+Y}(x) &= \int_{\mathbb{R}} f_Y(x)f_X(x-y)\,dy \\ &= \int_0^1 2y\mathbf{1}_{[0,1]}(x-y)\,dy. \end{aligned}$$

■

7.11 Let X, Y be two independent random variables with the same density function

$$f(x) = \frac{1}{x^2}1_{[1,\infty)}(x).$$

Let

$$U = XY \quad \text{and} \quad V = \frac{X}{Y}.$$

(a) Give the law of the vector (U, V).

(b) Give the marginal densities of U and V. Are they independent?

(c) Compute

$$\mathbf{E}\left(\frac{1}{\sqrt{UV}}\right).$$

Solution: The first part follows from Example 7.7. We obtain the density of (U, V) as

$$f_{(U,V)}(u, v) = \frac{1}{u^2}\frac{5}{4v}1_D(u, v)$$

with

$$D = \{(u, v) \in [1, \infty) \times [1, \infty) \mid \frac{u}{v} \geq 1\}.$$

We obtain the marginal densities as:

$$\begin{aligned} f_V(v) &= \int_{\mathbb{R}} f_{(U,V)}(u, v)\, du \\ &= \frac{5}{4v}1_{[1,\infty)}(v) \int_v^{\infty} \frac{1}{u^2}\, du \\ &= \frac{5}{4v}1_{[1,\infty)}(v)\frac{1}{4} = \frac{5}{4v^2}1_{[1,\infty)}(v), \end{aligned}$$

and similarly (left for the reader),

$$f_U(u) = \int_{\mathbb{R}} f_{(U,V)}(u, v)\, du.$$

■

7.12 Let (X, Y) be a random vector with joint density

$$f(x, y) = a^2 e^{-ay}\mathbf{1}_{\{0<x<y\}}.$$

(a) Prove that f is a density
(b) Compute the marginal distributions
(c) Find the density of the couple

$$(X, X - Y).$$

(d) Give the law of $X - Y$. Are the r.v.'s X and $X - Y$ independent?
(e) Find the density of the couple

$$(X, 2X - Y).$$

Solution:

$$\int_{\mathbb{R}^2} f(x,y)\,dxdy = a^2 \int_0^{\infty} dx \int_0^{x} dye^{-ay}$$
$$= a^2 \int_0^{\infty} dyye^{-ay}$$
$$= a^2 \frac{1}{a} \int_0^{\infty} e^{-ay} dy$$
$$= \left[-e^{-ay}\right]_{y=0}^{y=\infty} = 1.$$

The density of Y is

$$f_Y(y) = a^2 e^{-ay} y \mathbf{1}_{(0,\infty)}(y)$$

so $Y \sim \Gamma(2, a)$ and the density of Y is

$$f_X(x) = ae^{-ax}\mathbf{1}_{(0,\infty)}(x)$$

so $X \sim Exp(a) = \Gamma(1, a)$. Define the function

$$\varphi : D = \{= (x, y); 0 < x < y\} \to D' = (0, \infty) \times (-\infty, 0),$$

which is a bijection from D to D' with

$$\varphi^{-1}(u, v) = (u, u - v).$$

Since $detJ\varphi^{-1}(u, v) = -1$ the density of $(X, X - Y)$ is

$$g(u, v) = a^2\mathbf{1}_{(0,\infty)}(u)\mathbf{1}_{(-\infty,0)}(v)e^{-a(u-v)}.$$

The density of $X - Y$ is the second marginal density of the vector $(X, X - Y)$. It equals

$$f_{X-Y}(v) = a^2\mathbf{1}_{(-\infty,0)}(v)e^{av} \int_0^{\infty} e^{-au}du = a\mathbf{1}_{(-\infty,0)}(v)e^{av}.$$

We can see that

$$g(u, v) = f_X(u)f_{X-Y}(v),$$

so X and $X - Y$ are independent.

For the last part, define

$$\psi : D \to D'' = \{(u, v); u > v, u > 0\}.$$

This function ψ is a bijection from D to D'' and

$$\psi^{-1}(u, v) = (u, 2u, v).$$

Again $detJ\psi^{-1}(u, v) = -1$ and we obtain that the density of $(X, 2X - Y)$ is

$$h(u, v) = a^2 e^{-a(2u-v)} \mathbf{1}_{D''}(u, v).$$

■

7.13 Suppose that the random vectors (X, Y) have the joint density given by

$$f(x, y) = \mathbf{P}(X = x, Y = y) = \frac{k}{n(n+1)} \mathbf{1}_{\{1 \leq y \leq x \leq n\}}$$

(a) Find the constant k that makes f a proper distribution.
(b) Find the marginal distributions.
(c) Are X and Y independent?
(d) For $n = 5$ find the probabilities:

$$\mathbf{P}(X \leq 4, Y \leq 4), \qquad \mathbf{P}(X \leq 4), \qquad \mathbf{P}(Y > 3).$$

Solution: We need to have

$$\sum_{x,y} f(x, y) = 1$$

or equivalently

$$\begin{aligned} 1 = \sum_{y=1}^{n} \sum_{x=y}^{n} \frac{k}{n(n+1)} &= k \sum_{y=1}^{n} \sum_{x=y}^{n} \frac{1}{n(n+1)} \\ &= k \sum_{y=1}^{n} \frac{n - y - 1}{n(n+1)} \\ &= k \sum_{y=1}^{n} \left(\frac{1}{n} - \frac{y}{n(n+1)} \right) \\ &= k \left(1 - \frac{1}{n(n+1)} \frac{n(n+1)}{2} \right) \\ &= \frac{k}{2}, \end{aligned}$$

which implies $k = 2$.

(b) For every $x = 1, \ldots, n$, we have

$$\begin{aligned} \mathbf{P}(X = x) &= \sum_{y=1}^{x} \frac{2}{n(n+1)} \\ &= \frac{2x}{n(n+1)} \end{aligned}$$

and

$$\mathbf{P}(X = x) = 0 \qquad \text{if } x > n.$$

For every $y = 1, \ldots, n$, we obtain

$$\mathbf{P}(Y = y) = \sum_{x=y}^{n} \frac{2}{n(n+1)} = \frac{2(n-y+1)}{n(n+1)}$$

and

$$\mathbf{P}(Y = y) = 0 \text{ if } y > n.$$

(c) Clearly, in general

$$\mathbf{P}(X = x, Y = y) \neq \mathbf{P}(X = x)\mathbf{P}(Y = y),$$

which means that X and Y are not independent.

(d) We can write

$$\mathbf{P}(X \leq 4, Y \leq 3) = \sum_{y=1}^{3} \sum_{x=y}^{4} \frac{2}{5 \times 6} = \frac{3}{5}$$

and

$$\mathbf{P}(X \leq 4) = \sum_{x=1}^{4} \frac{2x}{5 \times 6} = \frac{2}{3}$$

and

$$\mathbf{P}(Y > 3) = 1 - \mathbf{P}(Y \leq 3) = 1 - \sum_{y=1}^{3} \frac{2(5-y+1)}{5 \times 6} = \frac{1}{5}.$$

■

Problems without Solution

7.14 Prove Proposition 7.26.

7.15 Suppose the random vector (X, Y) has the joint density

$$f(x, y) = \frac{1}{\pi} 1_{\Delta}(x, y)$$

where the set is the unit disk:

$$\Delta = \{(x, y) \in \mathbb{R}^2, x^2 + y^2 \leq 1\}.$$

(a) Calculate

$$\mathbf{P}\left(X \geq \frac{1}{\sqrt{2}}\right) \quad \text{and} \quad \mathbf{P}\left(Y \geq \frac{1}{\sqrt{2}}\right)$$

(b) Compute

$$\mathbf{P}\left(X \geq \frac{1}{\sqrt{2}}, Y \geq \frac{1}{\sqrt{2}}\right).$$

Are the r.v. X, Y independent?

(c) Compute $\mathbf{P}(X \geq Y)$.

7.16 Let X, Y be independent identically distributed random variables with a uniform distribution on $[0, 1]$. Denote by

$$Z = X + Y.$$

(a) Calculate $\mathbf{E}Z$.

(b) Find the density of Z.

(c) Show that for every $x \in (0, 1)$ the events

$$(Z > 1) \quad \text{and} \quad (-x < Z \leq 1 + x)$$

are independent.

7.17 Write down an expression for the density of the random variable defined in Example 7.6. Show that it defines a probability distribution.

7.18 Prove that the marginal distributions of the multinomial law are binomial distributions. See Example 7.6.

7.19 Let $C, \alpha > 0$. Let X be an r.v. with values in $\mathbb{N}^*$ such that

$$\mathbf{P}(X = k) = \frac{C}{k^\alpha}, \qquad \forall k \in \mathbb{N}^*.$$

(a) Find conditions on α and C such that the expression above defines a proper probability distribution.

(b) For what values of α the random variable X is integrable (has a finite expectation)?

7.20 Suppose $X \sim Exp(\theta)$ (exponentially distributed random variable with parameter $\theta > 0$) and let S be a uniformly distributed random variable on $\{-1, +1\}$. Assume S and X are independent and set

$$Y = XS.$$

(a) Show that Y follows a Laplace distribution with parameter θ. Recall that the Laplace law with parameter $\theta > 0$ has density

$$f(x) = \frac{\theta}{2} \exp(-\theta\,|x|).$$

Let Z be an r.v. with Laplace law with parameter θ. We further define random variables T and ζ by

$$T = \begin{cases} +1 & \text{if } Z \geq 0, \\ -1 & \text{if } Z < 0, \end{cases}$$

$$\zeta = |Z|\,.$$

(b) Show that T follows a discrete uniform distribution on $\{-1, +1\}$,

(c) Prove that ζ has an exponential distribution $Exp(\theta)$.

(d) Prove that the random variables T and ζ are independent.

7.21 We know that the random variables X and Y have a joint density given by the function $f(x, y)$. Assume that $\mathbf{P}(Y = 0) = 0$. Find the densities of the following variables in terms of the function f:

(a) $X + Y$

(b) $X - Y$

(c) XY

(d) $\frac{X}{Y}$

7.22 All children in Bulgaria are given IQ tests at ages 8 and 16. Let X be the IQ score at age 8 and let Y be the IQ score at age 16 for a randomly chosen Bulgarian 16-year-old. The joint distribution of X and Y can be described as follows. X is normal with mean 100 and standard deviation 15. Given that $X = x$, the conditional distribution of Y is normal with mean $0.8x + 30$ and standard deviation 9.

Among Bulgarian 16-year-olds with $Y = 120$, what fraction have $X \geq 120$?

7.23 Find a density function $f(x, y)$ such that if (X, Y) has density f, then $X^2 + Y^2$ is uniformly distributed on (0,10).

7.24 Let X be a unit exponential random variable (with density $f(x) = e^{-x}$, $x > 0$) and let Y be an independent $U[0, 1]$ random variable. Find the density of $T = Y/X$.

7.25 You have two opponents A and B with whom you **alternately** play games. Whenever you play A, you win with probability p_A; whenever you play B, you win with probability p_B, where $p_B > p_A$. If your objective is to *minimize the number of games you need to play to win two in a row*, should you start playing with A or with B?

7.26 Let X_1 and X_2 be independent, unit exponential random variables (so the common density is $f(x) = e^{-x}$, $x > 0$). Define

$$Y_1 = X_1 - X_2 \quad \text{and} \quad Y_2 = X_1/(X_1 - X_2).$$

Find the joint density of Y_1 and Y_2.

7.27 Let Z be an r.v. with an exponential distribution $Exp(a)$ with $a > 0$ and define

$$X = be^Z$$

with $b > 0$.

(a) Show that the density of X is given by

$$f(x) = \begin{cases} \frac{ab^a}{x^{a+1}}, & \text{if } x \geq b, \\ 0, & \text{otherwise.} \end{cases}$$

We will say that X follows a Paréto distribution denoted $P(a, b)$.

(b) For which values of a is the r.v. X integrable (i.e., has finite expectation)? In this case, compute its expectation.

(c) For what values of a is the r.v. X square integrable (i.e., X^2 has finite expectation)? In this case, compute the variance of X.

(d) Let Y, W be two independent random variables following Paréto distributions $P(a, 1)$ and $P(b, 1)$ respectively, with $a \neq b, a, b > 0$. Calculate the density of (U, V), where

$$U = YW \quad \text{and} \quad V = \frac{Y}{W}.$$

(e) Derive the marginal laws of U and V.

(f) Are U and V independent?

7.28 Let

$$f(x) = K\left(1 + xy(x^2 - y^2)\right)\mathbf{1}_{[-1,1]}(x)\mathbf{1}_{[-1,1]}(y).$$

(a) Find K such that f is a density.

(b) Compute the marginal densities.

(c) Compute $Cov(X, Y)$.

(d) Are X, Y independent?

(e) Let $U = XY$ and $V = \frac{X}{Y}$. Compute the density of the random vector (U, V).

7.29 Prove Proposition 7.2 on page 211.

7.30 Let (X, Y) be a Gaussian random vector such that X and Y are standard normal random variables $N(0, 1)$. Suppose that

$$Cov(X, Y) = \rho.$$

Let $\theta \in \mathbb{R}$. We set

$$U = X\cos\theta - Y\sin\theta, \qquad V = X\sin\theta + Y\cos\theta.$$

(a) Show that $|\rho| \leq 1$.

(b) Calculate $\mathbf{E}(U)$, $\mathbf{E}(V)$, $Var(U)$, $Var(V)$, and $Cov(U, V)$. What can be said about the vector (U, V) ?

(c) Suppose that $\rho \neq 0$. Are there values of θ such that U and V are independent?.

(d) Suppose $\rho = 0$. Give the marginal densities of U and V? Are U and V independent in this case?

7.31 Let X and Y be two independent exponentially distributed random variables with parameter $\lambda > 0$—that is, with density $f(x) = \lambda e^{-\lambda x}\mathbf{1}_{(0,+\infty)}(x)$. Let

$$S = X + Y \quad \text{and} \quad Q = \frac{Y}{X+Y}.$$

(a) Identify the law of X as a gamma law. Give its parameter values. Calculate the law of S.

(b) Give the density of the couple (S, Q) ? What is the density of Q?

(c) What can be said about the r.v.'s S and Q?

7.32 Let D be the set in $\mathbb{R}^2$ defined by $D = \{(x, y) \in \mathbb{R}^2 : 1 < y < x\}$. Let (X, Y) be a random vector with density f given by

$$f(x, y) = \frac{C}{x^3}\mathbf{1}_D(x, y).$$

(a) Draw the set D.

(b) Calculate the density of Y in terms of C and deduce the value of C.

(c) Calculate the density of X.

(d) Are X and Y independent? Do the r.v.'s X and Y have finite expectations?

(e) Let A be the subset of D defined by

$$A = \{(x, y) \in \mathbb{R}^2 : 1 < y < x - 1\}.$$

Draw an image of A and calculate the probability:

$$\mathbf{P}\big((X, Y) \in A\big).$$

7.33 Let (X, Y) be a vector with density f:

$$f(x, y) = \frac{C}{\sqrt{x}}\, e^{-y} 1_D(x, y),$$

where D is the domain

$$D = \{(x, y) \in \mathbb{R}^2 : x > 0,\ y > 0,\ y^2 > x\}.$$

(a) Compute the distribution of Y and then deduce C.
(b) Calculate the density of X.
(c) Are X and Y independent?
(d) Are the r.v.'s X and Z independent, where

$$Z = Y - \sqrt{X}?$$

(e) Identify the law of Z as a distribution we learned about.
(f) Define

$$T = \frac{X}{Y^2}.$$

Are T and Y independent?
(g) Calculate $\mathbf{E}(T)$.

7.34 We assume that the random vectors (X, Y) have the distribution

$$\mathbf{P}(X = i, Y = j) = \frac{\alpha}{(i + j + 1)!}$$

for every $(i, j) \in \mathbb{N}^2$.
(a) Explain why X and Y have the same law.
(b) Let $S = X + Y$. Show that

$$\mathbf{P}(S = k) = \frac{\alpha}{k!}$$

for all $k \in \mathbb{N}$.
(c) Derive the value of α which makes a proper density.
(d) Calculate $\mathbf{P}(X = 0)$. Are X and Y independent?
(e) Calculate $\mathbf{P}(X = Y)$ and derive $\mathbf{P}(X > Y)$.

CHAPTER EIGHT

Characteristic Function

8.1 Introduction/Purpose of the Chapter

The characteristic function of a random variable is a powerful tool for analyzing the distribution of sums of independent random variables. To some readers, characteristic functions may already be familiar in a different form: If a random variable is continuous and thus it has a probability density function $f(x)$, then its characteristic function is the Fourier transform of the function $f(x)$.

8.2 Vignette/Historical Notes

According to Kenney (1942) and (Todhunter, 1865, pp. 309–313), the first use of an analytic method substantially equivalent to the characteristic functions is due to Joseph Louis de Lagrange in his work *Réflexions sur la résolution algébrique des équations*, published around 1770 de Lagrange (1770). In this work, de Lagrange introduces a simple form of the Fourier transform. The Fourier transform and the Fourier series was properly introduced and refined by Joseph Fourier in his *Mémoire sur la propagation de la chaleur dans les corps solides*. He applies the series to find the solution for the heat equation.

The first general definition of the characteristic function is due to Pierre-Simon marquis de Laplace in his classic "*Théorie analytique des probabilités*" (Laplace, 1812, pp. 83–84). He first introduces the generating function as "fonction generatrice" (described in the following chapter on moment generating

---

Handbook of Probability, First Edition. Ionuţ Florescu and Ciprian Tudor.

function). Laplace is recognized with the introduction of the moment-generating function—or, in its more traditional name, Laplace transform. However, by setting $t^x = e^{isx}$ in the generating function, Laplace comes up with an equivalent form of the Fourier transform (the characteristic function) to which he does not assign a name. Perhaps this is done on purpose since Laplace was familiar with the Fourier transforms and he uses the function throughout his work to solve differential equations (which was exactly the purpose Fourier had in mind when he introduced it in the first place). However, there are two important differences between Laplace's approach and that of Fourier. First, the definition Laplace introduces can deal with discontinuities in the probability distributions; thus his definition (unlike Fourier) can deal with any random variables (not only continuous), a fact he explicitly recognized in his book. Second, he writes the inversion formula—what we now call the inverse Fourier transform, and once again this transform is written explicitly for discontinuous points as well. This creates a new way to solve differential equations without explicitly requiring that all functions be written as a fourier series (a requirement which is not possible because of those discontinuities).

The name characteristic function is due to Paul Levy in his book *Calcul des probabilités* Lévy (1925) who reintroduces the same function as Laplace. Since that time, the characteristic function is one of the primary tools of the probability theory. In fact an entire class of distributions (the Lévy distribution) is introduced through the characteristic function rather than via the distribution function.

8.3 Theory and Applications

8.3.1 DEFINITION AND BASIC PROPERTIES

Definition 8.1 *If $X : \Omega \to \mathbb{R}$ is a random variable, then its characteristic function*

$$\varphi_X : \mathbb{R} \to \mathbb{C}$$

is defined by

$$\varphi_X(\lambda) = \mathbf{E}(e^{i\lambda X}), \qquad \textit{for every } \lambda \in \mathbb{R}.$$

Note that the random variable $\mathbf{E}e^{i\lambda X}$ is complex-valued. In previous chapters we defined the expectation only for real-valued random variables. To take advantage of these definitions, recall that the complex exponential can be written as

$$e^{itx} = \cos(tx) + i\sin(tx)$$

Thus, we can write

$$\varphi_X(\lambda) = \mathbf{E}\cos(\lambda X) + i\mathbf{E}\sin(\lambda X),$$

where both expectations are now of real-valued random variables. This is the meaning of the expectation that appears in Definition 8.1.

Remark 8.2 *Since $|e^{itX(\omega)}| = 1$ for every $t \in \mathbb{R}$ and $\omega \in \Omega$ it is clear that the expectation introduced in Definition 8.1 exists and*

$$|\varphi_X(\lambda)| \leq 1.$$

This is easy to prove using Jensen inequality (see Appendix B) for the convex function $|x|$.

If X has a probability density $f_X(x)$, then the characteristic function reduces to

$$\varphi_X(\lambda) = \int_{\mathbb{R}} e^{i\lambda x} f_X(x)\, dx. \tag{8.1}$$

Formula (8.1) with $-\lambda$ replacing λ is known as the Fourier transform of f_X.

The Fourier transform is in fact defined for any integrable function f, not only for those which happen to be probability densities. Concretely, the Fourier transform of the function f_X is given by

$$\hat{f}(\lambda) = \int_{\mathbb{R}} e^{-i\lambda x} f_X(x)\, dx \tag{8.2}$$

for every $\lambda \in \mathbb{R}$, so

$$\varphi_X(\lambda) = \hat{f}(-\lambda).$$

The Fourier transform is well-defined for every function f integrable in the sense that

$$\int_{\mathbb{R}} |f(y)|\, dy < \infty$$

since $|e^{i\lambda x}| = 1$.

Remark 8.3 *In books and papers on analysis, pdes, and mathematical physics, slightly different definitions of the Fourier transform are often used—for example,*

$$\hat{f}(\lambda) = \frac{1}{\sqrt{2\pi}} \int_{\mathbb{R}} e^{i\lambda x} f_X(x)\, dx.$$

The constant $\sqrt{2\pi}$ is typically irrelevant.

We next list basic properties of the characteristic function.

Proposition 8.4 *Let X be a random variable and let φ_X denote its characteristic function. Then*

i. $\varphi_X(t) \leq 1$ *for every* $t \in \mathbb{R}$.
ii. $\varphi_X(0) = 1$.
iii. *For every* $t \in \mathbb{R}$ *we have*

$$\varphi_X(-t) = \varphi_{-X}(t) = \overline{\varphi_X(t)}.$$

Proof: The first two parts are obtained very fast. For every $t \in \mathbb{R}$

$$\begin{aligned}\varphi_X(t) &\leq |\varphi_X(t)| \leq \mathbf{E}|e^{itX}| \\ &= 1\end{aligned}$$

and

$$\varphi_X(0) = \mathbf{E}e^{i0X} = 1.$$

For the last part, obviously

$$\varphi_X(-t) = \mathbf{E}e^{-itX} = \varphi_{-X}(t) = \overline{\varphi_X(t)}.$$

■

Proposition 8.5 *The function φ_X is absolutely continuous.*

Proof: For every $h > 0$ we can write

$$\begin{aligned}&\left|\mathbf{E}e^{i(t+h)X} - \mathbf{E}e^{itX}\right| \leq \mathbf{E}\left|e^{i(t+h)X} - e^{itX}\right| \\ &= \mathbf{E}\left|e^{itX}\left(e^{ihX} - 1\right)\right| = \mathbf{E}\left[\left|e^{itX}\right|\left|\left(e^{ihX} - 1\right)\right|\right] \\ &= \mathbf{E}\left|e^{ihX} - 1\right|.\end{aligned}$$

Note that as $h \to 0$ the quantity $e^{ihX} - 1$ converges to zero for every ω. By the bounded convergence theorem, since

$$\left|e^{ihX} - 1\right| \leq 2$$

we can conclude that $\mathbf{E}\left|e^{ihX} - 1\right| \to 0$, and therefore $|\varphi_X(x+h) - \varphi(x)| \to 0$ as $h \to 0$ and thus the function φ_X is absolutely continuous. ■

Proposition 8.6 *If X, Y are two independent random variables, then*

$$\varphi_{X+Y}(\lambda) = \varphi_X(\lambda)\varphi_Y(\lambda).$$

Proof: Indeed,

$$\begin{aligned}\varphi_{X+Y}(\lambda) &= \mathbf{E}e^{i\lambda(X+Y)} \\ &= \mathbf{E}e^{i\lambda X}\mathbf{E}e^{i\lambda Y} = \varphi_X(\lambda)\varphi_Y(\lambda)\end{aligned}$$

for every $\lambda \in \mathbb{R}$, where we use the independence to be able to write the expectation of the product as product of expectations. ∎

Remark 8.7 *The result above extends easily to a finite sum of random variables; that is, if $X_1, \ldots, X_n$ are independent r.v.s, then*

$$\varphi_{X_1+\cdots+X_n}(\lambda) = \varphi_{X_1}(\lambda)\varphi_{X_1}(\lambda)\ldots\varphi_{X_n}(\lambda).$$

This can be proved by induction (we leave the proof to the reader).

In particular, if $X_1, \ldots, X_n$ are independent random variables with common distribution (i.i.d. random variables) and

$$X = X_1 + \cdots + X_n,$$

then

$$\varphi_X(t) = \left(\varphi_{X_i}(t)\right)^n$$

for every $i = 1, \ldots, n$.

The problem of existence of the derivative of the function φ_X is straightforward. The next theorem allows us to express the derivatives of the characteristic function in terms of the moments of random variable. Since $|\varphi_X| < 1$, we do not need any extra boundness assumption around 0 for the characteristic function. Instead, we require a finiteness condition on the moments of the random variable considered.

Theorem 8.8 *Suppose X is a random variable such that*

$$\mathbf{E}|X|^k < \infty.$$

Then for any j such that $0 \leq j \leq k$, the function φ_X has the j-th derivative given by

$$\varphi_X^{(j)}(t) = \mathbf{E}\left((iX)^j e^{itX}\right).$$

In particular,

$$\varphi_X^{(k)}(0) = i^j \mathbf{E}X^j,$$

and this expression connects the moments of the random variable with the characteristic function.

Proof: We will use an induction argument. The result is clear for $j = 0$. Suppose that the statement is true for $j - 1$ and denoted by

$$F(t) = (iX)^{j-1} e^{itX}.$$

Then

$$|F'(t)| = |(iX)^j e^{itX}| = |X|^j.$$

Since $\mathbf{E}|X|^k|$ is finite, we have that $\mathbf{E}|X|^j| < \infty$ for any $j \leq k$. Moreover,

$$\varphi_X^{(j)}(t) = \frac{d}{dt}\varphi_X^{(j-1)}(t) = \frac{d}{dt}\mathbf{E}(iX)^{j-1}e^{itX}$$
$$= \mathbf{E}\frac{d}{dt}(iX)^{j-1}e^{itX} = \mathbf{E}(iX)^j e^{itX};$$

thus the statement is true for step j.

Since the verification step $j = 0$ is true and for any $j - 1$ assumed true we showed that the statement is going to be true at j, then, by the induction argument, it follows that the statement is true for any $j \in \mathbb{N}$. ■

Remark 8.9 *As a particular case of Theorem 8.8, we get*

$$\mathbf{E}X = -i\varphi_X'(0)$$

and

$$\mathbf{E}X^2 = -\varphi_X''(0).$$

Proposition 8.10 *Let X be a random variable and let φ_X denote its characteristic function. For every a, $b \in \mathbb{R}$*

$$\varphi_{aX+b} = e^{itb}\varphi_X(at).$$

Proof: Clearly

$$\varphi_{aX+b}(t) = \mathbf{E}e^{it(aX+b)} = e^{itb}\mathbf{E}e^{itaX}$$
$$= e^{itb}\varphi_X(at).$$

■

8.3.2 THE RELATIONSHIP BETWEEN THE CHARACTERISTIC FUNCTION AND THE DISTRIBUTION

The characteristic function of a random variable uniquely characterizes the random variable (hence the name). If you recall, a random variable is uniquely characterized by its distribution. The clear consequence is that two random variables with the same characteristic function will have the same law (distribution). When the random variable has a density, this density can be recovered from the characteristic function. These statements will be proved next.

Theorem 8.11 **(Fourier inversion theorem)** *Let X be a continuous random variable. Suppose that the characteristic function of X, $\hat{f}_X$ given by (8.2) is integrable on the real line. Then, X admits a density f_X and f_X is continuous and bounded and given by*

$$f(x) = \frac{1}{2\pi}\int_{\mathbb{R}} \hat{f}(\lambda)e^{ix\lambda}d\lambda.$$

This theorem is the classic Fourier Inversion Theorem and it is proven in any functional analysis textbook (e.g., for a probability see Billingsley (1995)).

Remark 8.12 *Here is an example where the relationship does not hold. Suppose that X is distributed as an Exp(1). Then*

$$\varphi_X(t) = \frac{1}{1 - it}$$

and φ_X is not integrable since

$$|\varphi_X(t)| \sim \frac{1}{|t|}$$

as $|t| \to \infty$. This is consistent with the above theorem because the density of X is

$$f(x) = e^{-x} 1_{(0,\infty)}(x),$$

which is not continuous (at zero).

Clearly, the definition of the characteristic function (which is explicit) implies that two random variables with the same distribution have the same characteristic function. What is interesting and very useful in practice is that the reciprocal implication is also true. From Theorem 8.11, it follows immediately that if the random variable is continuous, the p.d.f. of the random variable $f(x)$ is uniquely determined by the characteristic function of the random variable.

However, this statement remains true even if the characteristic function is not integrable; in fact, it can be extended to random variables which are not necessarily continuous.

Theorem 8.13 (Uniqueness theorem) *If two random variables X, Y have the same characteristic function*

$$\varphi_X = \varphi_Y,$$

then they have the same distribution function (c.d.f.).

Proof: Let Z be a standard normal random variable ($Z \sim N(0, 1)$), independent of X and Y. Define two new random variables:

$$X_a = X + aZ,$$
$$Y_a = Y + aZ.$$

If the original variable X, Y have the same characteristic function, then these new variables X_a and Y_a will also have the same characteristic function. Indeed, for every $t \in \mathbb{R}$ we have

$$\begin{aligned}\varphi_{X_a}(t) &= \mathbf{E}e^{itX_a} = \mathbf{E}e^{itX+itaZ} = \mathbf{E}[e^{itX}]\mathbf{E}[e^{itaZ}] \\ &= \mathbf{E}e^{itX}e^{-\frac{a^2t^2}{2}} = \varphi_X(t)e^{-\frac{a^2t^2}{2}} \\ &= \varphi_Y(t)e^{-\frac{a^2t^2}{2}} \\ &= \varphi_{Y_a}(t),\end{aligned}$$

where we used the independence of Z from X and Y.

The difference now is that the random variables X_a and Y_a have an integrable characteristic function (this is easy to show and left as an exercise). Using Theorem 8.11 we obtain that X_a and Y_a have the same density. Therefore, by the definition we must have

$$\mathbf{E}g(X_a) = \mathbf{E}g(Y_a),$$

for any continuous bounded function $g : \mathbb{R} \to \mathbb{R}$. Now, we let $a \to 0$. From the Monotone Convergence Theorem we get

$$\mathbf{E}g(X) = \mathbf{E}g(Y) \tag{8.3}$$

for any continuous bounded function g. For an arbitrary $x \in \mathbb{R}$, let us choose

$$g(y) = \begin{cases} 1 & \text{if } y \in (-\infty, x), \\ 1 - n(y - x) & \text{if } y \in [x, x + \frac{1}{n}], \\ 0 & \text{if } y > x + \frac{1}{n}. \end{cases}$$

This function is continuous and bounded (plotted in Figure 8.1). Applying the result, we obtain

$$\mathbf{E}1_{(-\infty,x+\frac{1}{n})}(X) = \mathbf{E}1_{(-\infty,x+\frac{1}{n})}(Y), \qquad \forall x \in \mathbb{R}$$

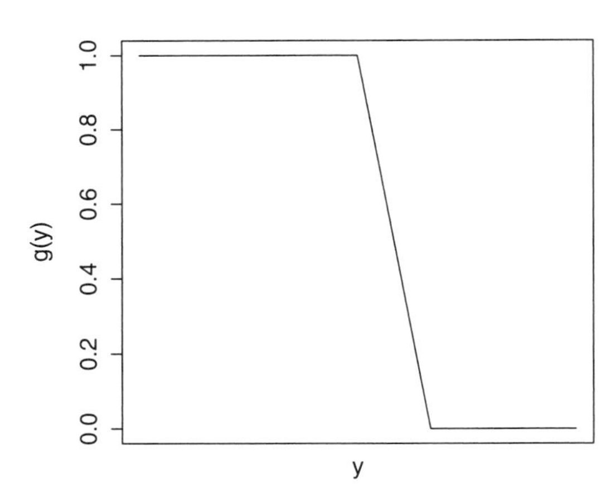

FIGURE 8.1 A plot of the function g used in Theorem 8.13.

Letting $n \to \infty$ and again applying the Monotone Convergence Theorem, we obtain

$$\mathbf{E}1_{(-\infty,x]}(X) = \mathbf{E}1_{(-\infty,x]}(Y)$$

for any x and that means that X, Y have the same c.d.f. ■

In the next section we will show that a random variable is $N(\mu, \sigma)$ distributed if and only if its characteristic function is

$$\varphi_X(t) = e^{i\mu t - \frac{\sigma^2 t^2}{2}}$$

for every $t \in \mathbb{R}$. This result can also be proven using that the sum of two independent r.v. normally distributed is normal (see exercise 8.2).

Theorem 8.14 (Uniqueness theorem for moments) *Let X, Y be two r.v. and denote by F_X, F_Y their cumulative distribution functions. If*

(i) *X and Y each have finite moments of all orders,*

(ii) *their moments are the same, that is, $\mathbf{E}X^k = \mathbf{E}Y^k := \alpha_k$ for all $k \geq 1$, and*

(iii) *the radius R of the power series $\sum_{k=1}^{\infty} \alpha_k \frac{u^k}{k!}$ is nonzero,*

then the two random variables have the same distribution,

$$F_X = F_Y.$$

Remark 8.15 *This theorem says that in principle two distributions are the same if all the moments identical. Recall that having the same distribution is also the same as having the same characteristic function.*

However, all conditions are needed. It is possible for two random variables to have the same moments but not the same distribution.

EXAMPLE 8.1 Identifying a Normal by calculating its moments

Here is an example of the applicability of the theorem. Let X be a random variable such that its moments

$$\alpha_k := \mathbf{E}X^k$$

satisfy

$$\alpha_k = \begin{cases} 1 \times 3 \times 5 \times (k-3) \times (k-1) = \frac{(2p)!}{2^p p!} & \text{if } k \text{ is even } = 2p, \\ 0 & \text{if } k \text{ is odd.} \end{cases}$$

Then, applying the theorem, the random variable X is $N(0, 1)$ distributed. Indeed, the distribution of a standard normal has these moments; and the series in the theorem,

$$\sum_{k=1}^{\infty} \alpha_k \frac{u^k}{k!},$$

has the radius of convergence $R = \infty$.

If the random variable is not continuous (or, in any other case), it is possible to state an inversion-type formula similar to the Fourier inversion theorem (Theorem 8.11) for the cumulative distribution function of a random variable.

Proposition 8.16 *i. Let X be a random variable with distribution function F_X, density f_X, and characteristic function φ_X. Suppose φ_X is integrable. Then*

$$F_X(b) - F_X(a) = \frac{1}{2\pi} \int_{\mathbb{R}} \frac{e^{-ita} - e^{-itb}}{it} \varphi_X(t)\, dt.$$

ii. Drop the assumption that φ_X is integrable. Then

$$F_X(b) - F_X(a) = \lim_{c \to 0} \frac{1}{2\pi} \int_{\mathbb{R}} \frac{e^{-ita} - e^{-itb}}{it} \varphi_X(t) e^{-\frac{c^2 t^2}{2}}\, dt.$$

Proof: Use Theorem 8.11. ■

If the random variable is discrete (the distribution is made of a points with corresponding probability mass function), an inversion formula can be found in Exercise 8.9.

Proposition 8.17 *Let X be an r.v. and let φ_X be its characteristic function. Then the law of X is symmetric if and only if φ_X is real-valued.*

Remark 8.18 *Here, "symmetric" means that the distribution is symmetric about zero: the distribution of X is symmetric if X and $-X$ have the same distribution. This is the same as saying that $\mathbf{P}(X \geq a) = \mathbf{P}(X \leq -a)$ for all a. In particular, if the random variable is continuous and has a density function f, it is symmetric if and only if f is an even function (recall that this means $f(-x) = f(x)$).*

Proof: Consider the case when X admits a continuous density f. In this case

$$\varphi_X(t) = \int_{\mathbb{R}} \cos(tx) f(x)\, dx + i \int_{\mathbb{R}} \sin(tx) f(x)\, dx.$$

If f is an even function, then

$$x \to \sin(tx)f(x)$$

is odd and consequently

$$\int_{\mathbb{R}} \sin(tx)f(x)\,dx = 0.$$

So φ_X is real-valued.

Conversely, if φ_X is real-valued, by Proposition 8.4,

$$\varphi_{-X}(t) = \varphi_X(-t) = \overline{\varphi_X(t)} = \varphi_X(t)$$

and the uniqueness theorem (Theorem 8.13) implies that X and $-X$ have the same distribution.

The general case is identical, but the integrals are expressed in terms of the c.d.f. $F(x)$, that is,

$$\int_{\mathbb{R}} \sin(tx)\,dF(x),$$

and

$$\int_{\mathbb{R}} \cos(tx)\,dF(x).$$

■

8.4 Calculation of the Characteristic Function for Commonly Encountered Distributions

In this section we provide examples of calculating characteristic functions for some distributions presented earlier.

8.4.1 BERNOULLI AND BINOMIAL

Proposition 8.19 *Suppose X is a Bernoulli random variable with probability of success p. Then*

$$\varphi_X(t) = 1 - p(1 - e^{it}).$$

Proof: Using the formula for the expectation of finite random variables, we obtain

$$\begin{aligned}\varphi_X(t) &= e^{it0}\mathbf{P}(X = 0) + e^{it1}\mathbf{P}(X = 1) = 1 - p + e^{it}p \\ &= 1 - p(1 - e^{it}).\end{aligned}$$

■

Proposition 8.20 *If Y follows a binomial distribution with parameters (n, p), then*

$$\varphi_Y(t) = \left(1 - p(1 - e^{it})\right)^n.$$

Proof: This is a consequence of the above result since Y can be written as

$$Y = X_1 + \cdots + X_n$$

where X_i are independent Bernoulli random variables with parameter p. Then just apply Remark 8.7. ■

8.4.2 UNIFORM DISTRIBUTION

Proposition 8.21 *If $X \sim U[0, 1]$, then*

$$\varphi_X(t) = \frac{e^{it} - 1}{it}.$$

More generally, if $X \sim U[a, b]$, then

$$\varphi_X(t) = \frac{e^{itb} - e^{ita}}{it(b - a)}.$$

Proof: Since the density of the law $U[0, 1]$ is $g(x) = 1_{[0,1]}(x)$, we get

$$\begin{aligned}\varphi_X(t) &= \int_0^1 e^{itx}\,dx = \frac{1}{it}\left[e^{itx}\right]_{x=0}^{x=1} \\ &= \frac{e^{it} - 1}{it}.\end{aligned}$$

The more general result is obtained by noting that $X \sim U[a, b]$ can be written as

$$X = a + (b - a)U,$$

with $U \sim U[0, 1]$. ■

Remark 8.22 *If $a = -b$ (i.e., X has a symmetric distribution $X \sim U[-b, b]$), then*

$$\begin{aligned}\varphi_X(t) &= \frac{e^{itb} - e^{-itb}}{it(b - (-b))} \\ &= \frac{\sin tb}{tb}\end{aligned}$$

and note that this is a real-valued function. This is in accord with Proposition 8.17.

8.4.3 NORMAL DISTRIBUTION

Proposition 8.23 *Suppose Z is a standard normal random variable. Then*

$$\varphi_Z(t) = e^{-\frac{t^2}{2}}.$$

Proof: We can use a direct "calculation":

$$\begin{aligned}
\mathbf{E}e^{itX} &= \frac{1}{\sqrt{2\pi}} \int_{\mathbb{R}} e^{itx-\frac{x^2}{2}} dx \\
&= \frac{1}{\sqrt{2\pi}} \int_{\mathbb{R}} e^{-(t^2+(x-it)^2)/2} dx \\
&= e^{-\frac{t^2}{2}} \frac{1}{\sqrt{2\pi}} \int_{\mathbb{R}} e^{-\frac{y^2}{2}} dy \\
&= e^{-\frac{t^2}{2}},
\end{aligned}$$

where we made the formal change of variables $y = x - it$. To mathematically justify this change of variables (which replace a real variable by a complex one!), we would need more complicated arguments from complex function theory. ■

Remark 8.24 *Even though the proof needs more arguments, the result is correct. A different proof of this result, based on differential equations, is given in Exercise 8.5.*

Proposition 8.25 *Suppose X is a normal random variable with expectation μ and variance σ^2, $X \sim N(\mu, \sigma^2)$. Then its characteristic function is*

$$\varphi_X(t) = e^{it\mu} e^{-\frac{\sigma^2 t^2}{2}} = e^{it\mu - \frac{\sigma^2 t^2}{2}}.$$

Proof: Since X can be written as $X = \mu + \sigma Z$, where Z is a standard normal random variable, the result is obtained using Propositions 8.23 and 8.10. ■

8.4.4 POISSON DISTRIBUTION

Proposition 8.26 *If X denotes a Poisson random variable with parameter $\lambda > 0$, then*

$$\varphi_X(t) = e^{-\lambda(1-e^{it})}.$$

Proof: Using the definition of the Poisson law, for every $\lambda \in \mathbb{R}$

$$\mathbf{E}e^{i\lambda X} = \sum_{k \geq 0} e^{i\lambda k} \frac{\lambda^k}{k!} e^{-\lambda}$$

$$= e^{-\lambda} \sum_{k \geq 0} \frac{(\lambda e^{it})^k}{k!}$$
$$= e^{-\lambda(1-e^{it})}.$$

■

8.4.5 GAMMA DISTRIBUTION

Proposition 8.27 *If $X \sim \Gamma(a, \lambda)$, then*

$$\varphi_X(t) = \left(\frac{1}{1 - \frac{it}{\lambda}}\right)^a.$$

Proof: Recall that the density of the gamma distribution is (see Section 5.4.4)

$$f(x) = \frac{\lambda^a}{\Gamma(a)} x^{a-1} e^{-\lambda x} 1_{(0,\infty)}(x).$$

We use the power series expansion of the exponential function

$$e^{itx} = \sum_{n \geq 0} \frac{(itx)^n}{n!}$$

and then (the interchange of the sum and of the integral can be rigourously argued)

$$\varphi_X(t) = \frac{\lambda^a}{\Gamma(a)} \sum_{n \geq 0} \int_0^\infty dx x^{a-1} e^{-\lambda x} \frac{(itx)^n}{n!}$$
$$= \sum_{n \geq 0} \frac{\lambda^a (it)^n}{\Gamma(a) n!} \int_0^\infty e^{-\lambda x} x^{a+n-1} dx$$
$$= \sum_{n \geq 0} \frac{\lambda^a (it)^n}{\Gamma(a) n!} \lambda^{-a-n} \int_0^\infty dy e^{-y} y^{a+n-1}$$
$$= \sum_{n \geq 0} \frac{\lambda^a (it)^n}{\Gamma(a) n!} \lambda^{-a-n} \Gamma(n+a)$$
$$= \sum_{n \geq 0} C_n^a \left(\frac{it}{\lambda}\right)^n$$
$$= \left(\frac{1}{1 - \frac{it}{\lambda}}\right)^a,$$

using the properties of the gamma function and the power series expansion of the function $(1-x)^a$. ■

Remark 8.28 *Recall that the distribution* $\Gamma(1, \lambda)$ *corresponds to the exponential distribution with parameter* $\lambda > 0$. *Therefore, if* $X \sim Exp(\lambda)$, *then*

$$\varphi_X(t) = \frac{1}{1 - \frac{it}{\lambda}}$$

for every $t \in \mathbb{R}$.

Remark 8.29 *Using Proposition 8.27, it is easy to check that if* $X \sim \Gamma(a, \lambda)$ *and* $Y \sim \Gamma(b, \lambda)$, *then*

$$X + Y \sim \Gamma(a + b, \lambda).$$

Indeed,

$$\begin{aligned}\varphi_{X+Y}(t) &= \varphi_X(t)\varphi_Y(t) \\ &= \left(\frac{1}{1 - \frac{it}{\lambda}}\right)^a \left(\frac{1}{1 - \frac{it}{\lambda}}\right)^b \\ &= \left(\frac{1}{1 - \frac{it}{\lambda}}\right)^{a+b},\end{aligned}$$

which means that $X + Y \sim \Gamma(a + b, \lambda)$. *This proof is much easier than a direct proof based on the density of the gamma distribution.*

8.4.6 CAUCHY DISTRIBUTION

Proposition 8.30 *Suppose X has a Cauchy distribution. Then*

$$\varphi_X(t) = e^{-|t|}.$$

Proof: We will use the integral formula:

$$\int_0^\infty \frac{\cos(tx)}{b^2 + x^2} dx = \frac{\pi}{2b} e^{-tb}, \qquad t \geq 0,$$

for every $b \in \mathbb{R}$. Note that the function under the integral is even, and so we have

$$\int_{\mathbb{R}} \frac{\cos(tx)}{b^2 + x^2} dx = \frac{\pi}{b} e^{-tb}, \qquad t \geq 0,$$

Recall that the density of the Cauchy distribution is

$$f(x) = \frac{1}{\pi} \frac{1}{1 + x^2}.$$

Since the imaginary part of φ_X vanishes (the law of X is symmetric) for $t \geq 0$, we have

$$\begin{aligned}\varphi_X(t) &= \frac{1}{\pi}\int_{\mathbb{R}} e^{itx}\frac{1}{1+x^2}dx \\ &= \frac{1}{\pi}\int_{\mathbb{R}} \cos(tx)\frac{1}{1+x^2} \\ &= e^{-t}.\end{aligned}$$

In a similar way for $t \leq 0$, we obtain

$$\varphi_X(t) = e^t,$$

which concludes the result. ■

8.4.7 LAPLACE DISTRIBUTION

Proposition 8.31 *Suppose X is Laplace distributed, that is, with density*

$$f_X(x) = \frac{1}{2}e^{-|x|}.$$

Then

$$\varphi_X(t) = \frac{1}{1+t^2}.$$

Proof: The formula follows from direct computation, using the fact that the law is symmetric and thus

$$\begin{aligned}\varphi_X(t) &= \frac{1}{2}\int_{\mathbb{R}} e^{itx-|x|}dx \\ &= \frac{1}{2}\int_{\mathbb{R}} \cos(tx)e^{-|x|}dx \\ &= \int_0^{\infty} \cos(tx)e^{-x}dx\end{aligned}$$

and finally using integration by parts two times. ■

Using the above result, we can prove a very interesting property of the Cauchy random variable.

Proposition 8.32 *Suppose $X_1, \ldots, X_n$ are independent Cauchy distributed r.v. Let $S_n = X_1 + \ldots + X_n$ and*

$$\bar{X}_n = \frac{S_n}{n}.$$

Then, the average $\bar{X}_n$ is Cauchy distributed for every n.

Proof: Following Proposition 8.30, we obtain the characteristic function of S_n as

$$\varphi_{S_n}(t) = e^{-n|t|}$$

and from Proposition 8.10 we obtain

$$\varphi_{\bar{X}_n}(t) = \varphi_{S_n}(\frac{t}{n}) = e^{-|t|}.$$

Therefore, $\bar{X}_n$ has a Cauchy law. ■

This property in fact leads to Levy distributions—a distribution defined entirely using its characteristic function. The final section in this chapter tells the story of this distribution.

8.4.8 STABLE DISTRIBUTIONS. LÉVY DISTRIBUTION

In finance, the continuously compounded return of an asset is one of the most studied objects. If the equity at time t is denoted with S_t, then the continuously compounded return over the period $[t, t + \Delta t]$ is defined as $R_t = \log S_{t+\Delta t} - \log S_t$. There are two major advantages to studying return versus the asset itself. First, in the regular Black Scholes model (Black and Scholes, 1973) the returns are independent, identically distributed as a Gaussian random variable. Second, the return over a larger period, say $[t, t + n\Delta t]$, is easily expressed as the sum of the returns calculated over the smaller periods $[t + i\Delta t, t + (i + 1)\Delta t]$. If we denote X_i the return over $[t + i\Delta t, t + (i + 1)\Delta t]$, then clearly

$$X(n\Delta t) = \log S_{t+n\Delta t} - \log S_t = \sum_{i=1}^{n} X_i.$$

Since the sum of i.i.d. Gaussians is a Gaussian, this model provides a very easy way to work with distribution.

It is well known today that the Black Scholes model is not a good fit for the asset prices. This is primarily due to the nature of the Gaussian distribution. When plotting histograms of the actually observed returns, the probability of large observations is much greater than that of a normal density (the leptokurtic property of the distribution). However, if we use a different density, we may lose the scaling property of the normal—that is, the ability of re-scaling the larger period returns to the same distribution. Is it possible to have a different distribution with this property?

Definition 8.33 (Stable distribution) *Consider the sum of n independent identically distributed random variables X_i, and denote it with*

$$X(n\Delta t) = X_1 + X_2 + X_3 + \cdots + X_n.$$

Since the variables are independent, the distribution of their sum may be obtained as the n-fold convolution,

$$P[X(n\Delta t)] = P(X_1) \otimes P(X_2) \cdots \otimes P(X_n).$$

The distribution of the X_i's is called ***stable*** *if the functional form of the $P[X(n\Delta t)]$ is the same as the functional form of $P[X(\Delta t)]$. More specifically, if for any $n \geq 2$, there exists a positive C_n and a D_n so that*

$$P[X(n\Delta t)] = P[C_n X + D_n],$$

where X has the same distribution as X_i for $i = 1, 2, \ldots, n$. If $D_n = 0$, then X is said to have a strictly stable distribution.

The writing in the definition is generic, but refer to the distribution as the c.d.f.

It can be shown (e.g., Samorodnitsky and Taqqu (1994)) that

$$C_n = n^{\frac{1}{\alpha}}$$

for some parameter α, $0 < \alpha \leq 2$. This α is an important characteristic of the processes of this type (which is in fact called α-stable), but we shall not talk about it here.

Paul Lévy (Lévy, 1925) and Aleksandr Khintchine (Khintchine and Lévy, 1936) found the most general form of the stable distributions. The general representation is through the characteristic function $\varphi(t)$ associated to the distribution.

It is very easy to understand why this form of characteristic function determine the stable distributions. The characteristic function of a convolution of independent random variables is simply the product of the respective characteristic functions; thus, owing to the special exponential form above, all stable distributions are closed under convolutions for a fixed value of α. Specifically, recall that (using our previous notation)

$$\varphi_{X(n\Delta t)}(t) = (\varphi(t))^n$$

and

$$\varphi_{C_n X + D_n}(t) = e^{iD_n t}\varphi(C_n t).$$

Thus, setting them equal in principle can provide the most general form for the stable distributions. It took the genius of Paul Lévy to actually see the answer.

The most common parametrization of Lévy distributions is given in the next definition.

Definition 8.34 (Lévy distribution) *A random variable X is said to have a Lévy distribution if its characteristic function is given by*

$$\ln(\varphi(t)) = \begin{cases} i\mu t - \gamma^\alpha |t|^\alpha \left[1 - i\beta \frac{t}{|t|} \tan(\frac{\pi\alpha}{2})\right] & \text{if } \alpha \in (0, 2] \setminus \{1\}, \\ i\mu t - \gamma |t| \left[1 + i\beta \frac{t}{|t|} \frac{2}{\pi} \ln|t|\right] & \text{if } \alpha = 1. \end{cases}$$

The parameters of this distribution are as follows:

- α *is called the stability exponent or the characteristic function.*
- γ *is a positive scaling factor.*
- μ *is a location parameter.*
- $\beta \in [-1, 1]$ *is a skewness (asymmetry) parameter.*

Remark 8.35 *All such distributions are heavy tailed (leptokurtic) and have no second moments for any* $\alpha < 2$. *When* $\alpha < 1$ *the distribution does not even have the first moment. Please note that neither third nor fourth moments-exist for these distributions, so the usual measures of skewness and kurtosis are undefined.*

The characterization above is due to (Khintchine and Lévy, 1936). If we take $\beta = 0$, *we obtain a (Lévy) symmetric alpha-stable distribution (Lévy, 1925). The distribution is symmetric about* μ.

8.4.8.1 Special Cases. In general, there is no formula for the p.d.f. $f(x)$ of a stable distribution. There are, however, three special cases which reduce to known characteristic functions (and therefore to known distributions).

1. When $\alpha = 2$ the distribution reduces to a Gaussian distribution with mean μ and variance $\sigma^2 = 2\gamma^2$. The skewness parameter β has no consequence.

$$f(x) = \frac{1}{\sqrt{4\pi\gamma}} e^{\frac{-(x-\mu)^2}{4\gamma^2}}.$$

2. When $\alpha = 1$ and $\beta = 0$, the distribution reduces to the Cauchy (Lorentz) distribution.

$$f(x) = \frac{1}{\pi} \frac{\gamma}{\gamma^2 + (x-\mu)^2}.$$

   This is the distribution of $\gamma X + \mu$, where X is the standard Cauchy we defined before ($\mu = 0.\ \gamma = 1$).

3. When $\alpha = \frac{1}{2}$ and $\beta = 1$ we obtain the Lévy–Smirnov distribution with location parameter μ and scale parameter γ.

$$f(x) = \sqrt{\frac{\gamma}{2\pi}} \frac{e^{-\gamma/2(x-\mu)}}{(x-\mu)^{\frac{3}{2}}} \qquad \text{if } x \geq \mu$$

8.4.9 TRUNCATED LÉVY FLIGHT DISTRIBUTION

The leptokurtic property of the stable distributions for $\alpha < 2$ is very desirable in finance. In practice, when working with daily and more frequently sampled returns, their marginal distribution has heavy tails. However, recall that the Lévy distribution in general does not have a second moment for any $\alpha < 2$; therefore, if distributed as Lévy, the returns have infinite variance. This is an issue when working with real data. In order to avoid this problem, Mantegna and Stanley (1994) consider a Lévy-type distribution truncated at some parameter l. That is, for any $x > l$ the density of the distribution, $f(x)$, equals zero. Clearly this distribution has finite variance.

This distribution was named the truncated Lévy flight (*TLF*):

$$T(x) = cP(x)\mathbf{1}_{(-l,l)}(x),$$

where $P(x)$ is any symmetric Lévy distribution (obtained when $\beta = 0$) and $\mathbf{1}_A(x)$ is the indicator function of the set A.

The truncation parameter l is quite crucial. Clearly, as $l \to \infty$, one obtains a regular Lévy distribution. However, the *TLF* distribution itself is not a stable distribution for any finite truncation level l. However, this distribution has finite variance; thus independent variables from this distribution satisfy a regular Central Limit Theorem (as we shall see in later chapters). If the parameter l is large the convergence may be very slow (Mantegna and Stanley, 1994). If the parameter l is small (so that the convergence is fast), the cut that it presents in its tails is very abrupt.

In order to have continuous tails, Koponen (1995) considered a *TLF* in which the cut function is a decreasing exponential characterized by a separate parameter l. The characteristic function of this distribution when $\alpha \neq 1$ can be expressed as

$$\varphi(t) = \\ \exp\left\{c_0 - c_1 \frac{(t^2 + 1/l^2)^{\frac{\alpha}{2}}}{\cos(\pi\alpha/2)} \cos(\alpha \arctan(l|t|))(1 + il|t|\beta \tan(t \arctan l|t|))\right\},$$

with c_1 a scale factor:

$$c_1 = \frac{2\pi \cos(\pi\alpha/2)}{\alpha\Gamma(\alpha)\sin(\pi\alpha)} At$$

and

$$c_0 = \frac{l^{-\alpha}}{\cos(\pi\alpha/2)} c_1 = \frac{2\pi}{\alpha\Gamma(\alpha)\sin(\pi\alpha)} Al^{-\alpha}t.$$

In the case of symmetric distributions we have $\beta = 0$. In this case, the variance of this distribution can be calculated from the characteristic function:

$$\sigma^2 = -\left.\frac{\partial^2\varphi(t)}{\partial t^2}\right|_{t=0} = \frac{2A\pi(1-\alpha)}{\Gamma(\alpha)\sin(\pi\alpha)} l^{2-\alpha}.$$

The remaining presentation is for this symmetric case of the *TLF*.

If we use time steps Δt apart, and $T = N\Delta t$, following the discussion about returns, at the end of each interval we must calculate the sum of N stochastic variables that are independent and identically distributed. Therefore, the new characteristic function will be

$$\varphi(t, N) = \varphi(t)^N = \exp\left\{c_0 N - c_1 \frac{N(t^2 + 1/l^2)^{\alpha/2}}{\cos(\pi\alpha/2)} \cos(\alpha \arctan(l|t|))\right\}.$$

The model can be improved by standardizing it. If the variance is given by

$$\sigma^2 = -\frac{\partial^2 \varphi(t)}{\partial t^2}|_{t=0},$$

we have that

$$-\frac{\partial^2 \varphi(t/\sigma)}{\partial t^2}|_{t=0} = -\frac{1}{\sigma^2}\frac{\partial^2 \varphi(t)}{\partial t^2}|_{t=0} = 1.$$

Therefore, a standardized model is

$$\ln \varphi_S(t) = \ln \varphi\left(\frac{t}{\sigma}\right) = c_0 - c_1 \frac{((t/\sigma)^2 + 1/l^2)^{\alpha/2}}{\cos(\pi\alpha/2)} \cos\left(\alpha \arctan\left(l\frac{|t|}{\sigma}\right)\right)$$

We leave it as an exercise to write the characteristic function in a more condensed form.

EXERCISES

Problems with Solution

8.1 Let Y be an uniform random variable on $\{-1, 1\}$. Show that

$$\varphi_Y(t) = \cos(t)$$

for every $t \in \mathbb{R}$.

Solution:

$$\begin{aligned}\mathbf{E}e^{itY} &= e^{it1}\mathbf{P}(Y = 1) + e^{-it}\mathbf{P}(X = -1)\\ &= \frac{1}{2}(e^{it} + e^{-it}) = \cos t\end{aligned}$$

because $e^{it} = \cos t + i \sin t$ and $e^{-it} = \cos t - i \sin t$. ■

8.2 Using characteristic functions, prove that if

$$X \sim N(a, \sigma_1^2)$$

and

$$Y \sim N(b, \sigma_2^2)$$

then

$$X + Y \sim N(a + b, \sigma_1^2 + \sigma_2)^2.$$

Solution: It holds

$$\begin{aligned}\varphi_{X+Y}(t) &= \varphi_X(t)\varphi_Y(t) = e^{ita - \frac{1}{2}\sigma_1^2 t^2} e^{itb - \frac{1}{2}\sigma_2^2 t^2} \\ &= e^{it(a+b) + \frac{1}{2}t^2(\sigma_1^2 + \sigma_2^2)}.\end{aligned}$$

This implies $X + Y \sim N(a + b, \sigma_1^2 + \sigma_2^2)$. Compare this proof with the proof given in Chapter 5. ■

8.3 Let X, Y be two independent random variables with uniform law $U[-1, 1]$.

(a) Give the density of $X + Y$.

(b) Calculate the characteristic function of $X + Y$.

(c) Show that the function

$$f(x) = \frac{1}{\pi} \left(\frac{\sin x}{x} \right)^2$$

is a density.

Proof: (a) Using Theorem 7.28, we find that

$$f_{X+Y}(x) = \frac{1}{4}(2 + x)\mathbf{1}_{(-2,0]}(x) + \frac{1}{4}(2 - x)\mathbf{1}_{(0,2)}(x).$$

(b) Using Remark 8.22, we obtain

$$\varphi_X(t) = \varphi_Y(t) = \frac{\sin t}{t}$$

and thus

$$\varphi_{X+Y}(t) = \varphi_X(t)\varphi_Y(t) = \left(\frac{\sin x}{x} \right)^2.$$

(c) We have, for every z,

$$\begin{aligned}\int_{\mathbb{R}} f(x) e^{-ixz} dx &= \frac{1}{\pi} \int_{\mathbb{R}} \varphi_{X+Y}(x) e^{-ixz} dx \\ &= 2 f_{X+Y}(z)\end{aligned}$$

and thus

$$\int_{\mathbb{R}} f(x)\,dx = 2f_{X+Y}(0) = 1.$$

■

8.4 Let (X, Y) be a random vector with joint density

$$f(x, y) = \frac{1}{4}\left[1 + xy(x^2 - y^2)\right] \mathbf{1}_{(|x|<1,\,|y|<1)}.$$

(a) Compute the marginal distributions of X and Y.
(b) Are X and Y independent?
(c) Compute the characteristic functions of X and Y.
(d) Compute the characteristic function of

$$Z = X + Y.$$

(e) What can you conclude?

Proof: (a) We have

$$\begin{aligned} f_X(x) &= \mathbf{1}_{(-1,1)}(x) \int_{-1}^{1} dy \frac{1}{4}\left[1 + xy(x^2 - y^2)\right] \\ &= \frac{1}{2}\mathbf{1}_{(-1,1)}(x) \end{aligned}$$

and similarly

$$f_Y(y) = \frac{1}{2}\mathbf{1}_{(-1,1)}(y).$$

So X and Y have uniform laws on $[-1, 1]$.
(b) From part a, we can easily see that X and Y are not independent.
(c) Using Remark 8.22, we obtain

$$\varphi_X(t) = \varphi_Y(t) = \frac{\sin t}{t}.$$

(d) After calculation, we get

$$\varphi_Z(t) = \left(\frac{\sin t}{t}\right)^2$$

for every t.
(e) We notice that

$$\varphi_Z(t) = \varphi_X(t)\varphi_Y(t)$$

for every t, but X and Y are not independent. This is a counterexample for the reciprocal of Theorem 8.6. ■

Problems without Solution

8.5 In this problem we shall derive the characteristic function of a normal using a differential equation approach.

(**a**) Let $f(x) = e^{-\frac{x^2}{2}}$. Show that the function f satisfies

$$\frac{d}{dx}f + xf = 0.$$

(**b**) Show that if $Z \sim N(0, 1)$, then its characteristic function φ_Z satisfies the same equation.

(**c**) Deduce that

$$\varphi_Z(t) = Ce^{-\frac{t^2}{2}}.$$

(**d**) Compute the constant C from the relation

$$C = \varphi_Z(0) = 1.$$

Hint: Part (a) is trivial since

$$\frac{d}{dx}\left(e^{-\frac{x^2}{2}}\right) = -xe^{-\frac{x^2}{2}}.$$

8.6 Compute the first four moments of a standard normally distributed random variable using the characteristic function.

8.7 Let X, Y be two independent random variables. Denote by φ_X, φ_Y their characteristic functions respectively. Then

$$\varphi_{XY}(t) = \mathbf{E}\varphi_X(tY) = \mathbf{E}\varphi_Y(tX) \qquad \text{for every } t \in \mathbb{R}.$$

8.8 Suppose the random variable X has characteristic function φ. Show that

$$\mathbf{P}(X = x) = \lim_{C\to\infty} \int_{-C}^{C} e^{itx}\varphi(t)\, dt.$$

8.9 Let X be an r.v. with characteristic function φ. Then for every $x \in \mathbb{R}$ we have

$$\mathbf{P}(X = x) = \lim_{T\to\infty} \frac{1}{2T} \int_{-T}^{T} e^{-itx}\varphi(t)\, dt.$$

Deduce that if $\varphi(t) \to 0$ as $|t| \to \infty$, then

$$\mathbf{P}(X = x) = 0 \qquad \text{for all } x \in \mathbb{R}.$$

8.10 Use exercise 8.9 to calculate $\mathbf{P}(X = k)$ in the case when X has a binomial distribution $B(n, p)$.

8.11 Let X be an r.v. with density

$$f(x) = (1 - |x|)\mathbf{1}_{[-1,1]}(x).$$

This X is said to have a triangular distribution, (see exercise 5.14).

(**a**) Compute the characteristic function of X

(**b**) Let X_1, X_2 be two independent random variable with uniform distribution on $[-\frac{1}{2}, \frac{1}{2}]$. Give the law of

$$S = X_1 + X_2.$$

8.12 Let Y be a random variable with density

$$f(y) = C\,e^{-|y|}.$$

Calculate C and show that the characteristic function of Y is

$$\Phi_Y(t) = \frac{1}{1+t^2}.$$

State the name of the distribution with this characteristic function and density.

8.13 Let X_i, $1 \le i \le 4$, be independent identically distributed $N(0, 1)$ random variables. Denote with

$$D = \det\begin{pmatrix} X_1 & X_2 \\ X_3 & X_4 \end{pmatrix},$$

the determinant of the matrix.

(**a**) Show that the characteristic function of the random variable X_1X_2 is

$$\varphi(t) = \frac{1}{\sqrt{1+t^2}}.$$

(**b**) Calculate the characteristic function of D and state the law of D.

8.14 Determine the formula for the distribution with characteristic function

$$\Phi(t) = \cos t$$

for every $t \in \mathbb{R}$.

8.15 Determine the distribution with characteristic function

$$\Phi(t) = \frac{t + \sin t}{2t}$$

for every $t \in \mathbb{R}$.

CHAPTER NINE

Moment-Generating Function

9.1 Introduction/Purpose of the Chapter

We chose to introduce the generating function and the moment-generating function after the chapter on characteristic function. This is contrary to the traditional and historical approach. In modern-day probability the characteristic functions are much more widespread simply because they exist and can be constructed for any random variable or vector regardless of its distribution. On the other hand, generating functions are defined only for positive discrete random variables. The moment-generating function does not exist for many discrete or continuous random variables. Despite this fact, when these functions exist, they are much easier to work with than the characteristic functions (recall that the characteristic function takes values in the complex plain). Thus, we feel that the knowledge of these functions cannot miss from the culture of anyone applying probability concepts.

9.2 Vignette/Historical Notes

According to Roy (2011), Euler found a generating function for the Bernoulli numbers around 1730's. That is, he defined numbers B_n such that

$$\frac{t}{e^t - 1} = \sum B_n \frac{t^n}{n!}.$$

Handbook of Probability, First Edition. Ionuţ Florescu and Ciprian Tudor.

Around the same time, DeMoivre independently introduced the method of generating functions in a general way (similar to our definition). He used this method to encode all binomial probabilities in a simple polynomial function. Laplace (1812) applied the Laplace transform to probability, in effect creating the moment-generating function.

9.3 Theory and Applications

9.3.1 GENERATING FUNCTIONS AND APPLICATIONS

First, we introduce a generating function for discrete probability distributions. This function in fact can be defined for any sequence of numbers.

Definition 9.1 (Regular generating function) *Let $a = \{a_i\}_{i\in\{0,1,2,\dots\}}$ be a sequence of real numbers. Define the generating function of the sequence a as*

$$G_a(s) = \sum_{i=0}^{\infty} a_i s^i.$$

The domain of definition of the function $G_a(s)$ is made of values $s \in \mathbb{R}$ for which the series converges.

Remark 9.2 *Please note that we can obtain all the terms of the sequence if we know the generating function. Specifically,*

$$a_i = \frac{G_a^{(i)}(0)}{i!},$$

where we denote with $G_a^{(i)}(x_0)$ the i-th derivative of the function G_a calculated at x_0.

EXAMPLE 9.1 A simple generating function

Take the sequence a as

$$a_n = (\cos\alpha + i\sin\alpha)^n = \cos(n\alpha) + i\sin(n\alpha) \qquad \text{(De Moivre's Theorem)}.$$

Then its generating function is

$$G_a(s) = \sum_{n=0}^{\infty} s^n \cdot a_n = \sum_{n=0}^{\infty} (s(\cos\alpha + i\sin\alpha))^n$$
$$= \frac{1}{1 - s(\cos\alpha + i\sin\alpha)},$$

provided that the power term converges to 0. This holds if $|s| < 1$ for all α; and obviously the radius of convergence may be increased, depending on the value of α. The DeMoivre theorem may be proven easily by taking derivatives of this expression and calculating them at $s = 0$.

Definition 9.3 *The convolution of two sequences $\{a_i\}$ and $\{b_i\}$ is the sequence $\{c_i\}$ with*

$$c_n = a_0 b_n + a_1 b_{n-1} + \cdots + a_n b_0.$$

We write $c = a * b$.

Lemma 9.4 *If two sequences a and b have generating functions G_a and G_b, then the generating function of the convolution $c = a * b$ is*

$$G_c(s) = G_a(s) G_b(s).$$

Proof: We have

$$G_c(s) = \sum_{n=0}^{\infty} c_n s^n = \sum_{n=0}^{\infty} s^n \sum_{k=0}^{n} a_k b_{n-k} = \sum_{n=0}^{\infty} \sum_{k=0}^{n} (a_k s^k)(b_{n-k} s^{n-k})$$
$$\overset{Fubini}{=} \sum_{k=0}^{\infty} a_k s^k \sum_{n=k}^{\infty} b_{n-k} s^{n-k} = G_a(s) G_b(s).$$

■

9.3.1.1 Generating Functions for Discrete Random Variables.

Definition 9.5 (Probability generating function(discrete)) *If X is a discrete random variable with outcomes $i \in \mathbb{N}$ and corresponding probabilities p_i, then we define*

$$G_X(s) = \mathbf{E}(s^x) = \sum_i s^i \mathbf{P}(X = i) = \sum_i s^i p_i.$$

This is the **probability-generating function** *of the discrete random variable X.*

Remark 9.6 *The first observation we want to make is that the definition is only for discrete random variables with positive integer values. However, the definition can be extended to any discrete random variable taking values in any countable set. Recall that the countable sets all have the same number of elements, so to use the definition we need to first relabel the outcomes of the random variable as $x_0 \to 0$, $x_1 \to 1$, $x_2 \to 2$, and so on, then the definition above will apply.*

Lemma 9.7 *Suppose that two* independent *discrete random variables X and Y have associated probability distributions $p = \{p_n\}$ and $q = \{q_n\}$ and generating* *functions* $G_X(s) = \sum_i s^i p_i$ *and* $G_Y(s) = \sum_i s^i q_i$ *respectively. Then the random variable* $X + Y$ *has distribution the convolution of the two probability sequences* $p * q$ *as in Definition 9.3 and its generating function is*

$$G_{X+Y}(s) = G_X(s)G_Y(s).$$

Proof: The proof is an exercise. Just use the preceding lemma and calculate

$$P(X + Y = n),$$

in terms of p_i and q_i. ■

This definition and the lemma above explain the main use of the generating functions. Recall that the probability distribution may be recovered entirely from the generating function. On the other hand, the generating function of a sum of two random variables is the product of individual generating functions. Thus, while the distribution of the sum using the convolution definition is complicated to calculate, calculating the generating function of the sum is a much simple process. Once calculated, the probability distribution of the sum may be obtained by taking derivatives of the generating function.

Proposition 9.8 (Properties of probability-generating function) *Let* $G_X(s)$, $G_Y(s)$ *be the generating functions of two discrete random variables* X *and* Y.

(i) *There exists* $R \geq 0$, *a radius of convergence such that the sum is convergent if* $|s| < R$ *and diverges if* $|s| > R$. *Therefore, the corresponding generating function* G_X *exists on the interval* $[-R, R]$.

(ii) *The generating function* $G_X(s)$ *is differentiable at any* s *within the domain* $|s| < R$.

(iii) *If two generating functions* $G_X(s) = G_Y(s) = G(s)$ *for all* $|s| < R$ *where* $R = \min\{R_X, R_Y\}$, *then* X *and* Y *have the same distribution. Furthermore, the common distribution may be calculated from the generating function using*

$$\mathbf{P}(X = n) = \frac{1}{n!} G^{(n)}(0).$$

(iv) *As particular cases of the definition, we readily obtain that*

$$G_X(0) = \mathbf{P}(X = 0),$$
$$G_X(1) = \sum_i p_i = 1.$$

The last relationship implies that the probability-generating function always exists for $s = 1$, *thus the radius of convergence* $R \geq 1$.

Proof: Exercise 9.8. ■

9.3.1.2 Generating Functions for Simple Distributions. Bellow we present the generating function for commonly encountered discrete random variables.

Once again we stress that the generating functions are very useful when working with discrete random variables (basically when they exist).

(i) Constant variable, identically equal to c (i.e., $\mathbf{P}(X = c) = 1$):

$$G(s) = \mathbf{E}(s^x) = s^c.$$

(ii) Bernoulli(p):

$$G(s) = \mathbf{E}(s^x) = 1 - p + ps.$$

(iii) Geometric(p):

$$\begin{aligned} G(s) &= \sum_{k=1}^{\infty} s^k (1-p)^{k-1} p \\ &= sp \cdot \sum_{k=1}^{\infty} (s(1-p))^{k-1} \\ &= \frac{sp}{1 - s(1-p)}. \end{aligned}$$

(iv) Poisson(λ):

$$G(s) = \sum_{k=1}^{\infty} s^k \cdot \frac{\lambda^k}{k!} e^{-\lambda} = e^{\lambda s} \cdot e^{-\lambda} = e^{\lambda(s-1)}.$$

Lemma 9.9 *If X has generating function $G(s)$, then:*

(i) $\mathbf{E}(X) = G'(1)$.

(ii) $\mathbf{E}[X(X-1)\cdots(X-k+1)] = G^{(k)}(1)$, *the k-th factorial moment of X.*

Proof: Again these results are easy to prove and are left as exercises. ■

9.3.1.3 Examples. In the next examples we show the utility of the generating functions.

EXAMPLE 9.2 Using generating function for Poisson distribution

Let X, Y be two independent Poisson random variables with parameters λ and μ. What is the distribution of $Z = X + Y$?

Solution: We could compute the convolution $f_Z = f_X * f_Y$ using the formulas in the previous section. However, calculating the distribution using generating functions is easier. We have

$$G_X(s) = e^{\lambda(s-1)},$$
$$G_Y(s) = e^{\mu(s-1)}.$$

By independence of X and Y and from the Lemma 9.7 above, we have

$$G_Z(s) = G_X(s)G_Y(s) = e^{\lambda(s-1)}e^{\mu(s-1)} = e^{(\lambda+\mu)(s-1)}.$$

But this is the generating function of a Poisson random variable. Identifying the parameters imply that Z has a Poisson$(\lambda + \mu)$ distribution. ■

EXAMPLE 9.3 Using generating function for Binomial distribution

We know that for a Binomial(n, p) random variable X may be written as $X = X_1 + \cdots + X_n$, where X_i's are independent Bernoulli(p) random variables. What is its generating function? What is the distribution of a sum of binomial random variables?

Solution: Again applying the Lemma 9.7, we obtain the generating function of a Binomial(n, p) as

$$G_X(s) = G_{X_1}(s)G_{X_2}(s)\cdots G_{X_n}(s) = (1 - p + sp)^n.$$

What if we have two independent binomially distributed random variables X Binomial(m, p) and Y Binomial(k, p)? Note that p the probability of success is identical. What is the distribution of $X + Y$ in this case?

$$G_{X+Y}(s) = (1 - p + sp)^m(1 - p + sp)^k = (1 - p + sp)^{m+k}.$$

We see that we obtain the generating function of a Binomial random variable with parameters $m + k$ and p.

This is of course expected. If X can be written and a sum of n Bernoulli(p) and Y can be written along with a sum of k Bernoulli(p), then of course $X + Y$ is a sum of $n + k$ Bernoulli(p) random variables. Even more intuitively, suppose X is the number of heads obtained in n tosses of a coin with probability p, and Y is the number of Heads obtained in k tosses of the same coin (separate tosses). Then of course $X + Y$ is the total number of heads in $n + k$ tosses. ■

Let us expand this results in which we presented the distribution of a sum of a fixed number of terms. Consider the case when the sum contains a random number of terms. The next theorem helps with such a situation.

Theorem 9.10 *Let $X_1, X_2, \ldots$ denote a sequence of i.i.d. random variables with generating function $G_X(s)$. Let N be an integer-valued random variable independent of the X_i's with generating function $G_N(s)$. Then the random variable*

$$S = X_1 + \cdots + X_N$$

has generating function:

$$G_S(s) = G_N(G_X(s))$$

Now let us give an example of application of the previous theorem.

EXAMPLE 9.4 A chicken and eggs problem

Suppose a hen lays N eggs. We assume that N has a Poisson distribution with parameter λ. Each egg then hatches with probability p independent of all other eggs. Let K be the number of chicken that hatch. Find $\mathbf{E}(K|N)$, $\mathbf{E}(N|K)$, and $\mathbf{E}(K)$. Furthermore, find the distribution of K.

Solution: We know that

$$f_N(n) = \frac{\lambda^n}{n!}e^{-\lambda},$$
$$f_{K|N}(k|n) = \binom{n}{k}p^k(1-p)^{n-k}.$$

The last distribution is the distribution of the number of eggs hatching if we know how many eggs were laid.

We may then calculate the conditional expectations. Let

$$\varphi(n) = \mathbf{E}(K|N=n) = \sum_k k \cdot \binom{n}{k}p^k(1-p)^{n-k} = np$$

since that is the formula for the expectation of a Binomial(n, p) random variable. Therefore, $\mathbf{E}(K|N) = \varphi(N) = Np$. Recall that $\mathbf{E}[\mathbf{E}[X|Y]] = \mathbf{E}[X]$. Using this relation, we get

$$\mathbf{E}[\mathbf{E}[K|N]] = \mathbf{E}[K] = \mathbf{E}[Np] = \lambda p.$$

To find $\mathbf{E}(N|K)$, we need to find $f_{N|K}$:

$$\begin{aligned} f_{N|K}(n|k) &= \mathbf{P}(N = n|K = k) = \frac{\mathbf{P}(N = n, K = k)}{\mathbf{P}(K = k)} \\ &= \frac{\mathbf{P}(K = k|N = n) \cdot \mathbf{P}(N = n)}{\mathbf{P}(K = k)} \\ &= \frac{\binom{n}{k} p^k (1-p)^{n-k} \frac{\lambda^n}{n!} e^{-\lambda}}{\sum_{m \geq k} \binom{m}{k} p^k (1-p)^{m-k} \frac{\lambda^m}{m!} e^{-\lambda}} \\ &= \frac{(q\lambda)^{n-k}}{(n-k)!} e^{-q\lambda} \end{aligned}$$

if $n \geq k$ and 0 otherwise, where $q = 1 - p$ and for the denominator we have used the following:

$$\begin{aligned} \mathbf{P}(K = k) &= \sum_{m=0}^{\infty} \mathbf{P}(K = k, N = m) = \sum_{m=0}^{\infty} \mathbf{P}(K = k|N = m)\mathbf{P}(N = m) \\ &= \sum_{m=k}^{\infty} \mathbf{P}(K = k|N = m)\mathbf{P}(N = m) \end{aligned}$$

since $\mathbf{P}(K = k|N = m) = 0$ for any $k > m$ (it is not possible to hatch more chicks than eggs laid).

Consequently, we immediately obtain

$$\begin{aligned} \mathbf{E}(N|K = k) &= \sum_{n \geq k} n \cdot \frac{(q\lambda)^{n-k}}{(n-k)!} e^{-q\lambda} \quad \text{change of var. } \overset{n-k=m}{=} \\ &= e^{-q\lambda} \sum_{m=0}^{\infty} (m+k) \frac{(q\lambda)^m}{m!} \\ &= e^{-q\lambda} \left(\sum_{m=0}^{\infty} m \frac{(q\lambda)^m}{m!} + k \sum_{m=0}^{\infty} \frac{(q\lambda)^m}{m!} \right) \\ &= e^{-q\lambda} \left(0 + q\lambda \sum_{m=1}^{\infty} \frac{(q\lambda)^{m-1}}{(m-1)!} + k e^{q\lambda} \right) = q\lambda + k, \end{aligned}$$

which gives us $\mathbf{E}(N|K) = q\lambda + K$.

Note that in the previous derivation we kind of obtained the distribution of K. We still have to calculate a sum to find the closed expression. Even though this is not that difficult, we will use generating functions to demonstrate how easy it is to calculate the distribution using them.

Let $K = X_1 + \cdots + X_N$, where the $X_i \sim$Bernoulli(p). Then using the the above theorem, we obtain

$$G_K(s) = G_N(G_X(s)),$$
$$G_N(s) = \sum_{n=0}^{\infty} s^n \frac{\lambda^n}{n!} e^{-\lambda} = e^{\lambda(s-1)},$$
$$G_X(s) = 1 - p + ps.$$

Using the above equations, we conclude that

$$G_K(s) = G_N(1 - p + ps) = e^{\lambda(1-p+ps-1)} = e^{\lambda p(s-1)}.$$

But since this is a generating function for a Poisson r.v., we immediately conclude that

$$K \sim \text{Poisson}(\lambda p).$$

■

EXAMPLE 9.5 A Quantum mechanics extension

In an experiment a laser beam emits photons with an intensity of λ photons/second. Each photon travels through a deflector that polarizes (spin up) each photon with probability p. A measuring device is placed after the deflector. This device measures the number of photons with a spin up (the rest are spin down). Let N denote the number of photons emitted by the laser bean in a second. Let K denote the number of photons measured by the capturing device. Assume that N is distributed as a Poisson random variable with mean λ. Assume that each photon is spinned independently of any other photon. Give the distribution of K and its parameters. In particular, calculate the average intensity recorded by the measuring device.

Solution: Exercise. Please read and understand the previous example. ■

9.3.2 MOMENT-GENERATING FUNCTIONS. RELATION WITH THE CHARACTERISTIC FUNCTIONS

Generating functions are very useful when dealing with discrete random variables. To be able to study any random variables, the idea is to make the transformation $s = e^t$ in $G(s) = \mathbf{E}(s^X)$. Thus, the following definition is introduced.

Definition 9.11 (Moment-generating function) *The moment-generating function of a random variable X is the function* $M : \mathbb{R} \to [0, \infty)$ *given* *by*

$$M_X(t) = \mathbf{E}(e^{tx}),$$

defined for values of t for which $M_X(t) < \infty$.

The moment-generating function is related to the Laplace transform of the distribution of the random variable X.

Definition 9.12 (Laplace transform of the function f) *Suppose we are given a positive function $f(x)$. Assume that*

$$\int_0^\infty e^{-t_0 x} f(x)\, dx < \infty$$

for some value of $x_0 > 0$. Then the **Laplace transform** *of f is defined as*

$$\mathcal{L}\{f\}(t) = \int_0^\infty e^{-tx} f(x)\, dx$$

for all $t > t_0$. If the random variable X has distribution function F, the following defines the **Laplace–Stieltjes transform***:*

$$\mathcal{L}_{LS}\{F\}(t) = \int_0^\infty e^{-tx} dF(x),$$

again with the same caution about the condition on the existence.

With the previous definitions it is easy to see that the moment-generating functions is a sum of two Laplace transforms (provided either side exists). Specifically, suppose the random variable X has density $f(x)$. Then,

$$\begin{aligned}
M_X(t) = \mathbf{E}(e^{tx}) &= \int_{-\infty}^\infty e^{tx} f(x)\, dx \\
&= \int_{-\infty}^0 e^{tx} f(x) dx + \int_0^\infty e^{tx} f(x)\, dx \\
&= \int_0^\infty e^{-tx} f(-x) dx + \int_0^\infty e^{-(-t)x} f(x)\, dx \\
&= \mathcal{L}\{f_-\}(t)\mathbf{1}_{(0,\infty)}(t) + \mathcal{L}\{f_+\}(t)\mathbf{1}_{(-\infty,0]}(t),
\end{aligned}$$

where

$$f_-(x) = \begin{cases} f(-x), & \text{if } x > 0, \\ 0, & \text{else} \end{cases}$$

and

$$f_+(x) = \begin{cases} f(x), & \text{if } x > 0, \\ 0, & \text{else.} \end{cases}$$

This result is generally needed to use the theorems related to Laplace and specially inverse Laplace transforms.

Proposition 9.13 (Properties) *Suppose that the moment-generating function of a random variable X, $M_X(t)$ is finite for all $t \in \mathbb{R}$.*

(i) *If the moment-generating function is derivable, then the first moment is $\mathbf{E}(X) = M'_X(0)$; in general, if the moment generating function is k times derivable, then $\mathbf{E}(X^k) = M_X^{(k)}(0)$.*

(ii) *More general, if the variable X is in L^p (the first p moments exist), then we may write (Taylor expansion of e^x):*

$$M(t) = \sum_{k=0}^{\infty} \frac{\mathbf{E}(X^k)}{k!} t^k.$$

(iii) *If X, Y are two independent random variables, then $M_{X+Y}(t) = M_X(t)M_Y(t)$.*

Remark 9.14 *It is the last property in the above proposition that makes it desirable to work with moment-generating functions. However, as mentioned these functions have a big disadvantage when compared with the characteristic functions. The expectation and thus the integrals involved may not be finite and therefore the function may not exist. Obviously, point iii in the Proposition 9.13 generalizes to any finite sum of independent random variables.*

Remark 9.15 *Another important fact concerning the moment-generating function is that, as in the case of the characteristic function, it characterizes the law of a random variable completely. See exercise 9.16.*

EXAMPLE 9.6 Calculating moments using the M.G.F.

Let X be an r.v. with moment-generating function

$$M_X(t) = \frac{1}{2}(1 + e^t).$$

Derive the expectation and the variance of X.

Solution: Since

$$M'_X(t) = \frac{1}{2}e^t,$$

we get for $t = 0$ and Proposition 9.13

$$\mathbf{E}X = M_X'(0) = \frac{1}{2}.$$

By differentiating M_X, twice, we obtain

$$M_X''(t) = \frac{1}{2}e^t$$

and

$$\mathbf{E}X^2 = M_X''(0) = \frac{1}{2}.$$

Thus

$$VarX = \frac{1}{2} - \frac{1}{4} = \frac{1}{4}.$$

■

Let us compute the moment-generating function for certain common probability distribution.

Proposition 9.16 *Suppose $X \sim N(0, 1)$. Then*

$$M_X(t) = e^{\frac{t^2}{2}}$$

for every $t \in \mathbb{R}$.

Proof: For every $t \in \mathbb{R}$, we have

$$\begin{aligned} M_X(t) &= \int_{-\infty}^{\infty} e^{tz} \frac{1}{\sqrt{2\pi}} e^{-\frac{z^2}{2}} dz \\ &= e^{\frac{t^2}{2}} \frac{1}{\sqrt{2\pi}} \int_{-\infty}^{\infty} e^{-\frac{(z-t)^2}{2}} dz \\ &= e^{\frac{t^2}{2}}. \end{aligned}$$

■

Proposition 9.17 *If $X \sim N(\mu, \sigma^2)$, prove that for every $t \in \mathbb{R}$, we have*

$$M_X(t) = e^{\mu t + \frac{t^2\sigma^2}{2}}.$$

Proof: See problem 9.3. ■

Proposition 9.18 *If $X \sim Exp(\lambda)$ with $\lambda > 0$, then*

$$M_X(t) = \frac{\lambda}{\lambda - t}.$$

Proof: See exercise 9.5. ■

Proposition 9.19 *If* $X \sim \Gamma(a, \lambda)$ *(gamma distribution with parameters* $a, \lambda > 0$*), then the moment-generating function* M_X *exists for* $t < \lambda$ *and in this case we have*

$$M_X(t) = \left(\frac{1}{1 - \frac{t}{\lambda}}\right)^a .$$

Proof: By the definition of the gamma density, we have

$$\begin{aligned} M_X(t) &= \int_0^\infty \frac{\lambda^a}{\Gamma(a)} e^{-\lambda x} x^{a-1} e^{tx} dx \\ &= \int_0^\infty \frac{\lambda^a}{\Gamma(a)} e^{-x(\lambda - t)} dx. \end{aligned}$$

Use the change of variables $y = x(\lambda - t)$ to conclude. ■

9.3.3 RELATIONSHIP WITH THE CHARACTERISTIC FUNCTION

If the moment-generating function of a random variable X exists, then we obviously have

$$\varphi_X(t) = M_X(it).$$

We recall that just as the moment-generating function is related to the Laplace transform, the characteristic function was related to the Fourier transform.

We note that the characteristic function is a better-behaved function than the moment-generating function. The drawback is that it takes complex values and we need to have a modicum of understanding of complex analysis to be able to work with it.

9.3.4 PROPERTIES OF THE MGF

The following theorems are very similar with the corresponding theorems for the characteristic function. This is why if the moment-generating function exists, it is typically easier to work with it than the characteristic function. We will state these theorems without proofs. The proofs are similar with the ones given for the characteristic function.

Theorem 9.20 (Uniqueness theorem) *If the MGF of* X *exists for any* t *in an open interval containing* 0*, then the MGF uniquely determines the c.d.f. of* X*. That is, no two different distributions can have the same values for the MGF's on an interval containing* 0.

The uniqueness theorem in fact comes from the uniqueness of the Laplace transform and inverse Laplace transform.

Theorem 9.21 (Continuity Theorem) *Let $X_1, X_2, \ldots$ a sequence of random variables with c.d.f.'s $F_{X_1}(x), F_{X_2}(x), \ldots$ and moment generating functions $M_{X_1}(t), M_{X_2}(t), \ldots$ (which are defined for all t). Suppose that*

$$M_{X_n}(t) \to M_X(t)$$

for all t as $n \to \infty$. Then, $M_X(t)$ is the moment generating function of some random variable X. If $F_X(x)$ denote the c.d.f. of X, then

$$F_n(x) \to F_X(x)$$

for all x continuity points of F_X

We note that the usual hypothesis of the theorem include the requirement that $M_X(t)$ be a moment-generating function. This is not needed in fact, the inversion theorem stated next makes this condition obsolete. The only requirement is that the MGF be defined at all x.

Recall that the MGF can be written as a sum of two Laplace transforms. The inversion theorem is stated in terms of the Laplace transform. This is in fact the only form in which it actually is useful.

Theorem 9.22 (Inversion theorem) *Let f be a function with its Laplace transform denoted with f^*, that is,*

$$f^*(t) = \mathcal{L}\{f\}(t) = \int_0^\infty e^{-tx} f(x)\, dx.$$

Then we have

$$f(x) = \frac{1}{2\pi i} \lim_{T\to\infty} \int_{c-iT}^{c+iT} e^{tx} f^*(t)\, dt,$$

where c is a constant chosen such that it is greater than the real part of all singularities of f^*. The formula above gives f as the inverse Laplace transform, $f = \mathcal{L}^{-1}\{f^*\}$.

Remark 9.23 *In practice, computing the complex integral is done using the Residuals Theorem (Cauchy residue theorem). If all singularities of f^* have negative real part or there is no singularity then c can be chosen 0. In this case the integral formula of the inverse Laplace transform becomes identical to the inverse Fourier transform (given in the chapter about the characteristic function, Chapter 8).*

If no complex analysis knowledge is available, the typical way of computing Laplace transform and inverse Laplace is through tables.

EXERCISES

Problems with Solution

9.1 Suppose that X has a discrete uniform distribution on $\{1, 2, \ldots, n\}$, that is,

$$\mathbf{P}(X = i) = \frac{1}{n}$$

for every $i = 1, \ldots, n$.

(a) Compute the moment-generating function of X.

(b) Deduce that

$$\mathbf{E}X = \frac{n+1}{2} \quad \text{and} \quad \mathbf{E}X^2 = \frac{(n+1)(2n+1)}{6}.$$

Solution: We have

$$\begin{aligned} M_X(t) &= \sum_{j=1}^{n} \frac{1}{n} e^{tj} \\ &= \frac{1}{n}(e^t + e^{2t} + \cdots + e^{nt}) \\ &= \frac{e^t(e^{nt} - 1)}{n(e^t - 1)}. \end{aligned}$$

By differentiating with respect to t,

$$\mathbf{E}X = M_X'(0) = \frac{1}{n}(1 + 2 + \cdots + n) = \frac{n+1}{2}$$

and

$$\mathbf{E}X^2 = M_X''(0) = \frac{1}{n}(1^2 + 2^2 + \cdots + n^2) = \frac{(n+1)(2n+1)}{6}.$$

■

9.2 Suppose X has uniform distribution $U[a, b]$ with $a < b$. Show that for every $t \in \mathbb{R} \setminus \{0\}$, we have

$$M_X(t) = \frac{e^{tb} - e^{ta}}{t(b-a)}$$

and $M_X(0) = 1$.

Solution: Clearly, $M_X(0) = 1$. For $t \neq 0$,

$$M_X(t) = \frac{1}{b-a}\int_a^b e^{tx}\,dx$$
$$= \frac{1}{b-a}\left[\frac{1}{t}e^{tx}\right]_{t=a}^{t=b}$$
$$= \frac{e^{tb} - e^{ta}}{t(b-a)}.$$

■

9.3 Prove Proposition 9.17.

Solution: For every t, we have

$$M_X(t) = \int_{-\infty}^{\infty} e^{ty}\frac{1}{\sqrt{2\pi}\sigma}e^{-\frac{(y-\mu)^2}{2\sigma^2}}$$
$$= e^{\mu t + \frac{t^2\sigma^2}{2}}\int_{-\infty}^{\infty}\frac{1}{\sqrt{2\pi}\sigma}e^{-\frac{1}{2\sigma^2}[(y-\mu)^2 - \sigma^2 t^2]}$$
$$= e^{\mu t + \frac{t^2\sigma^2}{2}}$$

since

$$\frac{1}{\sqrt{2\pi}\sigma}e^{-\frac{1}{2\sigma^2}[(y-\mu)^2 - \sigma^2 t^2]} = 1$$

(density of a normal law). ■

9.4 Use the moment-generating function of $X \sim N(\mu, \sigma^2)$ to prove that

$$\mathbf{E}X = \mu \quad \text{and} \quad Var\,X = \sigma^2.$$

Solution: From exercise 9.3, we get

$$M_X'(t) = e^{\mu t + \frac{t^2\sigma^2}{2}}(\mu + \sigma^2 t)$$

and thus

$$M_X'(0) = \mu.$$

Also,

$$M_X''(t) = e^{\mu t + \frac{t^2\sigma^2}{2}}\sigma^2 + e^{\mu t + \frac{t^2\sigma^2}{2}}(\mu + \sigma^2 t)^2$$

and thus $\mathbf{E}X^2 = M_X''(0) = \sigma^2 + \mu^2$. Then

$$VarX = \sigma^2 + \mu^2 - \sigma^2 = \sigma^2.$$

■

9.5 Prove Proposition 9.18.

Solution: We have

$$\begin{aligned} M_X(t) &= \int_0^\infty e^{tx} \lambda e^{-\lambda x} dx \\ &= \lambda \int_0^\infty e^{-(\lambda - t)x} dx \\ &= \frac{\lambda}{\lambda - t} \end{aligned}$$

for every $t \in \mathbb{R}$. ■

9.6 Suppose that X admits a moment-generating function M_X. Prove that

$$\mathbf{P}(X \geq x) \leq e^{-tx} M_X(t) \qquad \text{for every } t > 0$$

and

$$\mathbf{P}(X \leq x) \leq e^{-tx} M_X(t) \qquad \text{for every } t < 0.$$

(These are called the Chernoff bounds.)

Proof: One has

$$\mathbf{P}(X \geq x) = \mathbf{P}(e^{tX} \geq e^{tx}) \text{ if } t > 0$$

and

$$\mathbf{P}(X \leq x) = \mathbf{P}(e^{tX} \geq e^{tx}) \text{ if } t < 0.$$

We can use Markov's inequality to conclude. ■

Problems without Solution

9.7 If $X \sim Binom(n, p)$, use the moment generating function to get

$$\mathbf{E}X = np \quad \text{and} \quad VarX = np(1 - p).$$

9.8 Prove Proposition 9.8.

9.9 Show that

$$(\cos\theta + i\sin\theta)^n = \cos n\theta + i\sin n\theta$$

for any $n \in \mathbb{N}$ and $\theta \in [0, 2\pi]$, using generating functions.

9.10 Find the generating functions, both ordinary and moment-generating function, for the following discrete probability distributions.

(a) The distribution describing a fair coin.

(b) The distribution describing a fair die.

(c) The distribution describing a die that always comes up 3.

9.11 Let X, Y be two random variables such that

$$\mathbf{P}(X = 1) = \frac{1}{3}, \qquad \mathbf{P}(X = 4) = \frac{2}{3}$$

and

$$\mathbf{P}(Y = 2) = \frac{2}{3}, \qquad \mathbf{P}(Y = 5) = \frac{1}{5}.$$

(a) Show that X and Y have the same first and second moments, but not the same third and fourth moments.

(b) Calculate the moment generating functions of X and Y.

9.12 Let X be a discrete r.v. with values in $\{0, 1, \ldots, n\}$. Let M_X denote its moment-generating function. Find in terms of M_X the moment-generating functions of

(a) $-X$.

(b) $X + 1$.

(c) $3X$.

(d) $aX + b$.

9.13 Use Proposition 9.19 to show that $X \sim \Gamma(a, \lambda)$ and $Y \sim \Gamma(b, \lambda)$ are two independent random variables, then

$$X + Y \sim \Gamma(a + b, \lambda).$$

9.14 Suppose X has the density

$$f(x) = \frac{a}{x^{a+1}} 1_{(x>1)},$$

where $a > 0$. This means that X follows a Pareto distribution (power law).

(a) Show that $M_X(t) = \infty$ for every $t \in \mathbb{R}$.

(b) Show that

$$\mathbf{E}X^n < \infty$$

if and only if $a > n$.

9.15 Suppose $X \sim Poisson(a)$ with $a > 0$. Use exercise 9.6 to show that

$$\mathbf{P}(X \geq x) \leq e^{n-a} \left(\frac{a}{n}\right)^n .$$

9.16 Let X, Y be two random variables with moment-generating functions M_X and M_Y respectively. Show that X and Y have the same distribution if and only if

$$M_X(t) = M_Y(t)$$

for any $t \in \mathbb{R}$.

9.17 Let $X \sim Exp(\lambda)$. Calculate $\mathbf{E}X^3$ and $\mathbf{E}X^4$. Give a general formula for $\mathbf{E}X^n$.

9.18 Using the properties of the MGF, show that if X_i are i.i.d. $Exp(\lambda)$ random variables, then

$$S = X_1 + \cdots + X_n$$

is distributed as a gamma random variable with parameters n and λ.

9.19 Prove the following facts about Laplace transform:

(a)

$$\mathcal{L}\{e^{-\lambda x}\}(t) = \frac{1}{\lambda + t}$$

(b)

$$\mathcal{L}\{t^{n-1} e^{-\lambda x}\}(t) = \frac{\Gamma(n)}{(\lambda + t)^n},$$

where $\Gamma(n)$ denote the gamma function calculated at n.

9.20 Suppose f^* denotes the Laplace transform of the function f. Prove the following properties:

(a)

$$\mathcal{L}\{\int_0^x f(y)dy\}(t) = \frac{f^*(t)}{t}$$

(b)

$$\mathcal{L}\{f'(x)\}(t) = tf^*(t) - f(0+),$$

where $f(0+)$ denote the right limit of f at 0.

(c)

$$\mathcal{L}\{f^{(n)}(x)\}(t) = t^n f^*(t) - t^{n-1} f(0+) - t^{n-2} f'(0+) - \ldots f^{(n)}(0+).$$

9.21 Suppose that $X_1 \sim Exp(\lambda_1)$ and $X_2 \sim Exp(\lambda_2)$ are independent. Given that the inverse laplace transform of the function $\frac{1}{c+t}$ is e^{-cx} and that the inverse Laplace transform is linear, calculate the p.d.f. of $X_1 + X_2$.

9.22 Repeat the problem above with three exponentially distributed random variables, that is, $X_i \sim Exp(\lambda_i)$, $i = 1, 2, 3$. As a note, the distribution of a sum of n such independent exponentials is called the Erlang distribution and is heavily used in queuing theory.

CHAPTER TEN

Gaussian Random Vectors

10.1 Introduction/Purpose of the Chapter

Gaussian random variables and Gaussian random vectors (vectors whose components are jointly Gaussian, as defined in this chapter) play a central role in modeling real-life processes. Part of the reason for this is that noise like quantities encountered in many practical applications are reasonably modeled as Gaussian. Another reason is that Gaussian random variables and vectors turn out to be remarkably easy to work with (after an initial period of learning their features). Jointly Gaussian random variables are completely described by their means and covariances, which is part of the simplicity of working with them. Estimating these joint Gaussians means approximating only their means and covariances.

A third reason why Gaussian random variables and vectors are so important is that, in many cases, the performance measures we get for estimation and detection problems for the Gaussian case often bounds the performance for other random variables with the same means and covariances. For example, the minimum mean square estimator for Gaussian problems is the same as the linear least squares estimator for other problems with the same mean and covariance and, furthermore, has the same mean square performance. We will also find that this estimator has a simple expression as a linear function of the observations. Finally, we will find that the minimum mean square estimator for non-Gaussian problems always has a better performance than that for Gaussian problems with the same mean and covariance, but that the estimator is typically much more complex. The point of this example is that non-Gaussian problems are often more easily and more deeply understood if we first understand the corresponding Gaussian problem.

Handbook of Probability, First Edition. Ionuţ Florescu and Ciprian Tudor.

In this chapter, we develop the most important properties of Gaussian random variables and vectors, namely the moment-generating function, the moments, the joint densities, and the conditional probability densities.

10.2 Vignette/Historical Notes

The Gaussian distribution is named after Johann Carl Friedrich Gauss (30 April 1777–23 February 1855), one of the most famous mathematicians and physical scientists of the 18th century. Besides probability and statistics, Gauss contributed significantly to many other fields, including number theory, analysis, differential geometry, geodesy, geophysics, electrostatics, astronomy, and optics.

Gauss work on a theory of the motion of planetoids disturbed by large planets, eventually published in 1809 as *Theoria motus corporum coelestium in sectionibus conicis solem ambientum* (*Theory of motion of the celestial bodies moving in conic sections around the sun*), contained an influential treatment of the method of least squares, a procedure which is used today in all sciences to minimize the impact of measurement error. Gauss proved the method under the assumption of normally distributed errors, a methodology that today is the first step in analysis of errors produced by very complex processes.

With the proper formulation of the Brownian motion (Wiener process) by Norbert Wiener in the 1950s the Gaussian process became mainstream for the theory of stochastic processes. The chaos theory developed in the 1970s owes a lot of gratitude to Gaussian processes to be able to produce useful formulas and bounds. Most of the machine learning theory uses assumptions about errors behaving like Gaussian processes. For all these reasons, learning about them is a crucial skill to be acquired by any aspiring student in probability.

10.3 Theory and Applications

10.3.1 THE BASICS

Let us first recall that a random variable follows the normal law $N(\mu, \sigma^2)$ with $\sigma > 0$ if its density is given by

$$f(x) = \frac{1}{\sqrt{2\pi\sigma^2}} e^{-\frac{(x-\mu)^2}{2\sigma^2}}.$$

Before introducing the concept of Gaussian vector, let us list some properties of Gaussian random variables, which are already proven in the previous chapters of this book.

- The parameters μ and σ completely characterize the law of X.
- $X \sim N(\mu, \sigma^2)$ if and only if

$$X = \mu + \sigma Z,$$

where $Z \sim N(0, 1)$ is a standard normal random variable (a $N(0, 1)$ r.v.). See Remark 5.27.

- If $X \sim N(\mu, \sigma^2)$ and we denote $\mu_p = \mathbf{E}(X - \mathbf{E}X)^p = \mathbf{E}(X - \mu)^p$, the p-central moment, then

$$\mu_{2k+1} = 0$$

and

$$\mu_{2k} = \frac{(2k)!}{k!} \left(\frac{\sigma^2}{2} \right)^k .$$

Note in particular that the third central moment is zero, and the fourth central moment is equal to $3\sigma^4$. We point this out because these are crucial to describe the shape of distributions. The third standardized moment of a random variable X is defined as

$$\mathit{Skeweness}(X) = \mathbf{E}\left[\left(\frac{X - \mathbf{E}X}{\sqrt{V(X)}} \right)^3 \right] = \mathbf{E}\left[\left(\frac{X - \mu}{\sigma} \right)^3 \right]$$

and is a measure of the skew of a distribution (how much it deviates from being symmetric). This measure is typically compared to 0 which shows very little skew. A negative value of the measure indicates that the random variable is more likely.

The fourth standardized moment of a random variable X is defined as

$$\mathit{Kurtosis}(X) = \mathbf{E}\left[\left(\frac{X - \mathbf{E}X}{\sqrt{V(X)}} \right)^4 \right] = \mathbf{E}\left[\left(\frac{X - \mu}{\sigma} \right)^4 \right]$$

and is a measure of the tails of the distribution compared with the tails of a Gaussian random variable. Once again, recall that the kurtosis for a Gaussian is equal to 3.

A random variable with $\mathit{Kurtosis}(X) > 3$ is said to have a leptokurtic distribution—in essence the random variable takes large values more often than an equivalent normal variable. These random variables are crucial in finance where the observed behavior of returns calculated from price processes have this property. The Cauchy and t distributions are examples of distributions having this property.

A random variable with $\mathit{Kurtosis}(X) < 3$ is said to have a platykurtic distribution (extreme observations are less frequent than a corresponding normal). An example is the uniform distribution on a finite interval.

- If $X \sim N(\mu, \sigma^2)$, then

$$\mathbf{P}(a < X < b) = \Phi\left(\frac{b-\mu}{\sigma}\right) - \Phi\left(\frac{a-\mu}{\sigma}\right)$$

for every $a, b \in \mathbb{R}$, $a < b$, where

$$\Phi(x) = \frac{1}{2\pi}\int_{-\infty}^{x} e^{-\frac{u^2}{2}}\,du$$

is the cumulative distribution function of the standard normal $N(0, 1)$ (its values can be usually found in the table of the normal distribution).

- The sum of two independent normal random variables is normal; that is, if $X \sim N(\mu_1, \sigma_1^2)$, $Y \sim N(\mu_2, \sigma_2^2)$ and X, Y are independent, then

$$X + Y \sim N(\mu_1 + \mu_2, \sigma_1^2 + \sigma_2^2).$$

The result has been proven in Proposition 7.29. Clearly, the result can be easily extended by induction to finite sums of independent Gaussian random variables. That is, if $X_i \sim N(\mu_i, \sigma_i^2)$ for $i = 1, \ldots, n$, then

$$X_1 + \cdots + X_n \sim N\left(\sum_{i=1}^{n}\mu_i, \sum_{i=1}^{n}\sigma_i^2\right).$$

- The characteristic function of $X \sim N(\mu, \sigma^2)$ is

$$\varphi_X(t) = \exp\left(it\mu - \frac{\sigma^2 t^2}{2}\right) \tag{10.1}$$

for every $t \in \mathbb{R}$.

- The moment-generating function of $X \sim N(0, 1)$ is (see Proposition 9.16)

$$M_X(t) = e^{\frac{t^2}{2}}$$

for every $t \in \mathbb{R}$.

10.3.2 EQUIVALENT DEFINITIONS OF A GAUSSIAN VECTOR

Let us now define the Gaussian vector.

Definition 10.1 *A random vector $X = (X_1, \ldots, X_d)$ defined on the probability space $(\Omega, \mathcal{F}, \mathbf{P})$ is Gaussian if every linear combination of its components is a Gaussian random variable. That is, for every $\alpha_1, \ldots, \alpha_d \in \mathbb{R}$ the r.v.*

$$\alpha_1 X_1 + \cdots + \alpha_d X_d$$

is Gaussian.

The following two facts are immediate.

Proposition 10.2

1. *If $X = (X_1, \ldots, X_d)$, then X_i is a Gaussian random variable for every $i = 1, \ldots, d$.*
2. *If X_i is a Gaussian random variable for every $i = 1, \ldots, d$ and the r.v. X_i are mutually independent, then the vector*

$$X = (X_1, \ldots, X_d)$$

is a Gaussian vector.

Proof: 1. Indeed, this follows from the definition by choosing $\alpha_i = 1$ and $\alpha_j = 0$ for every $j \neq i, j = 1, \ldots, d$.

2. It is a consequence of the definition of the Gaussian vector and of the Proposition 7.29. We mention that the assumption that X_i are independent is crucial. In this chapter, we will see examples showing that it is possible to have random vectors with each component Gaussian which are not Gaussian. ■

Remark 10.3 *A standard normal Gaussian vector is a random vector*

$$X = (X_1, \ldots, X_d)$$

such that $X_i \sim N(0, 1)$ for every $i = 1, \ldots, d$ and X_i are independent. In this case the mean of X is $\mathbf{E}X = 0 \in \mathbb{R}^d$ and the covariance matrix is I_d, the unit matrix (matrix with 1 on the main diagonal and zero everywhere else). We denote

$$X \sim N(0, I_d).$$

As we mention above, the mean and the variance characterize the law of a Gaussian random variable (one-dimensional). In the multidimensional case, the mean (which is a vector) and the covariance matrix will completely determine the law of a Gaussian vector.

Recall that if $X = (X_1, \ldots, X_d)$ is a random vector, then

$$\mathbf{E}X = (\mathbf{E}X_1, \ldots, \mathbf{E}X_d) \in \mathbb{R}^d$$

and the covariance matrix of X, denoted by $\Lambda_X = (\Lambda_X(i, j))_{i,j=1,\ldots,d}$, is defined by

$$\Lambda_X(i, j) = Cov(X_i, X_j)$$

for every $i, j = 1, \ldots, d$.

Let us first remark that the mean and the covariance matrix of a Gaussian vector entirely characterize the first two moments (and thus the distribution) of every linear combination of the components of the vector. We recall the notation of a scalar product of two vectors:

$$\langle x, y\rangle = x^T y = \sum_{i=1}^{d} x_i y_i$$

if $x = (x_1, \ldots, x_d), y = (y_1, \ldots, y_d) \in \mathbb{R}^d$, and we denoted by x^T the transpose of the $d \times 1$ matrix x.

Proposition 10.4 *Let $X = (X_1, \ldots, X_d)$ be a d-dimensional Gaussian vector with mean vector μ and covariance matrix Λ_X. Let $\alpha = (\alpha_1, \ldots, \alpha_d) \in \mathbb{R}^d$. Define*

$$Y = \alpha_1 X_1 + \cdots + \alpha_d X_d,$$

a linear combination of the vector components. Then

$$Y \sim N\left(\langle\alpha, \mu\rangle, \langle\alpha, \Lambda_X \alpha\rangle\right).$$

Proof: It follows from Definition 10.1 that Y is a Gaussian r.v. (a linear combination of Gaussian random variables). It remains to compute its expectation and its variance. First

$$\begin{aligned}\mathbf{E}Y &= \alpha_1 \mathbf{E}X_1 + \ldots. + \alpha_d \mathbf{E}X_d \\ &= \langle\alpha, \mathbf{E}X\rangle = \langle\alpha, \mu\rangle\end{aligned}$$

and

$$\begin{aligned}VarY &= Cov\Big(\sum_{i=1}^{d} \alpha_i X_i, \sum_{j=1}^{d} \alpha_j X_j\Big) \\ &= \sum_{i,j=1}^{d} \alpha_i \alpha_j Cov(X_i, X_j) \\ &= \sum_{i=1}^{d} \alpha_i \left(\sum_{j=1}^{d} \alpha_j Cov(X_i, X_j)\right) \\ &= \langle\alpha, \Lambda_X \alpha\rangle\end{aligned}$$

by noticing that the components of the vector $\Lambda_X \alpha \in \mathbb{R}^d$ are

$$\left(\sum_{j=1}^{d} \alpha_j Cov(X_i, X_j)\right)_{i=1,\ldots,d}.$$

■

Using Proposition 10.4, it is possible to obtain the characteristic function of a Gaussian vector.

Theorem 10.5 *Let $X = (X_1, \ldots, X_d)$ be a Gaussian vector and denote by μ its expectation vector and by Λ_X its covariance matrix. Then the characteristic function of X is*

$$\varphi_X(u) = e^{i\langle \mu, u\rangle - \frac{1}{2}\langle u, \Lambda_X u\rangle} \tag{10.2}$$

for every $u \in \mathbb{R}^d$. Here we applied Proposition 8.23.

Proof: By definition, we have

$$\varphi_X(u) = \mathbf{E}\left(e^{i\langle X, u\rangle}\right).$$

Using Proposition 10.4, we have that, by Eq. (10.1),

$$Y \sim N(\langle \alpha, \mathbf{E}X\rangle, \langle \alpha, \Lambda_X \alpha\rangle),$$

where $Y = \langle X, u\rangle$. It suffices to note that

$$\begin{aligned} \varphi_X(u) &= \varphi_Y(1) = e^{i\mathbf{E}Y - \frac{1}{2}(V(Y))^2} \\ &= e^{i\langle \mu, u\rangle - \frac{1}{2}\langle u, \Lambda_X u\rangle} \end{aligned}$$

for every $u \in \mathbb{R}^d$. ∎

Remark 10.6 *Often in the literature, a Gaussian vector is defined through its characteristic function given in Theorem 10.5. That is, a random vector X is called a Gaussian random vector if its characteristic function is given by*

$$\varphi_X(u) = e^{i\langle \mu, u\rangle - \frac{1}{2}\langle u, \Sigma u\rangle}$$

for some vector μ and some symmetric positive definite matrix Σ.

Theorem 10.7 *If X is a d-dimensional Gaussian vector, then there exists a vector $m \in \mathbb{R}^d$ and a d-dimensional square matrix $A \in M_d(\mathbb{R})$ such that*

$$X =^{(d)} m + AN,$$

(where $=^{(d)}$ stands for the equality in distribution) where $N \sim N(0, I_d)$.

Proof: Let $A \in M_d(\mathbb{R})$ and define

$$Z = m + AN.$$

For every $u \in \mathbb{R}^d$ we have

$$\begin{aligned} \varphi_Z(u) &= \mathbf{E}e^{i\langle u, Z\rangle} \\ &= e^{i\langle m, u\rangle}\mathbf{E}e^{i\langle u, AN\rangle} \\ &= e^{i\langle m, u\rangle}\mathbf{E}e^{i\langle A^T u, N\rangle} \\ &= e^{i\langle m, u\rangle}\varphi_N(A^T u). \end{aligned}$$

Suppose that X is a Gaussian vector. Since Λ_X is symmetric and positive definite, there exists $A \in M_d(\mathbb{R})$ such that

$$\Lambda_X = AA^T.$$

Let $N_1 \sim N(0, I_d)$. We apply Theorem 10.5 and the beginning of this proof. It follows that X and $m + AN_1$ have the same characteristic function and therefore the same law. ■

The converse of Theorem 10.7 is also true.

Theorem 10.8 *If*

$$X = m + AN$$

with $m \in \mathbb{R}^d$, $A \in M_d(\mathbb{R})$, and $N \sim N(0, 1)$, then X is a Gaussian vector.

Proof: Let

$$a = (a_1, \dots, a_d) \in \mathbb{R}^d$$

and consider the linear combination

$$\begin{aligned} Y &= a_1X_1 + \cdots + a_dX_d \\ &= a_1(m_1 + (AN)_1) + \cdots + a_d(m_d + (AN)_d), \end{aligned}$$

where $(AN)_i$ is the component i of the vector $AN \in \mathbb{R}^d$, $i = 1, \dots, d$. Then, due to Proposition 10.4, we have

$$Y \sim N(a^T m, a^T AA^T a).$$

So every linear combination of the components of X is a Gaussian random variable, which means that X is a Gaussian vector. ■

By putting together the results in Theorems 10.5, 10.7, and 10.8, we obtain three alternative characterization of a Gaussian vector.

Theorem 10.9 *Let $X = (X_1, \dots, X_d)$ be a random vector with $\mu = \mathbf{E}X$ the mean vector and Λ_X the covariance matrix. Then the following are equivalent.*

1. *X is a Gaussian vector.*
2. *The characteristic function of X is given by*

$$\varphi_X(u) = e^{i\langle \mu, u\rangle - \frac{1}{2}\langle u, \Lambda_X u\rangle}$$

   for every $u \in \mathbb{R}^d$.
3. *There exist a vector $m \in \mathbb{R}^d$ and a matrix $A \in M_d(\mathbb{R})$ such that*

$$X =^{(d)} m + AZ$$

   (equality in distribution), where $Z \sim N(0, I_d)$ is a standard Gaussian vector.

Proof: The implication $1 \mapsto 2$ follows from Theorem 10.5. The implication $2 \mapsto 3$ is a consequence of the proof of Theorem 10.7. The implication $3 \mapsto 1$ has been showed in Theorem 10.8. ■

Remark 10.10 *Let us notice that the sum of two independent Gaussian vectors is a Gaussian vector. This follows easily from the definition of a Gaussian vector and the additivity property of the Gaussian random variables.*

EXAMPLE 10.1 A vector with all components Gaussian but which is not a Gaussian vector

Let us give an example of a vector with each component Gaussian which is not a Gaussian vector. Consider $N_1 \sim N(0, 1)$ and define the random variable N_2 by

$$N_2 = -N_1 \qquad \text{if} \quad N_1 \in [-a, a]$$

($a > 0$ is some fixed constant) and

$$N_1 = N_2 \qquad \text{otherwise.}$$

Let us show that N_2 is a Gaussian random variable. We compute its cumulative distribution function. It holds that

$$\begin{aligned} F_{N_2}(x) &= \mathbf{P}(N_2 < x) \\ &= \mathbf{P}\left((N_2 < x) \cap ((N_2 \in [-a, a]) \cup (N_2 \notin [-a, a]))\right) \\ &= \mathbf{P}\left((N_2 < x) \cap (N_2 \in [-a, a])\right) \\ &\quad + \mathbf{P}\left((N_2 < x) \cap (N_2 \notin [-a, a])\right) \\ &= \mathbf{P}\left((N_2 < x) \cap (N_1 \in [-a, a])\right) \\ &\quad + \mathbf{P}\left((N_2 < x) \cap (N_1 \notin [-a, a])\right) \\ &= \mathbf{P}(N_1 < x). \end{aligned}$$

As a consequence, N_2 has the same law as N_1, so

$$N_2 \sim N(0, 1).$$

However, the vector

$$(N_1, N_2)$$

is not a Gaussian vector. Indeed, the sum

$$S = N_1 + N_2$$

(which constitutes a linear combination of the components of (N_1, N_2)) is such that

$$\mathbf{P}(S = 0) = \mathbf{P}(N_1 + N_2 = 0) = \mathbf{P}(N_1 \in [-a, a]) > 0,$$

so S has strictly positive probability for a fixed value; thus it cannot be a random variable with a normal density (or any other continuous density).

Remark 10.11 *See also Example 10.5 for a situation when a non-Gaussian random vector has Gaussian marginals.*

10.3.3 UNCORRELATED COMPONENTS AND INDEPENDENCE

One of the most important properties of a Gaussian vector is the fact that its components are independent *if and only* if they are uncorrelated. One direction is always true for every random variable: If two random variables are independent, then they are uncorrelated. On the other hand, the converse is strongly related to the structure of the Gaussian vector, and it does not hold in general for other random variables. In Example 10.5 we show that that there exist Gaussian random variables (without the gaussian vector structure) which are uncorrelated but are not independent.

Theorem 10.12 *Let $X = (X_1, \ldots, X_d)$ be a d-dimensional Gaussian vector. Then for every $i, j = 1, \ldots, d$, $i \neq j$, the random variables X_i and X_j are independent if and only if*

$$Cov(X_i, X_j) = 0.$$

Proof: If X_i is independent by X_j, then clearly $Cov(X_i, X_j) = 0$, $i \neq j$.

Suppose that $Cov(X_i, X_j) = 0$, $i \neq j$. Denote by $\mu = \mathbf{E}X$ and $\Lambda_X = (\lambda_{i,j})_{i,j=1,\ldots,d}$ the mean vector and covariance matrix of the vector X. Since the individual covariances are zero, this matrix is a diagonal matrix and we have

$$\begin{aligned}
\varphi_X(u) &= \mathbf{E}e^{i\langle u, X\rangle} = e^{i\langle \mu, u\rangle} e^{-\frac{1}{2}\langle u, \Lambda_X u\rangle} \\
&= e^{i\sum_{j=1}^{d} u_j\mu_j - \frac{1}{2}\sum_{j=1}^{d} \lambda_{jj} u_j^2} \\
&= \prod_{j=1}^{d} \varphi_{X_j}(u_j)
\end{aligned}$$

for every $u = (u_1, \ldots, u_d) \in \mathbb{R}^d$. We used the fact that for every $j = 1, \ldots, d$, one has

$$X_j \sim N(\mu_j, \lambda_{jj}).$$

Since the characteristic function of the vector is the product of individual characteristic functions, the components of the vector X are independent. ∎

EXAMPLE 10.2

Let X, Y be two independent standard normal random variables and define

$$U = X + Y \quad \text{and} \quad V = X - Y.$$

Then U and V are independent.

Solution: Indeed, it is easy to see that (U, V) is a Gaussian vector (every linear combination of its components is a linear combination of X and Y and (X, Y) is a Gaussian vector). Moreover,

$$\begin{aligned} Cov(U, V) &= \mathbf{E}UV - \mathbf{E}U\mathbf{E}V \\ &= \mathbf{E}\left[(X + Y)(X - Y)\right] \\ &= \mathbf{E}(X^2 - Y^2) = 1 - 1 = 0 \end{aligned}$$

and the independence is obtained from Theorem 10.12. ∎

EXAMPLE 10.3

Let X_1, X_2, X_3, X_4 be four independent standard normal random variables. Define

$$\begin{aligned} Y_1 &= X_1 + X_2, \\ Y_2 &= \frac{1}{4}\left(X_1 - X_2 + X_3 + X_4\right), \\ Y_3 &= \frac{1}{2}(X_3 - X_4). \end{aligned}$$

We will show that (Y_1, Y_2, Y_3) is a Gaussian vector with independent components.

Solution: Let us first show that Y is a Gaussian vector. We note that

$$X = (X_1, X_2, X_3, X_4)$$

is a Gaussian vector with zero mean and covariance matrix I_4. Take $\alpha_1, \alpha_2, \alpha_3 \in \mathbb{R}$. Then

$$\begin{aligned}
\alpha_1 Y_1 + \alpha_2 Y_2 + \alpha_3 Y_3 \\
&= \left(\alpha_1 + \frac{1}{4}\alpha_2\right) X_1 + \left(\alpha_1 - \frac{1}{4}\alpha_2\right) X_2 \\
&\quad + \left(\frac{1}{4}\alpha_2 + \frac{1}{2}\alpha_3\right) X_3 + \left(\frac{1}{4}\alpha_2 - \frac{1}{2}\alpha_3\right) X_4,
\end{aligned}$$

and this is a Gaussian r.v. since X is a Gaussian vector. Moreover,

$$\begin{aligned}
Cov&(Y_1, Y_2) \\
&= \frac{1}{4}V(X_1) - \frac{1}{4}Cov(X_1, X_2) + \frac{1}{4}Cov(X_1, X_3) + \frac{1}{4}Cov(X_1, X_4) \\
&\quad - \frac{1}{4}Cov(X_2, X_1) - \frac{1}{4}VX_2 + \frac{1}{4}Cov(X_2, X_3) + \frac{1}{4}Cov(X_2, X_4) \\
&= \frac{1}{4}V(X_1) - \frac{1}{4}V(X_2) = 0.
\end{aligned}$$

In the same way, we have

$$Cov(Y_2, Y_3) = Cov(Y_1, Y_3) = 0,$$

so Y_1, Y_2, Y_3 are independent random variables. In fact, the vector Y is a Gaussian vector with $\mathbf{E}Y = 0$ and covariance matrix

$$\Lambda_Y = \begin{pmatrix} 2 & 0 & 0 \\ 0 & \frac{1}{4} & 0 \\ 0 & 0 & 1 \end{pmatrix}.$$

■

EXAMPLE 10.4

Let (X, Y) be a Gaussian couple with mean the zero vector such that $VarX = \sigma_X^2$ and $VarY = \sigma_Y^2$. Find a scalar α such that $X - \alpha Y$ and Y are independent random variables.

Solution: Since $X - \alpha Y$ and Y are Gaussian random variables, it suffices to impose the condition

$$\mathbf{E}(X - \alpha Y)Y = 0;$$

this implies

$$\alpha = \frac{\mathbf{E}XY}{\mathbf{E}Y^2} = \frac{Cov(X, Y)}{\sigma_Y^2} = \rho\frac{\sigma_X}{\sigma_Y},$$

where we denoted with

$$\rho := \frac{Cov(X, Y)}{\sigma_X \sigma_Y}$$

the correlation coefficient between X and Y. ■

EXAMPLE 10.5 Two uncorrelated normals which are *not* independent

Let $X \sim N(0, 1)$ and let ε be a random variable such that

$$\mathbf{P}(\varepsilon = 1) = \mathbf{P}(\varepsilon = -1) = \frac{1}{2}.$$

Suppose that X and ε are independent and define

$$Y = \varepsilon X.$$

Then, both of these random variables (X and Y) are normal, they are uncorrelated but they are not independent.

Solution: Let us show first that

$$Y \sim N(0, 1).$$

Indeed, by computing the cumulative distribution function of Y, we get for every $t \in \mathbb{R}$

$$\begin{aligned} F_Y(t) &= \mathbf{P}(Y \leq t) = \mathbf{P}(\varepsilon X \leq t) \\ &= \mathbf{P}(X \leq t, \varepsilon = 1) + \mathbf{P}(-X \leq t, \varepsilon = -1) \\ &= \mathbf{P}(X \leq t)\mathbf{P}(\varepsilon = 1) + \mathbf{P}(-X \leq t)\mathbf{P}(\varepsilon = -1) \\ &= \frac{1}{2}(\mathbf{P}(X \leq t) + \mathbf{P}(-X \leq t)) \\ &= \mathbf{P}(X \leq t) \end{aligned}$$

by using the fact that $-X \sim N(0, 1)$. So X, Y have the same law $N(0, 1)$. Let us show that X, Y are uncorrelated. Indeed,

$$\begin{aligned} Cov(XY) &= \mathbf{E}XY = \mathbf{E}(X^2\epsilon) \\ &= \mathbf{E}X^2\mathbf{E}\varepsilon = 0 \end{aligned}$$

since $\mathbf{E}\varepsilon = 0$. But it is easy to see that X and Y are not independent.

This example shows that it is possible to find two Gaussian uncorrelated r.v which are not independent. The reason why they are not independent is that the vector (X, Y) is not a Gaussian vector (see exercise 7.8). ■

A more general result can be stated as follows. The proof follows the arguments in the proof of Theorem 10.12.

Theorem 10.13 *Suppose $X \sim N(0, \Lambda_X)$ with $\Lambda_X \in M_n(\mathbb{R})$. Suppose that the components of X can be divided into two groups $(X_i)_{i\in I}$ and $(X_j)_{j\notin I}$, where $I \subset \{1, 2, \ldots, n\}$, and further suppose that*

$$Cov(X_i, X_j) = 0$$

for all $i \in I$ and $j \notin I$. Then the family $(X_i)_{i\in I}$ is independent of the family $(X_j)_{j\notin I}$.

Remark 10.14 *A Gaussian vector X is called degenerated if the covariance matrix Λ_X is not invertible (i.e.,* $\det \Lambda_X = 0$*). For example, the vector*

$$X = (X_1, X_1),$$

with X_1 a Gaussian r.v., is a degenerated Gaussian vector.

10.3.4 THE DENSITY OF A GAUSSIAN VECTOR

Let $X = (X_1, \ldots, X_d)$ be a Gaussian vector with independent components. Assume that for every $i = 1, \ldots, d$ we have

$$X_i \sim N(\mu_i, \sigma_i^2).$$

In this case it is easy to write the density of the vector X. It is, from Corollary 7.23,

$$f(x_1, \ldots, x_d) = \frac{1}{(2\pi)^{\frac{d}{2}}\sigma_1 \ldots \sigma_d} e^{-\left(\frac{(x_1-\mu_1)^2}{2\sigma_1^2}+\cdots+\frac{(x_d-\mu_d)^2}{2\sigma_d^2}\right)}. \tag{10.3}$$

In the case of a standard normal vector $X \sim N(0, I_d)$, we have

$$f(x_1, \ldots, x_d) = \frac{1}{(2\pi)^{\frac{d}{2}}} e^{-\frac{1}{2}(x_1^2+\cdots+x_d^2)}.$$

When the components of X are not independent, we have the following.

Theorem 10.15 *Let $X = (X_1, \ldots, X_d)$ be a Gaussian vector with $\mu =$ **E**X and covariance matrix denoted by Λ_X. Assume that Λ_X is invertible. Then X admits the following probability density function:*

$$f(x) = \frac{1}{(2\pi)^{\frac{d}{2}}(\det(\Lambda_X))^{\frac{1}{2}}} = e^{-\frac{1}{2}(x-\mu)^T\Lambda_X^{-1}(x-\mu)} \tag{10.4}$$

for any $x \in \mathbb{R}^d$.

Proof: From Theorem 10.7 we can write

$$X = AN + \mu$$

where

$$N \sim N(0, I_d) \quad \text{and} \quad \Lambda^X = AA^T.$$

We apply the change of variable formula (Theorem 7.24) to the function $h : \mathbb{R}^d \to \mathbb{R}^d$:

$$h(x) = Ax + \mu.$$

We have

$$h^{-1}(x) = A^{-1}(x - \mu).$$

We obtain

$$f_X(x) = f_N(h^{-1}(x))|\det J_{h^{-1}}(x)|$$

where f_N denotes the density of the vector $N \sim N(0, I_d)$. Since

$$\det J_{h^{-1}}(x) = \sqrt{\Lambda_X^{-1}} = \frac{1}{\sqrt{\Lambda_X}}$$

and

$$\|A^{-1}(x - \mu)\|^2 = (x - \mu)^T \Lambda_X^{-1}(x - \mu),$$

we obtain the conclusion of the theorem. ∎

Remark 10.16 *The above result applies only to non-degenerated Gaussian vectors. Indeed, if* $\det \Lambda_X = 0$, *the formula (10.4) has no sense. In fact, if X is a degenerated Gaussian vector, then X does not have a probability density function.*

However, in the case of degenerated Gaussian vectors, we have the property stated by the next proposition. We need to recall the notion of rank of a matrix. The rank of a matrix is the dimension of the vector space generated by its columns (or its rows). It is the largest possible dimension of a minor with determinant different from zero. A minor in a matrix can be constructed by eliminating any number of rows and columns. Obviously, if the d-dimensional matrix is invertible, then the rank of the matrix is d.

Proposition 10.17 *Let* $X \sim N(\mu, \Lambda_X)$ *in* $\mathbb{R}^d$ *and assume that*

$$rank(\Lambda_X) = k < d,$$

a degenerated random vector.

Then, there exists a vector space $H \subset \mathbb{R}^d$ of dimension $d - k$ such that $\langle a, X \rangle = a^T X$ is a constant random variable for every $a \in H$.

EXAMPLE 10.6

Let $X = (X_1, X_1)$ with X_1 Gaussian one-dimensional. Then clearly

$$rank \Lambda_X = 1 < 2.$$

Let

$$H = \{C = (c, -c), c \in \mathbb{R}\} \subset \mathbb{R}^2.$$

Then H is a subspace of $\mathbb{R}^2$ and $\langle C, X \rangle = C^T X = 0$ for every $C \in H$.

EXAMPLE 10.7

Let X be a two-dimensional Gaussian vector with zero mean and covariance matrix

$$\Lambda_X = \begin{pmatrix} 4 & 1 \\ 1 & 4 \end{pmatrix}.$$

Let us write the density of the vector X. First $det \Lambda_X = 15$ and

$$\Lambda_X^{-1} = \frac{1}{15} \begin{pmatrix} 4 & -1 \\ -1 & 4 \end{pmatrix}.$$

Therefore

$$\begin{aligned} f(x, y) &= \frac{1}{2\pi\sqrt{15}} e^{-\frac{1}{30}(x,y)^T \Lambda_X^{-1}(x,y)} \\ &= \frac{1}{2\pi\sqrt{15}} e^{-\frac{1}{30}(4x^2 - 2xy + 4y^2)}. \end{aligned}$$

Remark 10.18

1. *As we mentioned, we can see that μ and Λ_X completely characterize the law of a Gaussian vector. This is not true for other types of random vectors.*
2. *It is easy to see that in the case when the components of X are independent, formula (10.4) reduces to (10.3).*

In the case of a Gaussian vector of dimension 2 (a Gaussian couple), we have the following:

Proposition 10.19 *Let $X = (X_1, X_2)$ be a Gaussian couple with*

$$\mathbf{E}X = \mu = (\mu_1, \mu_2)$$

and

$$\Lambda_X = \begin{pmatrix} \sigma_1^2 & Cov(X_1, X_2) \\ Cov(X_1, X_2) & \sigma_2^2. \end{pmatrix} = \begin{pmatrix} \sigma_1^2 & \rho\sigma_1\sigma_2 \\ \rho\sigma_1\sigma_2 & \sigma_2^2. \end{pmatrix},$$

where we denoted the correlation,

$$\rho := \frac{Cov(X_1, X_2)}{\sigma_1\sigma_2}.$$

Assume $\rho^2 \neq 1$. Then the density of the vector X is

$$f_X(X_1, X_2) = \frac{1}{\sigma_1^2\sigma_2^2\sqrt{1-\rho^2}} e^{-\frac{1}{2(1-\rho^2)}\left[\frac{(x_1-\mu_1)^2}{\sigma_1^2} - 2\rho\frac{(x_1-\mu_1)(x_2-\mu_2)}{\sigma_1\sigma_2} + \frac{(x_2-\mu_2)^2}{\sigma_2^2}\right]}.$$

Proof: This follows from Theorem 10.4 since

$$\Lambda_X^{-1} = \frac{1}{\sigma_1^2\sigma_2^2(1-\rho^2)} \begin{pmatrix} \sigma_2^2 & -\rho\sigma_1\sigma_2 \\ -\rho\sigma_1\sigma_2 & \sigma_1^2. \end{pmatrix}.$$

■

10.3.5 COCHRAN'S THEOREM

Recall that if $X \sim N(0, 1)$, then $X^2 \sim \Gamma(\frac{1}{2}, \frac{1}{2})$, the gamma distribution with parameters $a = \frac{1}{2}$ and $\lambda = \frac{1}{2}$. This law is called the chi square distribution and is usually denoted by $\chi^2(1)$, with 1 denoting one degree of freedom. More generally, if $X_1, \ldots, X_d$ are independent standard normal random variables, then

$$\|X\|^2 := X_1^2 + \cdots + X_d^2$$

follows the law $\Gamma(\frac{d}{2}, \frac{1}{2})$ and this is called the chi square distribution with d degrees of freedom, denoted by $\chi^2(d)$.

This situation can be extended to nonstandard normal random variables.

Definition 10.20 *If $X = (X_1, \ldots, X_d)$ is a Gaussian vector with $\mathbf{E}X = \mu$ and $\Lambda_X = I_d$, then the law of $\|X\|^2$ is denoted by*

$$\chi^2(d, \|\mu\|^2)$$

and it is called the noncentral chi square distribution with d degrees of freedom and noncentrality parameter μ. When $\|\mu\| = 0$, then obviously $\chi^2(d, 0) = \chi^2(d)$.

Recall that the modulus (or the Euclidean norm) of a d-dimensional vector is

$$\|\mu\| = \sqrt{\mu_1^2 + \cdots + \mu_d^2}.$$

Remark 10.21 *The law of the random variable $\|X\|^2$ depends only on $\|\mu\|$ and d. Indeed, if $Y \in \mathbb{R}^d$ is such that*

$$Y \sim N(\mu', I_d)$$

with $\|\mu'\| = \|\mu\|$, then there exists an orthonormal matrix U such that

$$\mu = U\mu'.$$

Therefore,

$$UY \sim N(\mu, I_d)$$

and

$$\|Y\|^2 = \|UY\|^2 \sim \|X\|^2.$$

Remark 10.22 *We know that the density of the $\chi^2(d)$ law is*

$$f(x) = 2^{-\frac{d}{2}}\Gamma(\frac{d}{2})^{-1}e^{-\frac{x}{2}}x^{\frac{d}{2}-1}1_{(0,\infty)}(x).$$

In the case of the noncentral chi square distribution $\chi^2(d, a)$ we can prove that the density is

$$g(x) = \sum_{i=0}^{\infty} e^{-\frac{a}{2}} \left(\frac{a}{2}\right)^i \frac{1}{i!} f_{Y_{k+2i}}(x) \tag{10.5}$$

where Y_q denotes a random variable with distribution $\chi^2(q)$.

We can now state the Cochran theorem.

Theorem 10.23 (Cochran's theorem) *Assume*

$$\mathbb{R}^d = E_1 \oplus \cdots \oplus E_r,$$

where E_i, $i = 1, \ldots, r$ are orthogonal subspaces of $\mathbb{R}^d$ with dimension $d_1, \ldots, d_r$ respectively. Denote by

$$X_{E_i}, \qquad i = 1, \ldots, r$$

the orthogonal projection of X on the subspace E_i.

Then $X_{E_1}, \ldots, X_{E_r}$ are independent random vectors, $\|X_{E_1}\|^2, \ldots, \|X_{E_r}\|^2$ are also independent, and

$$\left(\|X_{E_1}\|^2, \ldots, \|X_{E_r}\|^2\right) \sim \left(\chi^2(d_1, \|\mu_{E_1}\|^2), \ldots, \chi^2(d_1, \|\mu_{E_r}\|^2)\right),$$

where μ_{E_i} is the projection of μ on E_i for every $i = 1, \ldots, r$.

Proof: Let

$$(e_{j_1}, \ldots, e_{j_{d_j}})$$

be an orthonormal basis of E_j. Then

$$X_{E_j} = \sum_{k=1}^{d_j} e_{j_k}(e_{j_k}X)^t.$$

The random vectors $(e_{j_k}X)^t$ are independent of distribution $N((e_{j_k}\mu)^t, 1)$, so the random vectors $X_{E_1}, \ldots, X_{E_r}$ are independent. To finish, it suffices to remark that

$$\|X_{E_j}\|^2 = \sum_{k=1}^{d_j}((e_{j_k}X)^t)^2$$

for every $j = 1, \ldots, d$. ■

Let us give an important application of the Cochran's theorem to Gaussian random vectors.

Proposition 10.24 *Let $X = (X_1, \ldots, X_n)$ denote a Gaussian vector with independent identically distributed $N(\mu, \sigma^2)$ components. Let us define*

$$\bar{X}_n = \frac{1}{n}\sum_{i=1}^{n} X_i$$

and

$$S_n^2 = \sum_{i=1}^{n}(X_i - \bar{X}_n)^2,$$

the sample mean and sample variance respectively.

Then

1. *$\bar{X}_n$ and S_n^2 are independent.*
2. *$\bar{X}_n \sim N(\mu, \frac{\sigma^2}{n})$.*
3. *$\frac{nS_n}{\sigma} \sim \chi^2(n-1)$.*

Proof: We set for every $i = 1, \ldots, n$

$$Y_i = \frac{X_i - m}{\sigma}.$$

Then Y_i are independent identically distributed $N(0, 1)$. We also set

$$e = (1, \ldots, 1) \in \mathbb{R}^n \quad \text{and} \quad E = Vect(e)$$

(by $Vect(e)$ we mean the vector space of $\mathbb{R}^n$ generated by the vector e). Then

$$\mathbb{R}^n = E \oplus E^{\perp}.$$

The projections of $Y = (Y_1, \ldots, Y_n)$ on E, $E^{\perp}$ are independent and given by

$$Y_E = \frac{1}{n} \sum_{i=1}^{n} Y_i \times e$$

and

$$Y_{E^{\perp}} = \left(Y_1 - \frac{1}{n} \sum_{i=1}^{n} Y_i, \ldots, Y_n - \frac{1}{n} \sum_{i=1}^{n} Y_i \right).$$

We therefore have

$$\frac{1}{\sigma}(\bar{X}_n - \mu) \times e = Y_E.$$

and

$$\frac{nS_n}{\sigma} = \| Y_{E^{\perp}} \|^2,$$

which gives the conclusion. ∎

10.3.6 MATRIX DIAGONALIZATION AND GAUSSIAN VECTORS

We start with some notion concerning eigenvalues and eigenvectors of a matrix.

Definition 10.25 *Let A be a matrix in $M_n(\mathbb{R})$. We say that λ is an eigenvalue of the matrix A if there exists a vector $u \in \mathbb{R}^n$, $u \neq 0$ such that*

$$Au = \lambda u.$$

In this case we will say that u is an eigenvector of the matrix A.

Remark 10.26 *For every vector $u \in \mathbb{R}^n$ we have $I_n u = u$ (where we denoted with I_n the identity matrix in $M_n(\mathbb{R})$). This implies that every vector $u \in \mathbb{R}^n$ is an eigenvector of the identity matrix I_n associated to the eigenvalue $\lambda = 1$.*

If $D = Diag(D_1, \ldots, D_n)$ is a diagonal matrix, then every D_i, $i = 1, \ldots, n$, is an eigenvalue of the matrix D and every vector $e_i = (0, 0, \ldots, 1, \ldots, 0)$ of the canonical basis of $\mathbb{R}^n$ is an eigenvector of D associated to the eigenvalue D_i.

Proposition 10.27 *If λ is an eigenvalue of the matrix $A \in M_n(\mathbb{R})$, then the set E_λ of the eigenvectors of A associated to λ is a vectorial subspace of $\mathbb{R}^n$.*

Proof: Let λ be an eigenvalue of A. If $u \in E_\lambda$ is a corresponding eigenvector, then for every $\alpha \in \mathbb{R}$, $\alpha \neq 0$ we have

$$A(\alpha u) = \alpha A u = \alpha \lambda u = \lambda(\alpha u)$$

so

$$\alpha u \in E_\lambda, \tag{10.6}$$

which shows that E_λ is closed under multiplication with scalars.

If v is another vector in E_λ, then

$$A(u + v) = Au + Av = \lambda u + \lambda v = \lambda(u + v)$$

so

$$u + v \in E_\lambda. \tag{10.7}$$

Relations (10.6) and (10.7) show that E_λ is a vector space in $\mathbb{R}^n$. ∎

Definition 10.28 *The vector space E_λ is called the eigenspace associated with the eigenvalue λ.*

Definition 10.29 *If $A \in M_n(\mathbb{R})$ is an n-dimensional matrix, we define*

$$KerA = \{u \in \mathbb{R}^n, Au = 0\},$$

the kernel or the null space of the matrix A.

Proposition 10.30 *Let $A \in M_n(\mathbb{R})$. Then the eigenvalues of A are solutions of the equation*

$$det(A - \lambda I_n) = 0,$$

and the eigenvectors associated with λ are the elements of $Ker(A - \lambda I_n)$.

Proof: If λ is an eigenvalue for A, there exists $u \neq 0$, $u \in \mathbb{R}^n$ such that $Au = \lambda u$; therefore

$$(A - \lambda I_n)u = 0.$$

This implies that the matrix $A - \lambda I_n$ is not invertible and thus $\det(A - \lambda I_n) = 0$.

Conversely, if $\det(A - \lambda I_n) = 0$, then $A - \lambda I_n$ is not invertible so there exists $u \in \mathbb{R}^n$ nonidentically zero such that $(A - \lambda I_n)u = 0$.

Finally, if λ is an eigenvalue of A, then the set of associated eigenvectors are the vectors $u \in \mathbb{R}^n$ satisfying $(A - \lambda I_n)u = 0$, which in fact is the definition of $Ker(A - \lambda I_n)$. ∎

EXAMPLE 10.8

Let

$$A = \begin{pmatrix} 1 & -3 \\ -2 & 2. \end{pmatrix},$$

then

$$A - \lambda I_2 = \begin{pmatrix} 1-\lambda & -3 \\ -2 & 2-\lambda \end{pmatrix}$$

and $\det(A - \lambda I_2) = \lambda^2 - 3\lambda - 4$. The solution of the equation $\det(A - \lambda I_2) =$ (which are the eigenvalues of A) is

$$\lambda_1 = -1 \quad \text{and} \quad \lambda_2 = 4.$$

Further,

$$\begin{aligned} E_{\lambda_1} &= \{(x, y) \in \mathbb{R}^2, A(x, y)^T = -(x, y)^T\} \\ &= \{(x, y) \in \mathbb{R}^2, 2x - 3y = 0\} \end{aligned}$$

and

$$\begin{aligned} E_{\lambda_2} &= \{(x, y) \in \mathbb{R}^2, A(x, y)^T = 4(x, y)^T\} \\ &= \{(x, y) \in \mathbb{R}^2, x + y = 0\}. \end{aligned}$$

Definition 10.31 *A matrix $A \in M_n(\mathbb{R})$ is diagonalisable if there exists a matrix $P \in M_n(\mathbb{R})$ invertible such that*

$$P^{-1}AP = D,$$

where D is a diagonal matrix.

In the case when A is diagonalizable, every column of the matrix P represents an eigenvector for A and the diagonal matrix D contains on its diagonal the eigenvalues of A. Each column i is an eigenvector for the eigenvalue i on the diagonal of D.

The following results apply to any random vector and not only the Gaussian random vectors. We shall review the requirement of a covariance matrix.

Definition 10.32 *A matrix A is called symmetric iff it is equal to its transpose $A = A^T$ or element-wise $a_{ij} = a_{ji}$ for all i, j. Note that from definition a symmetric matrix needs to be a square matrix (number of columns equal to the number of rows).*

A $d \times d$-dimensional matrix A is called positive definite if and only if

$$u^T A u > 0,$$

for any $u = \begin{pmatrix} u_1 \\ \vdots \\ u_d \end{pmatrix} \in \mathbb{R}^d,\ u \neq 0.$

The matrix A is called non-negative definite if and only if

$$u^T A u \geq 0.$$

Please note that $u^T Au$ is a number, thus its sign is unique. Further, we always consider vectors in $\mathbb{R}^d$ as matrices having dimension $d \times 1$.

Proposition 10.33 *Let* $X = \begin{pmatrix} X_1 \\ \vdots \\ X_d \end{pmatrix}$ *be a random vector with mean* μ *and covariance matrix* Λ_X. *Then the matrix* Λ_X *is symmetric and non-negative definite.*

Proof: To prove this proposition, let us first remark that the covariance matrix is a square matrix. Next, the element on row i column j is

$$\Lambda_X(i, j) = Cov(X_i, X_j) = \mathbf{E}[(X_i - \mu_i)(X_j - \mu_j)] = Cov(X_j, X_i) = \Lambda_X(j, i),$$

thus the matrix must be symmetric.

About the positive definiteness for any $u = \begin{pmatrix} u_1 \\ \vdots \\ u_d \end{pmatrix} \in \mathbb{R}^d$, we can construct the one-dimensional random variable: $u^T X$. Since this is a valid random variable, its variance must be non-negative. So let us calculate this variance:

$$Var(u^T X) = \mathbf{E}[(u^T X - u^T \mu)^2].$$

Since the number squared is one-dimensional, and a one-dimensional number is equal to its transpose, we may write

$$\begin{aligned} Var(u^T X) &= \mathbf{E}\left[(u^T X - u^T \mu)((u^T X - u^T \mu))^T\right] \\ &= \mathbf{E}\left[\left(u^T (X - \mu)\right)\left(u^T (X - \mu)\right)^T\right] \\ &= u^T \mathbf{E}\left[(X - \mu)(X - \mu))^T\right] u \\ &= u^T \Lambda_X u. \end{aligned}$$

Thus, the condition that variance is non-negative translates into the condition that the covariance matrix is non-negative definite. ■

Remark 10.34 *The distinction between non-negative and positive definite matrices is important for random variables. If it is possible to create a random variable which* *is identically zero, then its variance will be zero. If the components are independent, then the covariance matrix will be positive definite.*

As a simple example, consider the random vector $X = (X_1, -X_1)$ *for some random variable* X_1. *The covariance matrix of the vector will be non-negative definite since there exists the vector* $u = (1, 1) \in \mathbb{R}^2$ *and* $u^T X = 0$ *and thus its variance will be zero.*

Checking that a square matrix is positive definite can be complicated. However, there is an easy way to check involving the eigenvalues of the matrix.

Lemma 10.35 *A matrix* $A \in M_d(\mathbb{R})$ *is positive definite if all its eigenvalues are real and positive. A matrix is non-negative definite if all eigenvalues are non-negative.*

The following result is important because it can be applied to the covariance matrices.

Theorem 10.36 *Let A be a symmetric, positive definite* $n \times n$ *matrix. Then A is diagonalizable by an orthonormal matrix P. That is, there exists an orthonormal matrix P (i.e.,* $P^T = P^{-1}$*) such that*

$$A = P^T D P$$

with $D \in M_n(\mathbb{R})$ *a diagonal matrix.*

The above result says that every symmetric positive definite matrix is diagonalizable in an orthonormal basis. That is, it can be transformed by elementary transformations into a diagonal matrix.

Definition 10.37 *Two vectors* $x = (x_1, \ldots, x_n)$ *and* $y = (y_1, \ldots, y_n)$ *in* $\mathbb{R}^n$ *are called orthogonal if*

$$\langle x, y \rangle = \sum_{i=1}^{n} x_i y_i = 0.$$

A vector is called with norm 1 if $\|x\|^2 = \langle x, x \rangle = x_1^2 + \cdots + x_n^2 = 1$.

The vectors x and y are called orthonormal if they are orthogonal and $\|x\| = \|y\| = 1$.

Theorem 10.38 *Let X be a d-dimensional Gaussian vector with zero mean and covariance matrix* Λ_X. *Then there exists a matrix* $B \in M_d(\mathbb{R})$ *such that BX is a Gaussian vector with independent components.*

Proof: The proof follows by using Theorem 10.36. Since Λ_X is a symmetric matrix, it can be diagonalizable in an orthonormal basis. That is, there exists an orthogonal matrix B such that $B\Lambda_X B^T$ is diagonal. The covariance of BX is $B\Lambda_X B^T$, so the components of BX are independent Gaussian random variables. ∎

EXAMPLE 10.9

Let X be a centered Gaussian vector with

$$\Lambda_X = \begin{pmatrix} 3 & 2 \\ 2 & 3 \end{pmatrix}.$$

Note first that this matrix is symmetric. To be a valid covariance matrix, it needs to be positive definite, which can be checked by looking at the sign of the eigenvalues (all eigenvalues should be positive) so it can diagonalizable. Then two eigenvalues are the solution of the equation

$$det(\Lambda_X - \lambda I_2) = \begin{pmatrix} 3-\lambda & 2 \\ 2 & 3-\lambda \end{pmatrix} = \lambda^2 - 6\lambda + 5 = 0$$

and thus

$$\lambda_1 = 1 \quad \text{and} \quad \lambda_2 = 5.$$

The matrix therefore is positive definite and symmetric. To find the eigenvectors, we need to solve using the definition $\Lambda_X u = \lambda_i u$ for both $i = 1$ and $i = 2$. This gives

$$E_{\lambda_1} = \{(x, y) \in \mathbb{R}^2, x + y = 0\}$$

and

$$E_{\lambda_2} = \{(x, y) \in \mathbb{R}^2, x - y = 0\}.$$

The eigenspace E_{λ_1} is generated by the vector $(1, -1)$, while E_{λ_2} is generated by the vector $(1, 1)$. This two vectors are orthogonal but not orthonormal (their euclidian norm is not 1). To make them ortonormal, we normalize them. We define

$$e_1 = \frac{(1, -1)}{\|(1, -1)\|} = \left(\frac{1}{\sqrt{2}}, -\frac{1}{\sqrt{2}}\right)$$

and

$$e_2 = \frac{(1, 1)}{\|(1, 2)\|} = \left(\frac{1}{\sqrt{2}}, \frac{1}{\sqrt{2}}\right)$$

and therefore the matrix P given by

$$P = \begin{pmatrix} \frac{1}{\sqrt{2}} & \frac{1}{\sqrt{2}} \\ -\frac{1}{\sqrt{2}} & \frac{1}{\sqrt{2}}. \end{pmatrix}$$

Then it can be checked that

$$P^T \Lambda_X P = D,$$

where D is the diagonal matrix with the eigenvalues of Λ_X on the diagonal

$$D = \begin{pmatrix} 1 & 0 \\ 0 & 5 \end{pmatrix}.$$

Remark 10.39 *If the matrix Λ_X is small, then the decomposition shown above will work. However, if the dimensionality of the matrix is large while the methodology presented here will still work, it will become very tedious. In this case, one will enroll the use of a computer and reach the same decomposition using the methodology presented in Section 6.4.*

EXERCISES

Problems with Solution

10.1 Suppose

$$(X_1, X_2) \sim N(0, I_2)$$

and define

$$Y_1 = aX_1 + X_2 \quad \text{and} \quad Y_2 = X_1 + bX_2.$$

(a) Find $a, b \in \mathbb{R}$ such that $Y = (Y_1, Y_2)$ is a Gaussian vector with independent components.
(b) Write the density of the vector Y.

Solution: Y is clearly a Gaussian vector since $Y = AX$ with

$$A = \begin{pmatrix} a & 1 \\ 1 & b \end{pmatrix},$$

where each component of Y is a linear combination of the original independent components Gaussian vector. To have the components of Y independent we need to impose the condition

$$Cov(Y_1, Y_2) = 0.$$

But

$$\begin{aligned} Cov(Y_1, Y_2) &= aV(X_1) + abCov(X_1, X_2) + Cov(X_1, X_2) + bV(X_2) \\ &= aV(X_1) + bV(X_2) = a + b. \end{aligned}$$

Therefore if $a+b=0$, the components Y_1, Y_2 are independent. Since

$$Y_1 \sim N(0, a^2+1) \quad \text{and} \quad Y_2 \sim N(0, b^2+1)$$

we will have in this case $(a+b=0)$

$$f_Y(x,y) = \frac{1}{2\pi\sqrt{(a^2+1)(b^2+1)}} e^{-\left(\frac{x^2}{2(a^2+1)} + \frac{y^2}{2(b^2+1)}\right)}.$$

■

10.2 Let $X = (X_1, X_2, X_3)$ be a random vector in $\mathbb{R}^3$ with density

$$f(x_1, x_2, x_3) = k \exp\left\{ -\frac{1}{2}\left(3x_1^2 + 2x_2^2 + x_3^2 + 4x_1x_2 - 2x_1x_3 - 2x_2x_3\right)\right\}.$$

(**a**) Find the law of X. Derive k which makes the law a proper density function.

(**b**) Let

$$\widetilde{X}_2 = aX_1 + bX_2 \quad \text{and} \quad \widetilde{X}_3 = cX_1 + dX_2 + eX_3.$$

Find the parameters a, b, c, d, e such that the covariance matrix of $(X_1, \widetilde{X}_2, \widetilde{X}_3)$ is I_3.

(**c**) What can be said about the variables $X_1, \widetilde{X}_2$, and $\widetilde{X}_3$?

Solution: Note that

$$\left(3x_1^2 + 2x_2^2 + x_3^2 + 4x_1x_2 - 2x_1x_3 - 2x_2x_3\right) = (x_1, x_2, x_3) M \begin{pmatrix} x_1 \\ x_2 \\ x_3 \end{pmatrix},$$

where

$$M = \begin{pmatrix} 3 & 2 & -1 \\ 2 & 2 & -1 \\ -1 & -1 & 1 \end{pmatrix}.$$

To see this decomposition, please think about how the polynomial terms appear. It helps to note that the diagonal elements give the squares in a unique way and thus they are easy to recognize. For the off diagonal elements note that twice the element gives the coefficient (because the matrix is symmetric).

Once we write it in this form, we recognize the density of a Gaussian vector X with zero mean and covariance matrix

$$\Lambda_X = M^{-1}.$$

Consequently,

$$k = \frac{1}{(2\pi)^{\frac{3}{2}}\sqrt{\det(M^{-1})}} = \frac{\sqrt{\det M}}{(2\pi)^{\frac{3}{2}}}$$

To solve part (b), we need to impose that

$$Cov(X_1, \tilde{X}_2) = Cov(X_1, \tilde{X}_3) = Cov(\tilde{X}_2, \tilde{X}_3) = 0$$

and

$$VX_1 = V\tilde{X}_2 = V\tilde{X}_3 = 1.$$

To impose these conditions, we need to calculate the covariance matrix:

$$\Lambda_X = M^{-1}.$$

Thus we need to know how to invert a matrix or to use a software program to do so. Using R gives

$$\Lambda_X = M^{-1} = solve(M) = \begin{pmatrix} 1 & -1 & 0 \\ -1 & 2 & 1 \\ 0 & 1 & 2 \end{pmatrix}.$$

The command "*solve(M)*" is the R command to find the inverse of the matrix M. We can now read the covariances between the original vector components X_1, X_2, X_3.

Now, using the matrix above and the formulas for the vector, the conditions are

$$\begin{aligned}
Cov(X_1, \tilde{X}_2) &= aV(X_1) + bCov(X_1, X_2) = a - b = 0, \\
Cov(X_1, \tilde{X}_2) &= cV(X_1) + dCov(X_1, X_2) + eCov(X_1, X_3) = c - d = 0, \\
Cov(\tilde{X}_2, \tilde{X}_3) &= ac - ad - bc + 2bd + be = 0, \\
V(\tilde{X}_2) &= a^2 + 2b^2 - 2ab = 1, \\
V(\tilde{X}_3) &= c^2 + 2d^2 + 2e^2 - 2cd + 2de = 1.
\end{aligned}$$

$V(X_1)$ is already 1. Using $a = b$ and $c = d$ from the first two equations, the later equations become

$$\begin{aligned}
&a(c + e) = 0, \\
&a^2 = 1, \\
&c^2 + 2e^2 + 2ce = 1.
\end{aligned}$$

The first two equations are incompatible with $a = 0$, so we must have $c = -e$ and either $a = 1$ or $a = -1$. Using this in the last equation gives

$e = 1$. Thus the problem has more than one solution. Either of these vectors

$$(X_1, \widetilde{X}_2, \widetilde{X}_3) = (X_1, X_1 + X_2, -X_1 - X_2 + X_3),$$
$$(X_1, \widetilde{X}_2, \widetilde{X}_3) = (X_1, X_1 + X_2, X_1 + X_2 - X_3),$$
$$(X_1, \widetilde{X}_2, \widetilde{X}_3) = (X_1, -X_1 - X_2, -X_1 - X_2 + X_3),$$
$$(X_1, \widetilde{X}_2, \widetilde{X}_3) = (X_1, -X_1 - X_2, X_1 + X_2 - X_3)$$

will have the desired properties.

Finally, for part (c), since either one of these vectors is Gaussian by requiring that the covariance matrix is the identity, we found components which are mutually independent. ■

10.3 Let X, X, Z be independent standard normal random variables. Denote

$$U = X + Y + Z$$

and

$$V = (X - Y)^2 + (X - Z)^2 + (Y - Z)^2. \tag{10.8}$$

Show that U and V are independent.

Solution: Define

$$A = (X, Y, Z).$$

Clearly, A is a Gaussian vector with $\mathbf{E}A = 0 \in \mathbb{R}^3$ and covariance matrix

$$\Lambda_A = I_3$$

the identity matrix. It follows that the vector

$$B = (X + Y + Z, X - Y, X - Z, Y - Z)$$

is also a Gaussian vector (every linear combination of its components is a linear combination of X, Y, Z). We will show that the first component is independent of all the other three. To prove this since the vector is Gaussian, it suffices to show that the first component is uncorrelated with the other three. We have

$$Cov(X + Y + Z, X - Y) =$$
$$V(X) + Cov(Y, X) + Cov(Z, X) - Cov(X, Y) - V(Y) - Cov(Z, Y) = 0$$

and in a similar way

$$Cov(X + Y + Z, X - Z) = Cov(X + Y + Z, Y - Z) = 0.$$

Therefore the r.v. $X + Y + Z$ is independent of $X - Y, X - Z$, and $Y - Z$ respectively. By the associativity property of the independence, we have that $X+Y+Z$ is independent of $(X-Y, X-Z, Y-Z)$ and, thus, independent of V. ■

10.4 Let $X_1, \dots, X_n$ be independent $N(0, 1)$ distributed random variables. Let $a \in \mathbb{R}^n$. Give a necessary and sufficient condition on the vector a in order to have $X - \langle a, X\rangle a$ and $\langle a, X\rangle$ independent.

Solution: The vector

$$(X - \langle a, X\rangle a, \langle a, X\rangle)$$

is an $(n+1)$-dimensional Gaussian vector and for every $i = 1, \dots, n$ we obtain

$$\begin{aligned}
&Cov\left(X_i - \sum_{j=1}^n a_j X_j a_i, \sum_{k=1}^n a_k X_k\right) \\
&= \sum_{k=1}^n a_k Cov(X_i, X_k) - a_i \sum_{j,k=1}^n a_j a_k Cov(X_j, X_k) \\
&= a_i - a_i \sum_{j=1}^n a_j^2 \\
&= a_i(1 - \sum_{j=1}^n a_j^2).
\end{aligned}$$

Therefore if we impose the condition

$$\sum_{j=1}^n a_j^2 = 1,$$

then all covariances between $\langle a, X\rangle$ and the other terms of the vector will be zero. This will accomplish what is needed in the problem. ■

10.5 Let $X = (X_1, X_2, \dots, X_n)$ denote an n-dimensional random vector with independent components such that $X_i \sim N\left(\mu, \sigma^2\right)$ for every $i = 1, \dots, n$. Define

$$\overline{X}_n = \frac{1}{n}\sum_1^n X_i,$$

(a) Give the law of $\overline{X}_n$

(b) Let $a_1, \dots, a_n$ in $\mathbb{R}$. Give a necessary and sufficient condition (in terms of $a_1, \dots, a_n$) such that $\overline{X}_n$ and $a_1X_1 + \cdots + a_nX_n$ are independent.

(c) Deduce that the vector $(X_1 - \overline{X}_n, X_2 - \overline{X}_n, \ldots, X_n - \overline{X}_n)$ is independent of $\overline{X}_n$.

Solution: As a sum of independent normal random variables, $\bar{X}_n$ is a normal random variable. Its parameters can be easily calculated as mean μ and variance $\frac{\sigma^2}{n}$. So

$$\bar{X}_n \sim N\left(\mu, \frac{\sigma^2}{n}\right).$$

Note that the vector

$$(\bar{X}_n, a_1X_1 + \cdots + a_nX_n)$$

is a Gaussian random vector. Indeed, every linear combination of its components is a linear combination of the components of X, so it is a Gaussian random variable. Therefore, $\bar{X}_n$ and $a_1X_1 + \cdots + a_nX_n$ are independent if and only if they are uncorrelated; and after calculating the covariance, this is equivalent to

$$a_1 + \cdots + a_n = 0.$$

Hint for part (c): W_n is invariant by translation: $W_n(X) = W_n(X + a)$ if $X + a = (X_1 + a, \ldots, X_n + a)$ for $a \in \mathbb{R}$. Consider also Proposition 10.24. ■

10.6 Let (X, Y) be a Gaussian vector with mean 0 and covariance matrix

$$\Gamma = \begin{pmatrix} 1 & \rho \\ \rho & 1 \end{pmatrix}$$

with $\rho \in [-1, 1]$. What can be said about the random variables

$$X \quad \text{and} \quad Z = Y - \rho X?$$

Solution: Clearly, (X, Z) is a Gaussian vector as a linear transformation of a Gaussian vector. Since

$$Cov(X, Z) = \mathbf{E}XY - \rho\mathbf{E}X^2 = \rho - \rho = 0,$$

we note that the r.v.'s X and Z are independent. ■

10.7 Suppose $X \sim N(0, 1)$. Prove that for every $x > 0$ the following inequalities hold:

$$\frac{1}{\sqrt{2\pi}}e^{-\frac{x^2}{2}}\left(\frac{1}{x} - \frac{1}{x^3}\right) \leq \mathbf{P}(X \geq x) \leq \frac{1}{\sqrt{2\pi}}e^{-\frac{x^2}{2}}\frac{1}{x}.$$

Solution: Consider the following functions defined on $(0, \infty)$:

$$G_1(x) = \frac{1}{\sqrt{2\pi}} e^{-\frac{x^2}{2}} \left(\frac{1}{x} - \frac{1}{x^3} \right)$$

and

$$G_2(x) = \frac{1}{\sqrt{2\pi}} e^{-\frac{x^2}{2}} \frac{1}{x}.$$

We need to show that

$$G_1(x) < 1 - F(x) < G_2(x)$$

for every $x > 0$, where F is the c.d.f. of an $N(0, 1)$ distributed random variable.

Since the normal c.d.f. does not have a closed form, we look at the derivatives of these functions. We need to check that for $x > 0$:

$$G_1'(x) > -f(x) > G_2'(x),$$

where $f = \frac{1}{\sqrt{2\pi}} e^{-x^2/2}$ is the standard normal density. Therefore, integrating the respective positive functions on $(0, \infty)$, we obtain

$$G_1(x) - (1 - F(x)) < \lim_{x \to \infty} G_1(x) - (1 - F(x)) = 0$$

and

$$(1 - F(x)) - G_2(x) < \lim_{x \to \infty} (F(x) - G_2(x)) = 0.$$

■

Problems without Solution

10.8 Suppose

$$(X, Y) \sim N(0, I_2).$$

Show that XY has the same law as $\frac{1}{2}(X^2 - Y^2)$.

Hint: Use the polarization formula

$$XY = \frac{1}{4} \left((X + Y)^2 - (X - Y)^2 \right).$$

10.9 Prove the expression of the density function of the noncentral chi square distribution (10.5).

10.10 Let (X, Y) be a two-dimensional Gaussian vector with zero expectation and I_2 covariance matrix. Compute

$$\mathbf{E}\left[\max(X, Y)\right].$$

10.11 Let X, Y be two independent $N(0, 1)$ distributed random variables. Define

$$U = X^2 + Y^2 \quad \text{and} \quad Y = \frac{X^2}{U}.$$

(a) Prove that U and V are independent.
(b) Show that $U \sim Exp(\frac{1}{2})$.
(c) Show that the function

$$f(x) = \frac{1}{\pi} \frac{1}{\sqrt{x(1-x)}} \mathbf{1}_{(0,1)}(x)$$

defines a probability density.
(d) Show that V admits as density the function f above (this distribution is called the arcsin distribution).

10.12 Let $X_1, \ldots, X_n$ be i.i.d. $N(0, 1)$ random variables. Define

$$U = \left| \frac{1}{n} \sum_{i=1}^{n} X_i \right| \quad \text{and} \quad V = \frac{1}{n} \sum_{i=1}^{n} |X_i|.$$

Compare and calculate $\mathbf{E}U$ and $\mathbf{E}V$.

10.13 10.13 Let $(X_1, X_2) \sim \mathcal{N}_2(0, \Sigma)$ with

$$\Sigma = \begin{pmatrix} \sigma_1^2 & \rho\sigma_1\sigma_2 \\ \rho\sigma_1\sigma_2 & \sigma_2^2 \end{pmatrix} \quad \text{and} \quad \rho \in [-1, 1].$$

(a) Let (Y_1, Y_2) a standard Gaussian vector (i.e., Y_1, Y_2 are independent standard Gaussian random variables). Find a function

$$(Y_1, Y_2) \mapsto (X_1, X_2) = (aY_1, bY_1 + cY_2)$$

such that $(X_1, X_2) \sim \mathcal{N}_2(0, \Sigma)$.
(b) Let $\Theta \sim \mathcal{U}{\Leftarrow}(0, 2\pi])$ and $R^2 \sim Exp(\frac{1}{2})$ two independent random variables with the respective distributions. Let $R \geq 0$ be the square root of R^2.
Prove that the random variables $X = R\cos(\Theta)$ and $Y = R\sin(\Theta)$ are standard Gaussian and independent.

(c) Deduce that if U_1, U_2 are independent $\mathcal{U}([0, 1])$, then the r.v.s

$$X = \sqrt{-2 \ln U_1} \, \cos(2\pi U_2)$$

and

$$Y = \sqrt{-2 \ln U_1} \, \sin(2\pi U_2)$$

are standard Gaussian and independent.

(d) Consider U_1, U_2 independent with law $\mathcal{U}((0, 1])$. Construct from U_1, U_2 a vector (X_1, X_2) with law $\mathcal{N}_2(0, \Sigma)$.

10.14 Suppose $X \sim N(0, \Lambda)$ where

$$\Lambda_Y = \begin{pmatrix} \frac{7}{2} & \frac{1}{2} & -1 \\ \frac{1}{2} & \frac{1}{2} & 0 \\ -1 & 0 & \frac{1}{2} \end{pmatrix}.$$

Let

$$Y_1 = X_2 + X_3,$$
$$Y_2 = X_1 + X_3,$$
$$Y_3 = X_1 + X_2.$$

(a) Give the law of Y_i, $i = 1, 2, 3$.
(b) Write down the density of the vector

$$Y = (Y_1, Y_2, Y_3).$$

10.15 Let $X = (X_1, X_2, X_3)$ be a Gaussian vector with law $N_3(m, C)$ with density

$$f(x) = \frac{1}{\sqrt{120\pi^3}} e^{P(x)}$$

where

$$P(x) = -\frac{x_1^2}{6} - \frac{7x_2^2}{15} - \frac{3x_3^2}{10} - \frac{x_1 x_2}{3} + \frac{2x_2 x_3}{5} + \frac{x_1}{3} + \frac{2x_2}{15} + \frac{4x_3}{5} - \frac{13}{15}.$$

(a) Calculate m and C^{-1}.
(b) Calculate C and the marginal distributions of X_1, X_2, X_3.
(c) Give the law of

$$Y = \begin{pmatrix} 1 & -3 & 0 \\ 4 & 2 & 1 \end{pmatrix} X + \begin{pmatrix} 0 \\ 1 \end{pmatrix}.$$

10.16 Let (X, Y) be a normal random vector such that X and Y are standard normal $N(0, 1)$. Suppose that $Cov(X, Y) = \rho$. Let $\theta \in \mathbb{R}$ and put

$$U = X\cos\theta - Y\sin\theta, \qquad V = X\sin\theta + Y\cos\theta.$$

(a) Show that $|\rho| \leq 1$.
(b) Calculate $\mathbf{E}(U)$, $\mathbf{E}(V)$, $Var(U)$, $Var(V)$, and $Cov(U, V)$. What can we say about the vector (U, V) ?

Suppose $\rho \neq 0$. Do there exist values of θ such that U and V are independent?

(c) Assume $\rho = 0$. Give the laws of U and V?

(d) Are the r.v. U and V independent?

10.17 Let (X, Y, Z) be a Gaussian vector with mean $(1, 2, 3)$ and covariance matrix

$$\Gamma = \begin{pmatrix} 5 & -1 & -5 \\ -1 & 1 & 0 \\ -5 & 0 & 10 \end{pmatrix}.$$

Set

$$U = X - 3Y \quad \text{and} \quad V = X + 2Y.$$

(a) Give the law of the couple (U, V). What can be said about U and V? Write *without calculation* the density of (U, V).

(b) Find constants c and d such that the r.v. $W = Z + cU + dV$ is independent by (U, V).

(c) Write the covariance matrix of the vector (U, V, W).

10.18 If X is a standard normal random variable $N(0, 1)$, let φ denote its characteristic function and F its c.d.f. For every $p \geq 1$ integer, denote the p moment with

$$C(p) = \mathbf{E}(|X|^p).$$

Let $(Y_n\,,\ n \geq 1)$ be a sequence of independent r.v. with identical distribution $N(0, 1)$. For every $k \geq 1$ and $n \geq 1$ integers, let

$$X_k = \sum_{j=1}^{k} Y_j \quad \text{and} \quad S_n = \sum_{k=1}^{n} X_k.$$

(a) Give the distribution of X_k.

(b) Calculate $Cov\,(X_k, X_{k+1})$. Are the variables $(X_k,\ k \geq 1)$ independent?

(c) Show that $S_n = \sum_{k=1}^{n} (n + 1 - k)\, Y_k$. Deduce that for every integer $n \geq 1$ the r.v. S_n follows the law

$$N\left(0, \frac{1}{6} n\,(n + 1)\,(2n + 1)\right).$$

Hint: We recall the formula

$$\sum_{k=1}^{n} k^2 = \frac{n\,(n + 1)\,(2n + 1)}{6}$$

(d) For every integer $p \geq 1$, calculate $\mathbf{E}(|S_n|^p)$ in terms of $C\,(p)$.

(**e**) Let $\alpha > \frac{3}{2}$. Show that the sequence $(n^{-\alpha} S_n\,,\ n \geq 1)$ converges to 0 in L^2, that is,

$$\mathbf{E}\left[\left(\frac{S_n}{n^\alpha}\right)^2\right] \to 0,$$

as a sequence of numbers.

(**f**) Show that for every $\beta > 0$ and for every $p \geq 1$ integer, there exists a constant $K_p > 0$ such that

$$\mathbf{P}(n^{-\alpha}\,|S_n| \geq n^{-\beta}) \leq K_p\, n^{-p(\alpha-\frac{3}{2}-\beta)}.$$

(**g**) Show that the sequence $(n^{-\alpha} S_n\,,\ n \geq 1)$ converges almost surely to an r.v. and identify this limit.

(**h**) Calculate the characteristic function φ_{S_n} of S_n using φ.

(**i**) Let

$$T_n = n^{-\frac{3}{2}}\, S_n.$$

Deduce the expression of the characteristic function of T_n.

(**j**) Show that the sequence $(T_n, n \geq 1)$ converges in law to a limit and identify this limit.

10.19 Let X_1, X_2, and X_3 be three i.i.d. random variables where their distribution has zero mean and variance $\sigma^2 > 0$. Denote

$$Y_1 = X_1 - 2X_2 + 3X_3,$$
$$Y_2 = X_1 - X_2.$$

(**a**) Calculate the covariance matrix of the vector (X_1, Y_1, Y_2).

(**b**) Give an upper bound for

$$a = \mathbf{P}(|Y_1| \geq 10\,\sigma)$$

using Bienaymé–Tchebychev inequality in Appendix B.

(**c**) Suppose that X_1, X_2, and X_3 are Gaussian. In this case, give another upper bound for a and compare with the previous question.

(**d**) Give an upper bound for

$$b = \mathbf{P}(Y_1 < Y_2 + \sigma).$$

Hint: Use $\frac{10}{\sqrt{14}} \sim 2.67$ *and* $\frac{1}{\sqrt{10}} \sim 0.32$.

10.20 If $X \sim N(0, 1)$ and $Y \sim \chi^2(n, 1)$ and X, Y are independent, show that

$$Z = \frac{X}{\sqrt{\frac{Y}{n}}}$$

has a Student t distribution (t_n) with n degrees of freedom. Specifically, show that the probability density function of Z is given by

$$f(x) = \frac{\Gamma(\frac{n+1}{2})}{\sqrt{n\pi}\Gamma(\frac{n}{2})}\left(1+\frac{x^2}{n}\right)^{-\frac{n+1}{2}}.$$

10.21 Assume X and Y are two independent $N(0, 1)$ random variables. Find the law of

$$W = \frac{X+Y}{|X-Y|}.$$

Hint: We know (how to prove) that $X+Y$ and $X-Y$ are independent and each has $N(0, 2)$ distribution. Then, once we show that

$$Z = \frac{(X-Y)^2}{2}$$

is $\chi^2(1)$ distributed, we will obtain

$$W = \frac{\frac{X+Y}{\sqrt{2}}}{\sqrt{Z}},$$

which follows a Student distribution with one degree of freedom (see exercise 10.20).

10.22 Consider the matrix

$$A = \begin{pmatrix} 6 & 1 \\ 1 & 2. \end{pmatrix}.$$

(a) Find the eigenvalues of A.

(b) Find the eigenspaces associated with each eigenvalue.

(c) Diagonalize the matrix A.

(d) Let X be a Gaussian random vector with covariance matrix A and zero mean. Find a linear transformation that transforms X in a Gaussian vector with independent components.

10.23 Consider the matrix

$$A = \begin{pmatrix} 0 & 2 & 2 \\ 2 & 3 & 4 \\ 2 & 4 & 3 \end{pmatrix}.$$

(a) Check that $\lambda_1 = 8$ is an eigenvalue of A.

(b) Find the other eigenvalues of A.

(c) Find the eigenspaces associated with each eigenvalue.

(**d**) Diagonalize the matrix A.

(**e**) Can there exist a Gaussian random vector with covariance matrix A?

10.24 Let the matrix Σ be defined as

$$\Sigma = \begin{pmatrix} 2 & 2 & 3 \\ 2 & 4 & 4 \\ 3 & 4 & 9 \end{pmatrix}.$$

Check that the matrix is symmetric and positive definite. Find its eigenvalues and the associated eigenvectors. Now, let X be a Gaussian random vector with covariance matrix Σ and zero mean. Find a linear transformation that transforms X in a Gaussian vector with independent components.

CHAPTER ELEVEN

Convergence Types. Almost Sure Convergence. L^p-Convergence. Convergence in Probability

11.1 Introduction/Purpose of the Chapter

In probability theory, there exist several different notions of convergence of random variables. The convergence of sequences of random variables to some limit random variable is an important concept in probability theory, and it is a very important application to statistics and stochastic processes. The same concepts are known in more general mathematics as stochastic convergence, and they formalize the idea that a sequence of essentially random or unpredictable events can sometimes be expected to settle down into a behavior that is essentially unchanging when items far enough into the sequence are studied. The different notions of convergence relate to how such a behavior can be characterized. We can talk about a sequence that approaches a random variable exactly, with probability one and looking at the moments of the distribution. In this chapter we will talk about a notion of convergence defined purely by the distribution of random variables.

Handbook of Probability, First Edition. Ionuţ Florescu and Ciprian Tudor.

11.2 Vignette/Historical Notes

In their development of the calculus both Newton and Leibniz used "infinitesimals," quantities that are infinitely small and yet nonzero. They found it convenient to use these quantities in their computations and their derivations of results. Cauchy, Weierstrass, and Riemann reformulated Calculus in terms of limits rather than infinitesimals. Thus the need for these infinitely small (and nonexistent) quantities was removed and was replaced by a notion of quantities being "close" to others. The derivative and the integral were both reformulated in terms of limits of functions. In probability theory the notion of convergence is fundamental, and in fact the entire construction of the probability theory is based on limits. Historically, measure theory gave birth of the notions used in probability theory. However, because the space is finite (has probability one), the theorems are much more precise and more results may be generated for this simpler case. Without the convergence results in probability, there would be no statistics and the problem of estimating distributions and parameters by observing real-life data would be impossible to solve.

11.3 Theory and Applications: Types of Convergence

Let $(\Omega, \mathcal{F}, \mathbf{P})$ be a probability space and let $X_n : \Omega \to \mathbb{R}$ be a sequence of random variables and $X : \Omega \to \mathbb{R}$ be a target random variable. Throughout this chapter we shall take $n \in \mathbb{N}$. In the theory of the stochastic processes in continuous time the index may be $t \in (0, \infty)$. The notations and notions introduced here extend to that case as well.

11.3.1 TRADITIONAL DETERMINISTIC CONVERGENCE TYPES

Before we introduce notions taking advantage of the structure of the probability space, we would like to recall the more traditional Real analysis types of convergence. Any random variable is essentially a function from $\Omega \to \mathbb{R}$. The following two definitions are a reminder of these classic types of convergence.

Definition 11.1 (Uniform convergence) *The sequence of random variables $\{X_n\}_{n\in\mathbb{N}}$ is said to converge to X uniformly if and only if:*
for all $\varepsilon > 0$ there exists an $n_\varepsilon \in \mathbb{N}$ such that

$$|X_n(\omega) - X(\omega)| < \varepsilon, \qquad \forall n \geq n_\varepsilon, \forall \omega \in \Omega.$$

Note that in the previous definition the number n_ε is the same for all points $\omega \in \Omega$. Thus, it does not depend on ω; therefore the convergence takes place similarly throughout the entire space. There exists no point in Ω where the sequence is farther than the target at some n. This is why the convergence is called uniform—uniform on the whole Ω. This type is the most powerful convergence mode. Next, we drop the restriction that the convergence is uniform on the space.

Definition 11.2 (Pointwise convergence) *The sequence $\{X_n\}_n$ is said to converge to X pointwise if and only if:*
for any ω fixed, for all $\varepsilon > 0$ there exists $n_\varepsilon(\omega)$ such that

$$|X_n(\omega) - X(\omega)| < \varepsilon, \qquad \forall n \geq n_\varepsilon(\omega).$$

Note that as in the previous type, we have convergence for each point in Ω. However, for some points it may take a larger n to be close enough to the limit than for other points.

Both these definitions are related to the concept of *all the points*—this is in fact the reason why they are called *deterministic notions*. Specifically, if the convergence does not hold for a single point, then the convergence pointwise or uniform does not hold. In other words, one point *determines* the fate of the whole concept.

EXAMPLE 11.1

Let $\Omega = [0, 1]$ and define

$$X_n(\omega) = \left(1 + \frac{\omega}{n}\right)^n.$$

Prove that X_n converges pointwise on Ω as $n \to \infty$ to the r.v. X given by

$$X(\omega) = e^\omega.$$

Solution: This follows easily since for every $\omega \in \Omega = [0, 1]$

$$\lim_{n\to\infty} \left(1 + \frac{\omega}{n}\right)^n = e^\omega.$$

■

EXAMPLE 11.2

Let $\Omega = [0, 1]$ and define

$$X_n(\Omega) = \omega^n, \qquad \forall \omega.$$

Let $X(\omega) = 0$ for every $\omega \in \Omega$. Is the sequence X_n converging pointwise to X?

Solution: Note that for every $\omega \in [0, 1)$ we have

$$w^n \to 0$$

as $n \to \infty$. However, when $\omega = 1$ we obtain $X_n(\omega) = 1$, which converges to 1. Therefore, the sequence X_n does not converge pointwise to X. It converges pointwise to the r.v. Y given by

$$Y(\omega) = \begin{cases} 0, & \text{if } \omega \in [0, 1), \\ 1, & \text{if } \omega = 1. \end{cases}$$

■

11.3.2 CONVERGENCE OF MOMENTS OF AN r.v.—CONVERGENCE IN L^p

Unlike the classic mathematical world, in the real-world things do not happen strictly. Probability theory allows for this, and a statement does not need to hold at all points in the space in order to be true. If the points for which a statement occur very rarely then, in reality, we do not need to care about them.

In Real Analysis there is a notion of convergence which is applicable to Probability theory. That concept is looking at the convergence of integrals of functions instead of the convergence of the functions. The integrals may have a limit even though the functions themselves may not converge at some points. The integral of a random variable with respect to the probability density is called a moment in Probability Theory. If we look at the convergence of moments, we are referring to convergence in L^p.

Definition 11.3 (Convergence in L^p) *Let $p > 0$. The sequence of random variables $\{X_n\}_n$ is said to converge to X in L^p ($X_n \xrightarrow{L^p} X$ or $X_n \to X$ in L^p) if and only if*

$$\lim_{n\to\infty} \mathbf{E}|X_n - X|^p = 0.$$

The particular case when $p = 2$ is special. This case is known by several names such as convergence in L^2, convergence in the quadratic mean (or, simply quadratic convergence), convergence in mean squared, and so on.

Remark 11.4 *The concept of mean-square convergence (or convergence in mean-square) is based on the following intuition: Two random variables are "close to each other" if the square of their difference is, on average, small.*

Any L^p space is a complete normed vector space. This is interesting from the Real Analysis perspective. In probability theory the L^p-norm of a random variable

X defined on $(\Omega, \mathcal{F}, \mathbf{P})$ is

$$\|X\|_p = \left(\mathbf{E}\left[|X|^p\right]\right)^{\frac{1}{p}} = \left(\int_\Omega x^p dF(x)\right)^{\frac{1}{p}} = \left(\int_\Omega x^p f(x)\, dx\right)^{\frac{1}{p}},$$

with the last equality valid if the random variable X has a density $f(x)$. For our purposes the following result is important:

Proposition 11.5 *Let X denote a random variable on $(\Omega, \mathcal{F}, \mathbf{P})$. Then the sequence of norms $\|X\|_p$ is nondecreasing (increasing) in p. In other words, $\|X\|_{p_1} \leq \|X\|_{p_2}$ for any $p_1 < p_2$.*

This immediately implies that if a variable is in L^q for some q fixed, then it also is in any L^r with $r \leq q$. Therefore, as spaces of functions we have

$$L^1(\Omega) \supseteq L^2(\Omega) \supseteq L^3(\Omega) \ldots.$$

Proof: Let $p_1 < p_2$. Then the function $f(x) = |x|^{p_2/p_1}$ is convex on $[0, \infty)$. This is true since $p_2/p_1 > 1$, and we can apply Jensen's inequality (Lemma 14.5) for the convex function $f(x)$ and the non-negative r.v. $Y = |X|^{p_1}$. The inequality immediately yields the desired result. ■

Corollary 11.6 *If $X_n \xrightarrow{L^p} X$ and $p \geq q$ then $X_n \xrightarrow{L^q} X$.*

Proof: This follows from the inequality

$$\mathbf{E}\,|X_n - X|^q \leq \mathbf{E}\,|X_n - X|^p$$

(see Lemma 13.11). ■

11.3.3 ALMOST SURE (a.s.) CONVERGENCE

The next two definitions are convergence types on probability spaces. The concepts mimic the definitions of uniform and pointwise convergence. As we shall see, one of the concepts is superfluous in probability spaces (finite total measure spaces).

Definition 11.7 (Almost uniform convergence) *The sequence of random variables $\{X_n\}_n$ is said to converge to X almost uniformly ($X_n \xrightarrow{a.u.} X$ or $X_n \to X$ a.u.) if for all $\varepsilon > 0$ there exists a set $N \in \mathcal{F}$ with $\mathbf{P}(N) < \varepsilon$ and $\{X_n\}$ converges to X uniformly on N^c.*

This concept mimics the uniform convergence for probability spaces. The next concept presented—the almost sure convergence (or a.s. convergence)—is a corresponding type to the concept of pointwise convergence in the whole space. As we have seen, a sequence of random variables $X, X_2, \ldots, X_n, \ldots$ is pointwise convergent if and only if the sequence of real numbers $\{X_n(\omega)\}_n$ is convergent for all $\omega \in \Omega$. Achieving convergence for all the omega points in Omega is a very stringent requirement and does not use the probability at all. This requirement is weakened, by requiring only the convergence for the points which are relevant to the probability measure.

Definition 11.8 (Almost Sure Convergence) *The sequence of random variables* $\{X_n\}_n$ *is said to converge to* X *almost surely* ($X_n \xrightarrow{a.s.} X$ *or* $X_n \to X$ *a.s.) if for all* $\varepsilon > 0$ *there exists a set* $N \in \mathcal{F}$ *with* $\mathbf{P}(N) = 0$ *and* $\{X_n\}$ *converges to* X *pointwise on* N^c*, or written mathematically:*

$$\mathbf{P}\{\omega \in \Omega : |X_n(\omega) - X(\omega)| > \varepsilon, \forall n \geq n_\varepsilon(\omega)\} = \mathbf{P}(N) = 0.$$

An alternative way to write the a.s. convergence is

$$\mathbf{P}\{\omega \in \Omega : \lim_{n \to \infty} X_n(\omega) = X(\omega)\} = \mathbf{P}(N^c) = 1.$$

In other words, the convergence has to happen only for the points $\omega \in N^c$, since this set has probability 1. We do not care about the rest of the points in $N = \Omega \setminus N^c$ since the probability of this set is zero.

Remark 11.9 *It turns out that almost uniform and almost sure convergence, despite their apparent different forms, are completely equivalent on a finite measure space (probability space has total measure* 1*, thus finite). This is the reason why most books and papers never even mention almost uniform convergence. The following proposition (due to Egorov) has this result.*

Proposition 11.10 (Egorov Theorem) *If the space* Ω *has a finite measure, then the sequence* $\{X_n\}_n$ *converges almost uniformly to* X *if and only if it converges almost surely to* X.

The proof of the theorem uses measure theory, and it is not all that illustrative for our purpose. The interested reader is directed to Theorem 10.13 in Wheeden and Zygmund (1977).

11.3.3.1 Complete Probability Space. Here is the time to talk about a technical aspect of probability spaces.

It is possible to construct a sequence of random variables X_n which has a limit in the a.s. sense; however, the limiting variable is not a random variable itself. More specifically, it is possible that the limit is not $\mathcal{B}(\mathbb{R})$-measurable anymore. The main issue lies in the following problem. Suppose that a set A has probability 0. Then you would think that any set B included in A also has probability 0. This in fact would be true if any $B \subset A$ is in the σ-algebra $\mathcal{F}$. If it is not, then we cannot calculate its probability.

To avoid this technical problem, we assume that the probability space is complete (as defined next). And as a consequence, if it exists, the limit of random variables (in any sense) will always be a random variable.

Throughout this book we will always assume that the probability space we work with is complete.

Definition 11.11 (Complete probability space) *We say that the probability space $(\Omega, \mathcal{F}, \mathbf{P})$ is complete if any subset of a probability zero set in $\mathcal{F}$ is also in $\mathcal{F}$. Mathematically: If $N \in \mathcal{F}$ with $\mathbf{P}(N) = 0$, then for all $M \subset N$ we have $M \in \mathcal{F}$.*

The issue arises from the definition of a σ-algebra which does not require the inclusion of all subsets only of all the unions of sets already existing in the collection.

We can easily "complete" any probability space $(\Omega, \mathcal{F}, \mathbf{P})$ by simply adding to its σ-algebra all the sets included in sets of probability zero and setting the probability of any one of them equal to zero.

11.3.4 CONVERGENCE IN PROBABILITY

As we mentionned, different concepts of convergence are based on different ways of measuring the distance between two random variables (how "close to each other" two random variables are). Let us go now to the concept of convergence, in probability. In the definition of a.s. convergence, the limit was inside the probability. Suppose we want to take the limit out of the probability and instead look at the measure of N as $n \to \infty$. This means that instead of looking at N, we need to construct a specific set for every n. But then, the set also needs to depend on some small ε. Therefore we look at

$$N_n(\varepsilon) = \{\omega \in \Omega \mid |X_n(\omega) - X(\omega)| > \varepsilon\},$$

which expressed in words is: For some epsilon, this set is made of the points ω for which the sequence is far from the target. If we require that the probability of this set of points converge to zero for all ε, we obtain the definition of convergence in probability.

Definition 11.12 (Convergence in probability) *The sequence of random variables $\{X_n\}$ is said to converge to X in probability ($X_n \xrightarrow{P} X$ or $X_n \to X$ in probability) if and only if for all fixed $\varepsilon > 0$ the sets $N_\varepsilon(n) = \{\omega : |X_n(\omega) - X(\omega)| > \varepsilon\}$ have the property $\mathbf{P}(N_\varepsilon(n)) \to 0$ as $n \to \infty$, or*

$$\lim_{n\to\infty} \mathbf{P}\{\omega \in \Omega : |X_n(\omega) - X(\omega)| > \varepsilon\} = 0.$$

Please compare the definition of convergence in probability with the convergence a.s.. The only difference is the location of the limit (and accordingly n). However, that location makes a world of difference—the a.s. convergence is much more powerful than the convergence in probability (as we shall see next).

EXAMPLE 11.3

Let X be a Bernoulli(1/3) random variable, that is,

$$\mathbf{P}(X = 0) = \frac{2}{3} \quad \text{and} \quad \mathbf{P}(X = 1) = \frac{1}{3}.$$

For every $n \geq 1$, we define

$$X_n = \left(1 + \frac{1}{n}\right) X.$$

Then the sequence $(X_n)_{n \geq 1}$ converges in probability to the r.v. X.

Solution: Note that

$$|X_n - X| = \left|\frac{1}{n} X\right| = \frac{1}{n} X$$

because X is non-negative. Let $\varepsilon > 0$ fixed. Then

$$\begin{aligned}
\mathbf{P}\left(|X_n - X| > \varepsilon\right) &= \mathbf{P}\left(|X_n - X| > \varepsilon, X = 0\right) \\
&\quad + \mathbf{P}\left(|X_n - X| > \varepsilon, X = 1\right) \\
&= \mathbf{P}\left(\frac{1}{n} X > \varepsilon, X = 0\right) \\
&\quad + \mathbf{P}\left(\frac{1}{n} X > \varepsilon, X = 1\right).
\end{aligned}$$

Therefore, we get

$$\mathbf{P}\left(|X_n - X| > \varepsilon\right) = \frac{1}{3} \mathbf{1}_{\{n < \frac{1}{\varepsilon}\}},$$

which clearly converges to zero as $n \to \infty$. ■

Definition 11.13 (Bounded in probability, big $O_p(1)$ notation) *A sequence of random variables $(X_n)_{n \geq 1}$ is called bounded in probability if for every $\varepsilon > 0$ there exists an $M > 0$ such that*

$$\mathbf{P}\left(|X_n| < M\right) > 1 - \varepsilon$$

for every $n \geq 1$. We denote $O_p(1)$ a sequence bounded in probability. Similarly, if there exists a set of constants $a_n > 0$ such that for every every $\varepsilon > 0$ there exists an $M > 0$, and

$$\mathbf{P}\left(\left|\frac{X_n}{a_n}\right| < M\right) > 1 - \varepsilon,$$

then we say that X_n is bounded in probability by a_n and we denote $O_p(a_n)$.

Definition 11.14 (Convergent in probability, little $o_p(1)$ notation) *For a set of random variables $(X_n)_{n\geq 1}$ we use the notation $X_n = o_p(1)$ if for every positive ε we have*

$$\lim_{n\to\infty} \mathbf{P}(|X_n| > \varepsilon) = 0,$$

or equivalently $X_n \to 0$ in probability. Similar to the notation in the definition above, we denote $X_n = o_p(a_n)$ a random sequence such that $X_n/a_n \to 0$ in probability.

Remark 11.15 *The notions above are important to express the Central Limit Theorem (next chapter) in a concise manner. We shall see there that for a sequence $(X_n)_{n\geq 1}$ of i.i.d. random variables with mean μ we will have:*

$$\sqrt{n}(\bar{X}_n - \mu) = o_p(1),$$

where $\bar{X}_n = \frac{1}{n}\sum_{i=1}^{n} X_i$ is the average of the first n random variables.

Furthermore, if $X_n \to X$ in probability, we may write $|X_n - X| = o_p(1)$. And of course an o_p sequence is also O_p; that is, a sequence that converges in probability is bounded in probability (see exercise 11.12).

Remark 11.16 *Normally, in order to check the convergence in probability of X_n to X, we need to know the joint distribution of X_n and X. However, in the case when X is a constant, this is not needed (obviously). In fact, $X_n \to_n c$ if for every $\varepsilon > 0$ we have*

$$\mathbf{P}(|X_n - c| < \varepsilon) \to_{n\to\infty} 1.$$

Note that the probability only contains the law of X_n.

11.4 Relationships Between Types of Convergence

First the deterministic convergence types imply all of the probability types of convergence with the exception of the L^p convergence. That is natural since the expectation has first to be defined regardless of the deterministic convergence. As we saw already, almost uniform convergence and almost sure convergence are equivalent for finite measure spaces (thus in particular for probability spaces). We will not mention a.u. convergence from now on.

The relations between various types of convergence are depicted in Figure 11.1. The solid arrows denote that one type of convergence implies the other type. The dashed arrows imply the existence of a subsequence that is convergent in the stronger type.

In this section we prove the relations between these types of convergence.

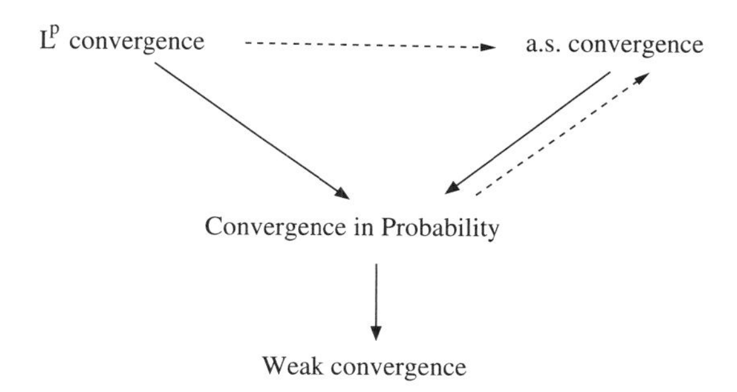

FIGURE 11.1 **Relations between different types of convergence. Solid arrows mean immediate implication. Dashed arrows mean that there exists a subsequence which is convergent in the stronger type.**

11.4.1 a.s. AND L^p

In general, L^p convergence and almost sure (a.s.) convergence are not related. The following example shows that a.s. convergence does not necessarily imply L^p convergence.

EXAMPLE 11.4

Let $(\Omega, \mathcal{F}, \mathbf{P}) = ((0, 1), \mathcal{B}((0, 1)), \lambda)$, where λ denotes the Lebesque measure. On this space, define the variables X_n as

$$X_n(\omega) = n\mathbf{1}_{(\frac{1}{n}, \frac{2}{n})}(\omega).$$

Then the variables constructed in this way have the property that $X_n \xrightarrow{a.s.} 0$ as $n \to \infty$ but $\mathbf{E}(X_n) = 1$ for all n. Therefore, $X_n \not\to X$ in L^1, thus it does not converge in any L^p with $p \geq 1$.

Solution: To show the a.s. convergence, let us start with some $\varepsilon > 0$ arbitrary. Then, for any number $\omega \in (0, 1)$ there exists a $n_\varepsilon(\omega)$ so that $\omega \notin (\frac{1}{n}, \frac{2}{n})$, $\forall n \geq n_\varepsilon(\omega)$. For example, take

$$n_\varepsilon(\omega) = \left[\frac{2}{\omega}\right],$$

where $[x]$ denotes the integer part of the real number x. Since $\omega \notin (\frac{1}{n}, \frac{2}{n})$, $\forall n \geq n_\varepsilon(\omega)$, we have that $X_n = 0$ and therefore $X_n(\omega) \longrightarrow 0$. Note that the convergence is, in fact, pointwise not only almost sure.

On the other hand,

$$\mathbf{E}[X_n] = \int_0^1 n\mathbf{1}_{(\frac{1}{n}, \frac{2}{n})}(\omega)\, d\lambda(\omega) = n\lambda\left(\frac{1}{n}, \frac{2}{n}\right) = 1, \qquad \forall n.$$

Thus, the limit of the expectations is also 1, which is different from $\mathbf{E}[\lim X_n] = \mathbf{E}[0] = 0$. Therefore, the sequence converges a.s. but not in L^1. ■

Remark 11.17 *The previous example may be read the other way as well to provide a counterexample to L^p does not imply a.s. convergence as well. Specifically, the sequence converges in L^1 to 0, but that does not imply that it converges a.s. to 0.*

So, are there conditions which being true would guarantee that a sequence converging almost surely would imply convergence in L^p to the same limit?

The answer is yes, and in fact we already know them. These conditions are the same conditions in the hypotheses of the Dominated Convergence Theorem (Theorem 13.7) and the Monotone Convergence Theorem (Theorem 13.4) in Appendix A (Chapter 13).

In fact, we have even more: If the sequence is dominated by an integrable random variable, then the convergence in probability (which is weaker than convergence a.s.) will imply convergence in L^p.

Theorem 11.18 *Suppose that $X_n \xrightarrow{P} X$ and assume there exists a random variable $Y \in L^p(\Omega)$ such that $|X_n| < Y$ a.s. for all n. Then $X_n \xrightarrow{L^p} X$.*

Proof: Since X_n converges in probability, by Theorem 11.19—which will be proven later in this chapter and does not use anything from this theorem—there exists a subsequence $(X_{n_k})_k$ that converges almost surely to the r.v. X as $k \to \infty$:

$$X_{n_k} \to_{k\to\infty} X, \qquad \text{a.s.} \tag{11.1}$$

Since X_n is dominated, we also have

$$\sup_n \mathbf{E}|X_n| \leq \mathbf{E}|Y|,$$

and from here it follows easily that

$$X \in L^1(\Omega) \quad \text{and} \quad |X| \leq Y.$$

Now, for a fixed $\epsilon > 0$

$$\begin{aligned}
\mathbf{E}\,|X_n - X| &= \int_{(|X_n - X| \leq \varepsilon)} |X_n - X|\, d\mathbf{P} + \int_{(|X_n - X| > \varepsilon)} |X_n - X|\, d\mathbf{P} \\
&\leq \varepsilon + \int_{(|X_n - X| > \varepsilon)} |X_n - X|\, d\mathbf{P} \\
&\leq \varepsilon + \int_{(|X_n - X| > \varepsilon)} 2Y d\mathbf{P}
\end{aligned}$$

in the least inequality we used

$$\sup_n |X_n - X| \leq 2Y.$$

Now let us pick an arbitrary $c > 0$. We can write

$$\mathbf{E}\,|X_n - X| \le \varepsilon + \int_{(|X_n - X|>\varepsilon)\cap(|Y|\le c)} 2Yd\mathbf{P} + \int_{(|X_n - X|>\varepsilon)\cap(|Y|>c)} 2Yd\mathbf{P}$$

$$\le \varepsilon + 2c\mathbf{P}(|X_n - X| > \epsilon) + 2\int_{\{|Y|\ge c\}} Yd\mathbf{P}.$$

Now look at the two final terms above. The sequence

$$\mathbf{1}_{\{|Y|\ge c\}}|Y|$$

converges to zero as $c \to \infty$ and since

$$\mathbf{E1}_{\{|Y|\ge c\}}|Y| \le \mathbf{E}Y < \infty$$

we apply the dominated convergence theorem and we obtain

$$\int_{\{|Y|\ge c\}} Yd\mathbf{P} \le \varepsilon$$

for a c large enough. The other term can be made as small as possible by the definition of convergence in probability, and therefore for any c there exists an n large enough so that

$$2c\mathbf{P}(|X_n - X| > \epsilon) \le \varepsilon.$$

Therefore the conclusion of the theorem follows. ■

The next example shows that a sequence converging in L^p does not necessarily converge almost surely. In fact, in this example the sequence also converges in probability beside converging in L^p. However, not even both types of convergence are enough for almost sure convergence.

EXAMPLE 11.5 A sequence converging in L^p but not a.s.

Let $(\Omega, \mathcal{F}, \mathbf{P}) = ((0, 1), \mathcal{B}((0, 1)), \lambda)$, where λ denotes the Lebesque measure. Define the following sequence:

$$X_1(\omega) = \omega + \mathbf{1}_{(0,1)}(\omega)$$
$$X_{2,1}(\omega) = \omega + \mathbf{1}_{(0,\frac{1}{2})}(\omega); X_{2,2}(\omega) = \omega + \mathbf{1}_{(\frac{1}{2},1)}(\omega)$$
$$X_{3,1}(\omega) = \omega + \mathbf{1}_{(0,\frac{1}{3})}(\omega); X_{3,2}(\omega) = \omega + \mathbf{1}_{(\frac{1}{3},\frac{2}{3})}(\omega); X_{3,3}(\omega) = \omega + \mathbf{1}_{(\frac{2}{3},1)}(\omega);$$
$$\vdots$$

and finally we can just re-index the sequence using integers as in $Y_1 = X_1$, $Y_2 = X_{2,1}$, $Y_3 = X_{2,2}$, $Y_4 = X_{3,1}$, and so on. We shall keep the notation using the X's since it is easier to follow.

To understand what is going on, plot a few of the random variables $X_{i,j}(\omega)$ for $\omega \in (0, 1)$.

Next, take $X(\omega) = \omega$ (the candidate limit variable). Then,

$$\mathbf{P}\{\omega : |X_{n,k}(\omega) - X(\omega)| > \varepsilon\} = \mathbf{P}\left\{\omega \text{ is in some interval of length } \frac{1}{n}\right\} = \frac{1}{n}$$

Therefore,

$$\lim_{n\to\infty} \mathbf{P}\{\omega : |X_{n,k}(\omega) - X(\omega)| > \varepsilon\} = 0,$$

and thus $X_n \xrightarrow{P} X$. Furthermore, for any k we have

$$\mathbf{E}[|X_{n,k} - X|^p] = \mathbf{E}\left[\left(\mathbf{1}_{\left(\frac{k}{n}, \frac{k+1}{n}\right)}\right)^p\right] = \mathbf{E}\left[\mathbf{1}_{\left(\frac{k}{n}, \frac{k+1}{n}\right)}\right] = 1/n,$$

and again taking $n \to \infty$ this shows that $X_n \xrightarrow{L^p} X$. Therefore, X_n converge in both probability and L^p for any p.

However, $X_{n,k} \not\to X$ almost surely. To understand why, just take any ω and any n. Due to the way we constructed the sequence, there exists an $m > n$ and a k such that

$$X_{m,k}(\omega) - X = 1$$

for that particular pair. The sequence at any ω alternates between many ω and a $\omega + 1$.

EXAMPLE 11.6

Let $(X_n)_{n\geq 1}$ be a sequence of independent random variables, each uniformly distributed on $(-\frac{1}{n}, \frac{1}{n})$, or

$$X_n \sim U\left(-\frac{1}{n}, \frac{1}{n}\right).$$

Find the limit in probability, if it exists, of the sequence X_n.

Solution: Let us write the density of the random variable X_n:

$$f_{X_n}(x) = \frac{n}{2} \qquad \text{if } x \in \left(-\frac{1}{n}, \frac{1}{n}\right)$$

and

$$f_{X_n}(x) = 0 \qquad \text{otherwise.}$$

Since this density seems to be concentrating around $x = 0$ as $n \to \infty$ it seems reasonable to think that X_n converges to zero in probability. Let us try and prove this claim using the definition. Let $\varepsilon > 0$. Then

$$\begin{aligned}
\mathbf{P}((|X_n - X| \geq \varepsilon) &= \mathbf{P}(|X_n| \geq \varepsilon) \\
&= 1 - \mathbf{P}(-\varepsilon \leq X_n \leq \varepsilon) \\
&= 1 - \int_{-\varepsilon}^{\varepsilon} f_{X_n}(x)\, dx \\
&= 1 - \int_{-\varepsilon \vee -\frac{1}{n}}^{\varepsilon \wedge \frac{1}{n}} \frac{n}{2}\, dx,
\end{aligned}$$

where we used the notation $-\varepsilon \vee -\frac{1}{n} = \max(-\varepsilon, -\frac{1}{n})$, and the similar one for the minimum. Since for any ϵ for n big we have $\varepsilon > \frac{1}{n}$, we can write

$$\begin{aligned}
\lim_n \mathbf{P}((|X_n - X| \geq \varepsilon) &= 1 - \lim_n \int_{-\frac{1}{n}}^{\frac{1}{n}} \frac{n}{2}\, dx \\
&= 1 - 1 = 0,
\end{aligned}$$

which shows that the sequence converges in probability to 0. ■

11.4.2 PROBABILITY AND a.s./L^p

Recall Example 11.5. Over there we presented a sequence that converged in probability but not almost surely. The next theorem says that even though the convergence in probability does not imply convergence a.s., we can always extract a subsequence that will converge almost surely to the same limit. Note that, in Example 11.5, if we omit the terms in the sequence with values $\omega + 1$, the resulting subsequence converges almost surely.

Theorem 11.19 (Relation between a.s. convergence and convergence in probability) *We have the following relations:*

1. *If $X_n \xrightarrow{a.s.} X$, then $X_n \xrightarrow{P} X$.*
2. *If $X_n \xrightarrow{P} X$, then there exists a subsequence n_k such that*

$$X_{n_k} \xrightarrow{a.s.} X, \qquad \text{as } k \to \infty.$$

Proof: 1. Let

$$N^c = \{\omega : \lim_{n\to\infty} |X_n(\omega) - X(\omega)| = 0\}.$$

From the definition of almost sure convergence, we know that $\mathbf{P}(N) = 0$. Now take an $\varepsilon > 0$ arbitrary and consider

$$N_\varepsilon(n) = \{\omega : |X_n(\omega) - X(\omega)| \geq \varepsilon\}.$$

Now let

$$M_k = \left(\bigcup_{n\geq k} N_\varepsilon(n)\right)^c = \bigcap_{n\geq k} N_\varepsilon(n)^c \tag{11.2}$$

using De Morgan's law. We have the following:

- The sets M_k are increasing since

  $$M_k = N_\varepsilon(k)^c \cap M_{k+1},$$

  which implies $M_k \subseteq M_{k+1}$.
- If we take an $\omega \in M_k$ and we look at the way it is defined, we have that for all $n \geq k$ ω is in $N_\varepsilon(n)^c$, or in other words:

  $$|X_n(\omega) - X(\omega)| < \varepsilon.$$

  By definition, this means that the sequence is convergent at ω; therefore

  $$M_k \subseteq N^c, \ \forall k,$$

  thus

  $$\bigcup_k M_k \subseteq N^c.$$

- We leave it as an easy exercise to take an $\omega \in N^c$ and to show that there must exist a k_0 such that $\omega \in M_{k_0}$. Therefore, we also obtain

  $$N^c \subseteq \bigcup_k M_k.$$

Using the double inclusion, we have

$$\bigcup_k M_k = N^c$$

and therefore

$$\mathbf{P}\left(\bigcup_k M_k\right) = 1,$$

using the hypothesis.

Since the sets M_k are increasing using the monotone convergence property of probability, we get that $\mathbf{P}(M_k) \to 1$ when $k \to \infty$. Looking at the definition of M_k in (11.2), this clearly implies that

$$\mathbf{P}\left(\bigcup_{n\geq k} N_\varepsilon(n)\right) \to 0, \qquad \text{as } k \to \infty;$$

therefore $\mathbf{P}\left(N_\varepsilon(k)\right) \to 0$, as $k \to \infty$, which is just the definition of the convergence in probability.

2. For the second part of the theorem we will use the Borel–Cantelli lemmas (Lemma 2.34). Let us take ε in the definition of convergence in probability of the form $\varepsilon_k > 0$ and make it to go to zero when $k \to \infty$.

By the definition of convergence in probability for every such ε_k, we can find an n_k such that

$$\mathbf{P}\{\omega : |X_n(\omega) - X(\omega)| > \varepsilon_k\} < \frac{1}{2^k},$$

for every $n \geq n_k$. We start with $m_1 = n_1$ and an easy iterative process will construct $m_k = \min(m_{k-1}, n_k)$, so that the subsequence becomes increasing while still maintaining the above, desired property. Denote

$$N_k = \{\omega : \left|X_{m_k}(\omega) - X(\omega)\right| > \varepsilon_k\}.$$

Then, from the above we have

$$\mathbf{P}(N_k) < \frac{1}{2^k},$$

which implies

$$\sum_k \mathbf{P}(N_k) < \sum_k \frac{1}{2^k} < \infty.$$

Therefore, applying the first Borel–Cantelli lemma to the N_k sets, the probability that N_k occurs infinitely often is zero.

This means that with probability one, N_k^c will happen eventually. Or, in words, the set of ω for which there exists a k_0 and $\left|X_{m_k}(\omega) - X(\omega)\right| < \varepsilon_k$ for all $k \geq k_0$ has probability 1. Therefore, at each such ω the subsequence is convergent, and thus the set

$$\{\omega : X_{m_k}(\omega) \to X(\omega)\}$$

has probability 1. But this is exactly what we need to prove to show that the subsequence converges a.s. ■

In general convergence in probability does not imply a.s. convergence. A counterexample was provided already in Example 11.5 but here is another variation of that example (written as an exercise).

EXAMPLE 11.7 Counterexample. $\xrightarrow{p}$ implies $\xrightarrow{a.s.}$

You may construct your own counterexample. For instance, take $\Omega = (0, 1)$ with the Borel sets defining the σ-algebra and with the Lebesque measure (which is a probability measure for this Ω). For every $n \in \mathbb{N}$ and $1 \leq m \leq 2^n$, we now construct

$$X_{n,m}(\omega) = \mathbf{1}_{\left[\frac{m-1}{2^n}, \frac{m}{2^n}\right]}(\omega).$$

We again form a single subscript sequence by taking $Y_1 = X_{0,1}$, $Y_2 = X_{1,1}$, $Y_3 = X_{1,2}$, $Y_4 = X_{2,1}$, $Y_5 = X_{2,2}$, $Y_6 = X_{2,3}$, $Y_7 = X_{2,4}$, and so on. Once again plot the graph of these variables on a piece of paper for a better understanding of what is going on.

The example asks that you prove that the sequence $\{Y_k\}$ has the property that $Y_k \xrightarrow{p} 0$ but $Y_k \not\to Y$ a.s. In fact, this sequence does not converge at any $\omega \in \Omega$.

Proposition 11.20 *If $X_n \xrightarrow{L^p(\Omega)} X$ for some p then $X_n \xrightarrow{P} X$.*

Proof: This proposition, which says that convergence in L^p implies convergence in probability, is an easy application of the Markov inequality (Proposition 14.1). In that inequality we take $g(x) = |x|^p$, and the random variable as $X_n - X$, for the particular p for which we have convergence. We obtain

$$\mathbf{P}\left(|X_n - X|^p > \varepsilon\right) \leq \varepsilon^{-p}\mathbf{E}|X_n - X|^p.$$

Therefore, if $X_n \xrightarrow{L^p(\Omega)} X$, then we necessarily have $X_n \xrightarrow{P} X$ as well. ∎

Remark 11.21 *The converse of the previous result is not true in general. Consider the probability ensemble of exercise 11.23.*

Specifically, take

$$X_n(\omega) = n\mathbf{1}_{[0, \frac{1}{n}]}(\omega).$$

Show that $X_n \xrightarrow{P} X$ but $X_n \not\to X$ in any L^p with $p \geq 1$.

Yet another example of a sequence of random variable that converges in probability but not in L^p may be found in exercise 11.3.

Finally, here is a criterion for convergence in probability using expectations.

Theorem 11.22 $X_n \xrightarrow{P} 0$ *if and only if* $\mathbf{E}\left(\frac{|X_n|}{1+|X_n|}\right) \to 0$.

Proof: Left as an exercise (exercise 11.13). ∎

Next we look at properties of the convergence in probability. The next proposition says that convergence in probability is stable under summation and multiplication.

Proposition 11.23 *If*

$$X_n \xrightarrow{P} X \text{ and } Y_n \xrightarrow{P} Y$$

as $n \to \infty$, then

$$X_n + Y_n \xrightarrow{P} X + Y$$

and

$$X_n Y_n \xrightarrow{P} XY. \tag{11.3}$$

Proof: Let us prove the first assertion. The second part is left as an exercise.
For every $\varepsilon > 0$ we can write

$$\begin{aligned}\mathbf{P}\left(|(X_n + Y_n) - (X + Y)| > \varepsilon\right) &= \mathbf{P}\left(|(X_n - X) + (Y_n - Y)| > \varepsilon\right)\\ &\leq \mathbf{P}\left(|X_n - X| + |Y_n - Y| \geq \varepsilon\right)\\ &\leq \mathbf{P}\left(|X_n - X| \geq \frac{\varepsilon}{2}\right) + \mathbf{P}\left(|Y_n - Y| \geq \frac{\varepsilon}{2}\right)\end{aligned}$$

and both terms on the right-hand side converge to zero as $n \to \infty$. ■

Another important property of the convergence in probability is that it is stable when we apply continuous functions.

Proposition 11.24 *Let X_n be a sequence of random variables taking values in a subset V of $\mathbb{R}$, or $X_n : \Omega \to V$, where V is an open subset of $\mathbb{R}$. Let $f : V \to \mathbb{R}$ be a continuous function. If $X_n \xrightarrow{P} X$ as $n \to \infty$, then*

$$f(X_n) \xrightarrow{P} f(X).$$

Proof: We will sketch the proof in the case when $\bar{V}$ is a compact set. Since f is continuous, f is uniformly continuous on V. So, for every $\varepsilon > 0$ there exists $\alpha > 0$ such that for any x, y with the property $|x - y| \leq \alpha$ we have

$$|f(x) - f(y)| \leq \varepsilon.$$

As a consequence

$$\{|f(X_n) - f(X)| \geq \varepsilon\} \subset \{|X_n - X| \geq \varepsilon\}$$

and therefore

$$\mathbf{P}\{|f(X_n) - f(X)| \geq \varepsilon\} \leq \mathbf{P}\{|X_n - X| \geq \varepsilon\}.$$

But, since X_n converges to X in probability, we have

$$\mathbf{P}\{|X_n - X| \geq \varepsilon\} \to 0$$

and clearly

$$\mathbf{P}\{|f(X_n) - f(X)| \geq \varepsilon\} \to 0.$$

■

Remark 11.25 *Sometimes, it is possible to prove the convergence of $f(X_n)$ to $f(X)$ without assuming the continuity of f. Indeed, consider the following function:*

$$f : \mathbb{R} \setminus \{0\} \to \mathbb{R}$$

given by $f(x) = \frac{1}{x}$. Note that the function is not continuous at $x = 0$. However, if $X_n \xrightarrow{P} X$ and $\mathbf{P}(X \neq 0) = 1$ (to avoid problems with the random variable $f(X)$), we have that

$$\frac{1}{X_n} \xrightarrow{P} \frac{1}{X}.$$

Remark 11.26 *We further note that the convergence in probability*

$$X_n \xrightarrow{P} c$$

when c is a constant does not necessarily imply either

$$\mathbf{E}X_n \to c$$

or

$$Var\, X_n \to 0.$$

Indeed, to show a counterexample for the remark above, consider the sequence of random variables

$$X_n = \begin{cases} n, & \text{with probability } \frac{1}{n}, \\ 0, & \text{with probability } 1 - \frac{1}{n}. \end{cases}$$

Then $X_n \to 0$ in probability (we already showed this in exercise 11.23), but

$$\mathbf{E}X_n = 1,$$

and

$$Var\, X_n = n - 1.$$

11.4.2.1 Back to Almost Sure versus L^p Convergence. Let us return to the earlier relations for a moment. We have already shown (by providing counterexamples) that neither necessarily implies the other one. In either of these examples the limit implied did not exist.

If both limits exist, then they necessarily must be the same.

Proposition 11.27 *If $X_n \xrightarrow{L^p(\Omega)} X$ and $X_n \xrightarrow{a.s} Y$, then*

$$X = Y \text{ a.s.}.$$

Proof: (Sketch) We have already proven that both types of convergence imply convergence in probability. The proof then ends by showing a.s. the uniqueness of a limit in probability. ∎

11.4.3 UNIFORM INTEGRABILITY

We have seen that convergence a.s. and convergence in L^p are generally not compatible. However, we will give an integrability condition that will make all the convergence types equivalent. In practice, it is a very desirable property that the moments converge when $X_n \xrightarrow{a.s.} X$.

Definition 11.28 (Uniform integrability criterion) *A collection of random variables $\{X_n\}_{n\in\mathcal{I}}$ is called uniform integrable (U.I.) if*

$$\lim_{M\to\infty} \sup_{n\in\mathcal{I}} \mathbf{E}\left[|X_n|\mathbf{1}_{\{|X_n|>M\}}\right] = 0.$$

In other words, the tails of the expectation converge to 0 uniformly for all the family.

In general, the criterion for uniform integrability is hard to verify. The following theorem contains two necessary and sufficient conditions for uniform integrability (U.I.).

Theorem 11.29 (Conditions for U.I.) *A family of r.v.'s $\{X_n\}_{n\in\mathcal{I}}$ is uniformly integrable if and only if one of the following two conditions are true:*

1. **Bounded and Absolutely Continuous Family.** $\mathbf{E}[|X_n|]$ *is bounded for every $n \in \mathcal{I}$ and for any $\varepsilon > 0$, there exists a $\delta(\varepsilon) > 0$ such that if $A \in \mathcal{F}$ is any set with $\mathbf{P}(A) < \delta(\varepsilon)$, then we must have $\mathbf{E}\left[|X_n|\mathbf{1}_A\right] < \varepsilon$.*
2. **de la Vallée–Poussin Theorem.** *There exists some increasing function f, $f \geq 0$, such that $\lim_{x\to\infty} \frac{f(x)}{x} = \infty$ with the property that $\mathbf{E}f(|X_n|) \leq C$ for some constant C and all n.*

The proof of this theorem is tedious, and we skip it here. For the interested reader, a proof of the first part may be found in Theorem 4.5.3 in Chung (2000), and a proof for the second part appears in Theorem 7.10.3 in Grimmett and Stirzaker (2001).

Here is the main result using the U.I. concept.

Theorem 11.30 *Suppose that $X_n \xrightarrow{P} X$ and for a fixed $p > 0$, $X_n \in L^p(\Omega)$. Then the following three statements are equivalent:*

(i) *The family* $\{|X_n|^p\}_{n\in\mathbb{N}}$ *is U.I.*

(ii) $X_n \xrightarrow{L^p} X$.

(iii) $\mathbf{E}[X_n^p] \to \mathbf{E}[X^p] < \infty$.

The theorem shows that for a uniformly integrable family, all types of convergence are equivalent. The proof is once again skipped and we refer the interested reader to Theorem 4.5.4 in Chung (2000).

EXAMPLE 11.8 Examples of U.I. families

1. Any integrable random variable $X \in L^1$ is U.I.

   Clearly, $\mathbf{E}|X| < \infty$ implies that $\mathbf{E}\left[|X|\mathbf{1}_{\{|X|>M\}}\right] \xrightarrow{M\to\infty} 0$.
2. Suppose the family $\{X_n\}$ is bounded by an integrable random variable, $|X_n| \le Y$ and $Y \in L^1$. Then X_n is a U.I. family.

   Indeed, from the boundness we have

$$\mathbf{E}\left[|X_n|\mathbf{1}_{\{|X_n|>M\}}\right] \le \mathbf{E}\left[Y\mathbf{1}_{\{|Y|>M\}}\right],$$

   which does not depend on n and converges to 0 when $M \to \infty$ as in the previous example.
3. Any finite collection of random variables in L^1 is U.I.

   This is just an application of the previous point. Suppose that

$$\{X_1, X_2, \ldots, X_n\}$$

   is the finite collection of integrable r.v.'s. Then we can find a random variable that dominates them, for example

$$Y = |X_1| + |X_2| + \cdots + |X_n|.$$

   Applying the previous part concludes.
4. The family

$$\{a_n Y\}$$

   is U.I., where $Y \in L^1$ and the sequence $a_n \in [-1, 1]$ of nonrandom constants.
5. Any bounded collection of integrable r.v.'s is U.I.

We conclude the chapter with an example of a family which is not U.I.

EXAMPLE 11.9 A family of r.v.'s which is *not* uniformly integrable

Consider the probability space of all infinite sequences of coin tosses. Assume that the coin is fair—that is, the Ω space of all infinite sequences of any combination of heads and tails. Consider any σ-algebra and probability on this space.

Define a sequence of random variables:

$$X_n = \inf\{i : \text{such that } i > n \text{ and toss } i \text{ is a } H\},$$

in other words, the first toss after the nth toss where we obtain a head. Let $M > 0$ an integer. Using the way we defined the random variable, there exists another integer n greater than M and for this n we have

$$X_n > n \geq M.$$

Therefore, for this n we have

$$\mathbf{E}\left[|X_n|\mathbf{1}_{\{|X_n|>M\}}\right] = \mathbf{E}[X_n] > n,$$

since the set is always true. This implies by definition that the $\{X_n\}$ family is not U.I.

EXERCISES

Problems with Solution

11.1 Let $\Omega = [0, 1]$ and define for every $n \geq 1$

$$X_n(\omega) = \frac{\omega}{4n}.$$

Find the pointwise limit of the sequence X_n.

Solution: For all $\omega \in \Omega = [0, 1]$, we have

$$\lim_n X_n(\omega) = \lim_n \frac{\omega}{4N} = 0,$$

so the sequence X_n converges pointwise to 0. ■

11.2 Let p_n be a sequence of numbers in $[0, 1]$ such that

$$\lim_{n\to\infty} p_n = 0.$$

Let X_n be a sequence of independent r.v. with Bernoulli law with parameter p_n. That is,

$$\mathbf{P}(X_n = 1) = p_n \quad \text{and} \quad \mathbf{P}(X_n = 0) = 1 - p_n.$$

(a) Prove that X_n converges to zero in probability.

(b) Under which condition on $\sum_n p_n$ does the sequence X_n converge almost surely to zero?

Solution: (a) We have for every $\varepsilon > 0$

$$\mathbf{P}(X_n \geq \varepsilon) \leq p_n \to 0$$

as $n \to \infty$.

(b) Set

$$\Lambda_n = \{\omega : X_n(\omega) = 1\}.$$

Then the set

$$\Lambda = \limsup_n \Lambda_n = \cap_n \cup_{k \geq n} \Lambda_k$$

contains the points $\omega \in \Omega$ for which $X_n(\omega)$ does not converge to zero. Using Borel–Cantelli, we obtain

$$\mathbf{P}(\Lambda) = 0 \quad \text{if and only if} \quad \sum_n \mathbf{P}(\Lambda_n) < \infty.$$

Thus, X_n converges to zero almost surely if and only if

$$\sum_n p_n < \infty.$$

■

11.3 Consider X a continuous r.v. with density function

$$f(x) = \frac{\log a}{x(\log x)^2} \mathbf{1}_{(x \geq a)}$$

with $a > 1$ a constant. Let

$$X_n = \frac{X}{n}, \qquad n \geq 1.$$

(a) Show that the sequence $(X_n)_{n \geq 1}$ converges to zero in probability.

(b) Show that this sequence does not converge to zero in L^p, for any $p > 0$.

Solution: Fix $\varepsilon > 0$. Then

$$\begin{aligned}
\mathbf{P}(|X_n| > \varepsilon) &= \mathbf{P}(X > n\varepsilon) \\
&= \int_{n\varepsilon}^{\infty} \frac{\log a}{x(\log x)^2}\, dx \\
&= \frac{\log a}{\log(n\varepsilon)} \\
&\to_{n\to\infty} 0,
\end{aligned}$$

so the conclusion of part (a) is obtained.

For part (b), for every $p > 0$, we obtain

$$\begin{aligned}
\mathbf{E}|X|^p &= \mathbf{E}\left(\frac{X}{n}\right)^p \\
&= \frac{\log a}{n^p} \int_a^{\infty} \frac{1}{x^{1-p}(\log x)^2}\, dx = \infty,
\end{aligned}$$

so X_n does not converge to zero in L^p. ■

11.4 Let X_n be a sequence of r.v.s iid with law $Exp(\lambda)$. Set

$$Y = \limsup \frac{X_n}{\ln n}.$$

(a) Show that

$$\mathbf{P}\left(\limsup_n \left(\frac{X_n}{\ln n} \geq \frac{1}{\lambda}\right)\right) \leq \mathbf{P}\left(\limsup_n \frac{X_n}{\ln n} \geq \frac{1}{\lambda}\right).$$

(b) Show that

$$\mathbf{P}\left(\limsup \left(\frac{X_n}{\ln n} \geq \frac{1}{\lambda}\right)\right) = 1.$$

(c) Deduce that $\mathbf{P}(Y \geq \frac{1}{\lambda}) = 1$.

(d) Show that for every $\varepsilon > 0$, we obtain

$$\mathbf{P}\left(\limsup_n \frac{X_n}{\ln n} > \frac{1+\varepsilon}{\lambda}\right) \leq \mathbf{P}\left(\limsup_n \left(\frac{X_n}{\ln n} > \frac{1+\varepsilon}{\lambda}\right)\right).$$

(e) Deduce that

$$\mathbf{P}\left(Y = \frac{1}{\lambda}\right) = 1.$$

(f) Show that $\frac{X_n}{\ln n}$ converges to zero in probability. Does the sequence converge almost surely to zero?

Solution: (**a**) This part follows from the following fact: If $(x_n)_n$ is a sequence of real numbers, then

$$x_n \geq 0 \qquad \text{for an infinite number of } n \Rightarrow \limsup_n x_n \geq 0.$$

(**b**) We have

$$\mathbf{P}\left(\frac{X_n}{\ln n} \geq \frac{1}{\lambda}\right) = \mathbf{P}\left(X_n \geq \frac{\ln n}{\lambda}\right) = e^{-\ln n} = \frac{1}{n}.$$

Therefore

$$\sum_n \mathbf{P}\left(\frac{X_n}{\ln n} \geq \frac{1}{\lambda}\right) = \sum_n \frac{1}{n} = \infty.$$

Since the random variables X_n are independent, the Borel–Cantelli lemma implies that

$$\mathbf{P}\left(\limsup_n \frac{X_n}{\ln n} \geq \frac{1}{\lambda}\right) = 1$$

and from part (a) we have

$$\mathbf{P}\left(Y \geq \frac{1}{\lambda}\right) = 1.$$

(**c**) Follows from the following fact: If $(x_n)_n$ is a sequence of real numbers, then

$$\limsup_n x_n > 0 \Rightarrow x_n > 0 \qquad \text{for an infinite number of } n.$$

(**d**) Now

$$\mathbf{P}\left(\frac{X_n}{\ln n} > \frac{1+\varepsilon}{\lambda}\right) = n^{-(1+\varepsilon)}$$

and

$$\sum_n \mathbf{P}\left(\frac{X_n}{\ln n} > \frac{1+\varepsilon}{\lambda}\right) = \sum_n n^{-(1+\varepsilon)} < \infty.$$

The Borel–Cantelli lemma says that

$$\mathbf{P}\left(\limsup_n \frac{X_n}{\ln n} > \frac{1+\varepsilon}{\lambda}\right) = 0$$

and from part (c) we have

$$\mathbf{P}\left(Y > \frac{1+\varepsilon}{\lambda}\right) = 0.$$

(e) Letting $\varepsilon \to 0$, we obtain

$$\mathbf{P}\left(Y > \frac{1}{\lambda}\right) = \lim_{\varepsilon \to 0} \mathbf{P}\left(Y > \frac{1+\varepsilon}{\lambda}\right) = 0.$$

Hence

$$\mathbf{P}\left(Y = \frac{1}{\lambda}\right) = \mathbf{P}\left(Y \geq \frac{1}{\lambda}\right) - \mathbf{P}\left(Y > \frac{1}{\lambda}\right) = 1 - 0 = 1.$$

(f) We have

$$\mathbf{P}\left(\frac{X_n}{\ln n} \geq a\right) = \mathbf{P}\left(X_1 \geq a \ln n\right) \to 0.$$

■

11.5 Let $f : \mathbb{R} \to \mathbb{R}$ be such that

$$|f(x) - f(y)| \leq C|x - y|, \qquad \forall x, y \in \mathbb{R}$$

for some $C > 0$.

(a) If $X_n \to X$ almost surely, prove that

$$\mathbf{E}f(X_n) \to \mathbf{E}f(X) \qquad \text{as } n \to \infty.$$

(b) Let $\varepsilon > 0$. If $X_n \to X$ in probability, show that

$$\mathbf{P}\left(|f(X_n) - f(X)| \geq \varepsilon\right) \to 0.$$

Solution: (a) We have

$$\begin{aligned} |\mathbf{E}f(X_n) - \mathbf{E}f(X)| &\leq \mathbf{E}|f(X_n) - f(X)| \\ &\leq C\mathbf{E}\inf(1, |X_n - X|). \end{aligned}$$

For almost all $\omega \in \Omega$ we obtain

$$\inf(1, |X_n - X|) \to 0$$

and for every ω we have

$$\inf(1, |X_n - X|) \leq 1.$$

The dominated convergence theorem implies that

$$\mathbf{E}\inf(1, |X_n - X|) \to 0,$$

so $|\mathbf{E}f(X_n) - \mathbf{E}f(X)| \to 0$.

(**b**) By Chebyshev's inequality we have

$$\mathbf{P}\left(|f(X_n) - f(X)| \geq \varepsilon\right) \leq \frac{1}{\varepsilon}\mathbf{E}|f(X_n) - f(X)| \to 0.$$

■

11.6 Suppose that $X_n \to X$ in probability and assume that

$$\sup_n \mathbf{E}|X_n|^q < K$$

for some $q > 1$, where $K > 0$ is a constant.

(**a**) Prove that the family $(X_n)_{n\geq 1}$ is uniformly integrable.

(**b**) Show that X_n converges to X in L^p for every $0 < p < q$.

Solution: Let $a > 0$ such that $|X_n| > a$. Then

$$|X_n|^{p-q} < a^{p-q}.$$

This implies

$$\begin{aligned}\mathbf{E}|X_n|^p \mathbf{1}_{(X_n > a)} &\leq a^{p-q}\mathbf{E}|X_n|^q \\ &\leq Ba^{p-q},\end{aligned}$$

where $B = \sup_n \mathbf{E}|X_n|^q$. Because $p - q < 0$ we will have

$$\sup_n \mathbf{E}|X_n|^p \mathbf{1}_{(X_n > a)} \to 0,$$

as $a \to \infty$. That means that the sequence $(X_n)_n$ is uniformly integrable.

Part (b) follows from Theorem 11.30. ■

11.7 Let X be a continuous random variable with density f. Suppose that f_n, for $n \geq 1$ is a sequence of densities such that

$$f_n(x) \to f(x)$$

for every $x \in \mathbb{R}$. Define, for every $n \geq 1$,

$$X_n = \frac{f_n(X)}{f(X)}.$$

(**a**) Prove that the sequence X_n, $n \geq 1$ converges to 1 almost surely.

(**b**) Prove that the sequence X_n, $n \geq 1$ converges to 1 in L^1.

Solution: The first point is obvious because

$$\frac{f_n(x)}{f(x)} \to 1$$

for every $x \in \mathbb{R}$ as long as $f(x) > 0$. Hence

$$\begin{aligned}\mathbf{P}(X_n \to 1) &\geq \mathbf{P}(f(X) > 0)\\ &= 1 - \mathbf{P}(f(X) = 0)\end{aligned}$$

and

$$\mathbf{P}(f(X) = 0) = \int_{f(x)=0} f(x)\,dx = 0.$$

The almost sure convergence implies the convergence in probability. On the other hand, we have

$$\begin{aligned}\mathbf{E}|X_n| &= \mathbf{E}\frac{f_n(X)}{f(X)}\\ &= \int_{\mathbb{R}} \frac{f_n(x)}{f(x)} f(x)\,dx\\ &= \int_{\mathbb{R}} f_n(x)\,dx\\ &= 1\end{aligned}$$

and we can conclude the L^1 convergence using the Theorem 11.30. ■

Problems without Solution

11.8 Prove the convergence in (11.3).

11.9 Prove the convergence in Remark 11.25.

Hint: The general result in Proposition 11.24 cannot be applied, so this convergence should be proven starting from scratch.

11.10 Let U be an uniformly distributed r.v. on $[0, 1]$. Set

$$V = U - 1 \text{ and } W = -U.$$

(a) Calculate the density of V.

(b) Compare the laws of V and W.

(c) Show that for every $p \geq 1$ integer we have

$$\mathbf{E}(|V|^p) < +\infty$$

and compute $\mathbf{E}(V^p)$.

11.11 Let $(V_n, n \geq 1)$ be a sequence of i.i.d. random variables uniformly distributed on the interval $(-1, 0)$. For every integer $n \geq 1$, define the following sequences of random variables:

$$X_n = \frac{V_1^3 + V_2^3 + \cdots + V_n^3}{n},$$

$$Y_n = \frac{V_1^2 + V_2^2 + \cdots + V_n^2}{n},$$

and

$$Z_n = \frac{V_1^3 + V_2^3 + \cdots + V_n^3}{V_1^2 + V_2^2 + \cdots + V_n^2}.$$

(a) Show that the sequences $(X_n, n \geq 1)$ and $(Y_n, n \geq 1)$ converge almost surely.

(b) Show that the sequence $(Z_n, n \geq 1)$ converge almost surely. Identify the limit.

(c) Does the sequence $(Z_n, n \geq 1)$ converge in L^1? If the answer is yes, identify the limit.

11.12 Prove that a sequence that converges in probability is bounded in probability (see Definition 11.13).

11.13 Prove Theorem 11.22.

11.14 Let $(X_n , n \geq 1)$ denote a sequence of i.i.d. random variables with mean 0 and variance 1. Let $(Z_n , n \geq 1)$ be given by

$$Z_n = \prod_{k=1}^{n} (a + bX_k) .$$

(a) Calculate $\mathbf{E}(Z_n)$ and $Var(Z_n)$.

(b) Show that

$$Z_n \xrightarrow{L^2} 0$$

if $a^2 + b^2 < 1$.

11.15 *Maximum and minimum of uniform laws*

Let $a < b$ be two constants and let $(X_n , n \geq 1)$ denote a sequence of i.i.d. random variables with common distribution $U([a, b])$. For $n \geq 1$ we denote

$$M_n = \max_{1 \leq i \leq n} X_i,$$

Let $0 < \varepsilon < 1$.

(a) Calculate

$$\mathbf{P}(M_n \leq b - \varepsilon) .$$

(b) Deduce that

$$M_n \xrightarrow{P} b.$$

(c) Show that

$$\sum_{n\geq 1} \mathbf{P}\left(|M_n - b| \geq n^{-\alpha}\right) < +\infty$$

for every $\alpha \in (0, 1)$.

(d) Deduce that

$$M_n \xrightarrow{a.s.} b.$$

(e) Calculate the density of M_n.

(f) Does the sequence (M_n) converge in L^1?

11.16 Using the same notations as in the problem above, let

$$I_n = \min_{1\leq i\leq n} X_i.$$

Study the convergence of the sequence (I_n) in probability, almost surely, and L^1.

11.17 Let U and V two arbitrary random variables. Prove that

$$\{|U + V| > \varepsilon\} \subset \{|U| > \varepsilon/2\} \cup \{|V| > \varepsilon/2\}.$$

11.18 Use the previous problem and derive that if (X_n) and (Y_n) are two sequences such that $X_n \xrightarrow{P} X$ and $Y_n \xrightarrow{P} Y$, then

$$X_n + Y_n \xrightarrow{P} X + Y.$$

11.19 *Moving average sequence*

Let $(\varepsilon_n\,,\ n \geq 0)$ a sequence of i.i.d. random variables with mean 0 and variance σ^2. For every $a \in \mathbb{R}$, we consider the sequence $(Y_n\,,\ n \geq 1)$ (called the *moving average* of X_n) defined by

$$Y_n = \varepsilon_n + a\varepsilon_{n-1},$$

for every $n \geq 1$.

(a) Compute

$$Var(Y_n),$$
$$Cov(Y_n, Y_{n+1}),$$

and a general formula for

$$Cov(Y_n, Y_{n+k}), \qquad \text{where } |k| \geq 2.$$

(b) Using the previous part derive the value of

$$Var(\overline{Y_n}),$$

where $\overline{Y_n} = \sum_i Y_i/n$ denotes the average of the first n values.

(c) Does the sequence $\overline{Y_n}$ converge in L^2 ?
(d) Does it converge in probability?

Suppose that the fourth moment of the original random variables is finite, that is,

$$\mu_4 = \mathbf{E}(\varepsilon_i^4) < \infty.$$

Define the sequence

$$C_n = \frac{1}{n}\sum_{i=1}^{n} Z_i, \qquad \text{where } Z_i = Y_i Y_{i+1}$$

(e) Calculate the variance of Z_i. What is

$$Cov(Z_i, Z_j)$$

if $\left|i - j\right| > 2$?
(f) Calculate $\mathbf{E}(C_n)$ and $Var(C_n)$.
(g) Prove that the sequence (C_n) converges in L^2.

11.20 *Estimation of the mean of X when X is observed with probability p.*
Let $(X_n\,,\ n \geq 1)$ be a sequence of i.i.d. square integrable random variables. Denote

$$m = \mathbf{E}(X_1)$$

and assume that

$$Var(X_1) > 0.$$

Let $(Y_n\,,\ n \geq 1)$ be i.i.d. with Bernoulli law with parameter p, for some $0 < p < 1$, and assume that the sequence (Y_n) is independent of (X_n). Set

$$Z_n = X_n Y_n,$$

$$A_n = \frac{1}{n}\sum_{1}^{n} X_i Y_i,$$

and

$$B_n = \frac{1}{n}\sum_{1}^{n} Y_i.$$

(a) Calculate $\mathbf{E}(Z_n)$ and $Var(Z_n)$.
(b) Show that $Var(Z_n) > 0$.

(c) Prove that the sequences (A_n) and (B_n) converge in probability.

(d) Deduce that

$$R_n = \frac{\sum_1^n X_i Y_i}{\sum_1^n Y_i} \xrightarrow{P} m.$$

Interpret the result.

11.21 Assume $(X_n\ ,\ n \geq 0)$ are i.i.d. random variables such that

$$\mathbf{E}(X_1) = m$$

and

$$\mathbf{E}(X_1^2) = \mu_2 > 0.$$

Furthermore, assume that the fourth moment of the distribution is finite. Denote by

$$\mu_4 = \mathbf{E}(X_1^4) < +\infty.$$

(a) Prove that

$$\frac{1}{n}\left(X_1^2 + \cdots + X_n^2\right) \xrightarrow{P} \mu_2$$

(b) Prove that

$$\frac{1}{n}(X_1X_2 + \cdots + X_{n-1}X_n) \xrightarrow{P} m^2.$$

(c) Find the limit in probability of the sequence

$$R_n = \frac{X_1X_2 + \cdots + X_{n-1}X_n}{X_1^2 + \cdots + X_n^2}.$$

11.22 Suppose we know that there exists a constant γ such that

$$\sum_{k=1}^{n} \frac{1}{k} - \ln(n) - \gamma \to 0$$

when $n \to +\infty$. This constant is called the Euler constant.

Let $(X_n\ ,\ n \geq 1)$ be a sequence of independent random variables such that for every integer $n \geq 1$, the distribution of X_n is given by

$$\mathbf{P}\left(X_n = -\frac{1}{\sqrt{n}}\right) = \mathbf{P}\left(X_n = \frac{1}{\sqrt{n}}\right) = \frac{1}{2}.$$

Let

$$S_n = \sum_{k=1}^{n} X_k$$

the sum of the random variables for every $n \geq 1$.

(a) Calculate

$$\mathbf{P}(|X_n| \geq n^{-\alpha})$$

in terms of $\alpha > 0$.

(b) Does the sequence $(X_n \, , \; n \geq 1)$ converge almost surely?

(c) Does the sequence $(X_n \, , \; n \geq 1)$ converge in L^2?

(d) Calculate $\mathbf{E}(S_n^2)$. Is the sequence $\mathbf{E}(S_n^2)$ convergent?

(e) Does the sequence $(S_n \, , \; n \geq 1)$ converge in L^2?

(f) Does the sequence

$$\left(\frac{S_n}{n} \, , \; n \geq 1\right)$$

converge in L^2?

(g) Does the sequence

$$\left(\frac{S_n}{n} \, , \; n \geq 1\right)$$

converge almost surely?

11.23 Consider the sequence of random variables

$$X_n = \begin{cases} n, & \text{with probability } \frac{1}{n}, \\ 0, & \text{with probability } 1 - \frac{1}{n}. \end{cases}$$

Show that $X_n \to 0$ in probability.

11.24 Show that if $X_n \xrightarrow{L^p} X$, then $\mathbf{E}|X_n|^p \to \mathbf{E}|X|^p$.

Hint: The $\| \cdot \|_p$ is a proper norm (recall the properties of a norm).

11.25 Let $(X_n)_{n \geq 1}$ such that

$$\mathbf{P}(X_n = -n) = \frac{1}{2n^2},$$

$$\mathbf{P}(X_n = n) = \frac{1}{2n^2},$$

and

$$\mathbf{P}(X_n = 0) = 1 - \frac{1}{n^2}.$$

(a) Calculate $\mu = \mathbf{E}X_n$.

(b) Calculate $Var\, X_n$ for every $n \geq 1$.

(c) Study the convergence in probability of the sequence $(X_n)_{n\geq 1}$.

(d) Study the convergence of the sequence $(\mathbf{E}\,|X_n - \mu|)_{n\geq 1}$.

11.26 Show that if $f : \mathbb{R} \to \mathbb{R}$ is a continuous function and $X_n \xrightarrow{a.s.} X$, then $f(X_n) \xrightarrow{a.s.} f(X)$ as well.

11.27 Let $U \sim U[0, 1]$ and define

$$U_n = U 1_{[\frac{1}{n}, 1]}(U).$$

Show that the sequence $(U_n)_n$ converges almost surely to U.

11.28 Define

$$f_n(x) = \frac{n}{\pi(1 + n^2 x^2)}$$

for $x \in \mathbb{R}$.

(a) Show that f_n is a density.

(b) Let X_n be an r.v. with density f_n. Compute $\mathbf{E}X_n$ and $Var\, X_n$.

(c) Show that the sequence X_n, $n \geq 1$ converges in probability to zero.

CHAPTER TWELVE

Limit Theorems

12.1 Introduction/Purpose of the Chapter

The notions in the current chapter could be presented together with the previous chapter concerning types of convergence. We decided to separate this chapter and present it on its own for two reasons. First, the convergence in distributions (weak convergence) presented here is special, and it is characteristic for the probability theory. Second, the two big theorems related to convergence in distribution (the law of large numbers and the central limit theorem) are the basis of statistics and stochastic processes, and we believe they deserve to be presented in their own chapter.

The present chapter is dedicated to the following problem. Suppose we have an experiment repeating itself with minor changes in the surrounding conditions. Can we say something about the underlying parameters of the experiment? Can we gather quantities such as the average or variance and use them to describe the conditions of the experiment? The notions in this chapter are some of the most useful in probability, but they have to be presented in the last chapter since they use all the other probability concepts presented throughout the book.

12.2 Vignette/Historical Notes

The Law of Large Numbers was first proved by the Swiss mathematician Jacob (Jacques) Bernoulli. Bernoulli was a Swiss mathematician, the first in the Bernoulli

Handbook of Probability, First Edition. Ionuţ Florescu and Ciprian Tudor.

family, a family of famous scientists of the eighteenth century. Jacob Bernoulli's most original work was *Ars Conjectandi*, published in Basel in 1713, eight years after his death. The work was incomplete at the time of his death, but it still was a work of the greatest significance in the development of the Theory of Probability. The fourth part of his work presents the statement and the proof of the law of large numbers in a simplified form of what we now call Bernoulli trials (or *Bernoulli*(p) random variables). In modern notation he has an event that occurs with probability p but he does not know p. He wants to estimate p by the sample proportion $\hat{p}$ of the times the event occurs when the experiment is repeated a number of times. Bernoulli discusses in detail the problem of estimating, by this method, the proportion of white balls in an urn that contains an unknown number of white and black balls. Repeatedly, he draws samples of n balls with replacement (i.e., each ball drawn is recorded and then replaced in the urn for the next draw). He estimates p by the sample proportion $\hat{p}$ obtained at each sequence of n drawn balls. He shows that, by choosing n large enough, he can obtain any desired accuracy and reliability of the estimate. He also expands the applicability of his theorem to estimates of probability of dying from a particular disease, of different kinds of weather occurring, and more. When writing about the number of trials which are necessary to make a judgment, Bernoulli observes that the "man on the street" believes the "law of averages."

Further, it is clear that to make proper judgements, it is not enough to use one or two trials, but rather a great number of trials is required. And sometimes the stupidest man "by some instinct of nature per se and by no previous instruction (this is truly amazing)" knows for sure that the more observations of this sort that are taken, the less the danger will be of straying from the mark (*Bernoulli*).

But he goes on to say that he must contemplate another possibility. Something further must be contemplated here which perhaps no one has thought about until now. It certainly remains to be inquired whether after the number of observations has been increased, the probability is increased of attaining the true ratio between the number of cases in which some event can happen and in which it cannot happen, so that this probability finally exceeds any given degree of certainty; or whether the problem has, so to speak, its own asymptote—that is, whether some degree of certainty is given which one can never exceed.[1] Of course, here he talks about the consistency of the estimator—a concept proven by the central limit theorem.

Bernoulli recognized the importance of this theorem, writing: "Therefore, this is the problem which I now set forth and make known after I have already pondered over it for twenty years. Both its novelty and its very great usefulness, coupled with its just as great difficulty, can exceed in weight and value all the remaining chapters of this thesis."

The Central Limit Theorem for the Bernoulli trials was first proved by Abraham de Moivre (1667–1754). De Moivre was a French mathematician who lived

[1] *Source*: Grinstead and Snell (1997).

most of his life in England.[2] De Moivre pioneered the modern approach to the Theory of Probability, in his work *The Doctrine of Chance: A Method of Calculating the Probabilities of Events in Play* in the year 1718. A Latin version of the book had been presented to the Royal Society and published in the *Philosophical Transactions* in 1711. The definition of statistical independence appears in this book for the first time. *The Doctrine of Chance* appeared in new expanded editions in 1718, 1738, and 1756. The birthday problem appeared in the 1738 edition, the gambler's ruin problem in the 1756 edition. The 1756 edition of *The Doctrine of Chance* contained what is probably de Moivre's most significant contribution to probability, namely the approximation of the binomial distribution (Bernoulli trials) with the normal distribution. This result is now known in most probability textbooks as "The First Central Limit Theorem" (presented and proven in this chapter). He understood the notion of standard deviation and is the first to write the normal integral (and the distribution density). In *Miscellanea Analytica* (1730) he derives Stirling's formula (wrongly attributed to Stirling) which he uses in his proof of the Central Limit Theorem. In the second edition of the book in 1738 de Moivre gives credit to Stirling for an improvement to the formula. De Moivre wrote:

> I desisted in proceeding farther till my worthy and learned friend Mr James Stirling, who had applied after me to that inquiry, [discovered that $c = \sqrt{2}$].

De Moivre also investigated mortality statistics and the foundation of the theory of annuities. In 1724 he published one of the first statistical applications to finance *Annuities on Lives*, based on population data for the city of Breslau. In fact, in *A History of the Mathematical Theory of Probability* (London, 1865), Isaac Todhunter says that probability:

> ... owes more to [de Moivre] than any other mathematician, with the single exception of Laplace.

De Moivre died in poverty. He did not hold a university position despite his influential friends Leibnitz, Newton, and Halley, and his main income came from tutoring.

De Moivre, like Cardan (Girolamo Cardano), predicted the day of his own death. He discovered that he was sleeping 15 minutes longer each night and summing the arithmetic progression, calculated that he would die on the day when he slept for 24 hours. He was right!

---

[2] A Protestant, he was incarcerated between 18 and 21 years of age. At age 21 he was pushed to leave France after Louis XIV revoked the Edict of Nantes in 1685, leading to the expulsion of the Huguenots.

12.3 Theory and Applications

12.3.1 WEAK CONVERGENCE

Let us now discuss the weakest form of convergence, the convergence in distribution, aptly named as such. It is in fact implied by all other types of convergence.

Definition 12.1 (Convergence in distribution (weak convergence)) *Consider a sequence of random variables $(X_n)_{n\geq 1}$ defined on probability spaces $(\Omega_n, \mathcal{F}_n, \mathbf{P}_n)$ (which might be all different) and a random variable X, defined on $(\Omega, \mathcal{F}, \mathbf{P})$. Let $F_n(x)$ and $F(x)$ be the corresponding distribution functions of the random variables.*

The sequence $(X_n)_{n\geq 1}$ is said to converge to X in distribution (written $X_n \xrightarrow{\mathcal{D}} X$ or $F_n \Rightarrow F$) if for every point x at which F is continuous, we have

$$\lim_{n\to\infty} F_n(x) = F(x).$$

Remark 12.2 *This type of convergence is different from the types presented in the previous chapter. Since we only need the convergence of the distribution functions of the random variables, each of the random variables X_n may live on a different probability space $(\Omega_n, \mathcal{F}_n, \mathbf{P}_n)$ and yet the convergence is valid. As functions, each distribution $F_n : \mathbb{R} \to [0, 1]$, thus regardless of the underlying probability spaces the distribution functions all live in the same space.*

However, if $(\Omega_n, \mathcal{F}_n, \mathbf{P}_n) = (\Omega, \mathcal{F}, \mathbf{P})$ for all n, then the variables are all defined on the same probability space. Only in this case we may talk about relations with the rest of convergence types.

Remark 12.3 *There are many notations for weak convergence which are used interchangeably in various books. We mention $X_n \xrightarrow{\mathcal{L}} X$ (convergence in law), $X_n \Rightarrow X$, $X_n \xrightarrow{Distrib.} X$, $X_n \xrightarrow{d} X$.*

EXAMPLE 12.1

Why do we require x to be a continuity point of F? The simple answer is that if x is a discontinuity point, the convergence may not happen at x even though we might have convergence everywhere else around it. We present a simple example that illustrates this fact.

Let X_n be a $1/n$Bernoulli$(1/n)$ random variable. That is, the random variable X_n takes value $1/n$ with probability $1/n$ and the value 0 with probability $1 - 1/n$. Its distribution function is

$$F_n(t) = \begin{cases} 0, & \text{if } t < \frac{1}{n}, \\ 1, & \text{if } t \geq \frac{1}{n}. \end{cases}$$

Looking at this, it makes sense to say that the limit is $X = 0$ with probability 1 which has distribution function

$$F(t) = \begin{cases} 0, & \text{if } t < 0, \\ 1, & \text{if } t \geq 0. \end{cases}$$

Recall that any distribution function must be right continuous (Proposition 3.9). So according to our definition the convergence is valid.

However, look at the discontinuity point of F. We have $F(0) = 1 \neq \lim_{n\to\infty} F_n(0) = 0$. So if we insisted on convergence holding at all the points, we would not have it. This is why one excludes these points from the definition.

There exists a quantity which we know how to calculate and where the isolated points do not matter. That is the integral of a function. That is why we have an alternate definition for convergence in distribution given by the next theorem. We note that the definition above and the next theorem applies to random vectors as well (X_n, X taking values in $\mathbb{R}^d$).

Theorem 12.4 (The characterization theorem for weak convergence) *Let X_n defined on probability spaces $(\Omega_n, \mathcal{F}_n, \mathbf{P}_n)$ and X, defined on $(\Omega, \mathcal{F}, \mathbf{P})$. Then*

$$X_n \xrightarrow{\mathcal{D}} X$$

if and only if for any bounded, continuous function ϕ defined on the range of X we have:

$$\mathbf{E}[\phi(X_n)] \to_{n\to\infty} \mathbf{E}[\phi(X)], \qquad \text{as } n \to \infty,$$

or equivalently:

$$\int \phi(t)\, dF_n(t) \to_{n\to\infty} \int \phi(t)\, dF(t).$$

A function ϕ with the above properties is also called a test function. In the case when the variables have densities $f_n(x)$ and $f(x)$ respectively then convergence in distribution is equivalent with

$$\int \phi(t) f_n(t)\, dt \to \int \phi(t) f(t)\, dt.$$

Proof: We prove here one implication. The other implication, which is much simpler, uses the boundness of the c.d.f. or p.d.f. and the bounded convergence theorem. Suppose that for any bounded, continuous function ϕ defined on the range of X we have

$$\mathbf{E}[\phi(X_n)] \to \mathbf{E}[\phi(X)], \qquad \text{as } n \to \infty.$$

For every $\varepsilon > 0$ consider the function

$$\Phi_{t,\varepsilon} : \mathbb{R} \to \mathbb{R}$$

given by

$$\Phi_{t,\varepsilon}(u) = \begin{cases} 1, & u \le t, \\ 0, & u \ge t + \varepsilon, \\ 1 - \frac{u-t}{\varepsilon}, & t < u < t + \varepsilon. \end{cases}$$

Then $\Phi_{t,\varepsilon}$ is continuous and bounded and

$$\lim_n \mathbf{E}\Phi_{t,\varepsilon}(X_n) = \mathbf{E}\Phi_{t,\varepsilon}(X).$$

On the other hand, since

$$1_{(-\infty,t]} \le \Phi_{t,\varepsilon} \le 1_{(-\infty,t+\varepsilon]}$$

we have for every n

$$F_{X_n}(t) \le \mathbf{E}\Phi_{t,\varepsilon}(X_n)$$

and

$$\mathbf{E}\Phi_{t,\varepsilon}(X) \le F_X(t).$$

Therefore,

$$\limsup_n F_{X_n}(t) \le F_X(t)$$

for every $t \in \mathbb{R}$. Next, using the function

$$g_{t,\varepsilon}(u) = \begin{cases} 0, & u \le t - \varepsilon, \\ 1, & u \ge t, \\ 1 - \frac{u-t+\varepsilon}{\varepsilon}, & t - \varepsilon < u < t, \end{cases}$$

one can similarly prove that

$$\liminf_n F_{X_n}(t) > F_X(t - \varepsilon).$$

■

Letting $\varepsilon \to 0$, we obtain

$$\lim_n F_{X_n}(t) = F_X(t)$$

for every t such that F_X is continuous in t.

Corollary 12.5 *As an immediate consequence of the previous theorem and the fact that composition of continuous functions is continuous, we have the following: If* $X_n \xrightarrow{D} X$ *and* g *is a continuous function, then* $g(X_n) \xrightarrow{D} g(X)$.

Remark 12.6 *The continuity theorem below (Theorem 12.7) gives a criterion for weak convergence using characteristic functions. The result of the theorem is very useful and explains the main reason why we use characteristic function.*

Theorem 12.7 (Levy's continuity theorem) *Let* X_n, $n \geq 1$ *and* X *be random variables with characteristic functions* φ_n, $n \geq 1$, *and* φ, *respectively. Thus:*

1. *If* $X_n \to X$ *in distribution as* $n \to \infty$, *then*

$$\varphi_n(t) \longrightarrow \varphi(t)$$

   for every $t \in \mathbb{R}$ *as* $n \to \infty$.

2. *If* $\varphi_n(t) \longrightarrow \varphi(t)$ *for every* $t \in \mathbb{R}$ *and* φ *is continuous at zero, then* $X_n \xrightarrow{D} X$ *in distribution as* $n \to \infty$.

Proof: *Part 1 follows directly from the de definition of convergence in distribution since the function* e^{itx} *is a bounded continuous function of* x *for every* t. *The proof of the second part utilizes the Levy inversion formula (Proposition 8.16). We omit the details.* ■

The following proposition formally states that the convergence in probability (and thus all the others) will imply convergence in distribution. That is perhaps, the reason for the name "weak convergence."

Proposition 12.8 *Suppose that the sequence of random variables* X_n *and the random variable* X *are defined on the same probability space* $(\Omega, \mathcal{F}, \mathbf{P})$.
If $X_n \xrightarrow{P} X$, *then* $X_n \xrightarrow{\mathcal{D}} X$.

Proof: Suppose that for every $\varepsilon > 0$ we have

$$\mathbf{P}[|X_n - X| > \varepsilon] = \mathbf{P}\{\omega : |X_n(\omega) - X(\omega)| > \varepsilon\} \longrightarrow 0, \qquad \text{as } n \to \infty.$$

following the definition of convergence in probability.

Let $F_n(x)$ and $F(x)$ the distribution functions of X_n and X. Then we may write

$$\begin{aligned}
F_n(x) &= \mathbf{P}[X_n \le x] = \mathbf{P}[X_n \le x, X \le x + \varepsilon] \\
&\quad + \mathbf{P}[X_n \le x, X > x + \varepsilon] \\
&\le \mathbf{P}[X \le x + \varepsilon] + \mathbf{P}[X_n - X \le x - X, X - x > \varepsilon] \\
&= F(x + \varepsilon) + \mathbf{P}[X_n - X \le x - X, x - X < -\varepsilon] \\
&\le F(x + \varepsilon) + \mathbf{P}[X_n - X < -\varepsilon] \\
&\le F(x + \varepsilon) + \mathbf{P}[X_n - X < -\varepsilon] + \mathbf{P}[X_n - X > \varepsilon] \\
&= F(x + \varepsilon) + \mathbf{P}[|X_n - X| > \varepsilon].
\end{aligned}$$

We may repeat the line of reasoning above by starting with F and replacing x above with $x - \varepsilon$, that is,

$$\begin{aligned}
F(x - \varepsilon) &= \mathbf{P}[X \le x - \varepsilon] \\
&= \mathbf{P}[X \le x - \varepsilon, X_n \le x] + \mathbf{P}[X \le x - \varepsilon, X_n > x] \\
&\le \mathbf{P}[X_n \le x] + \mathbf{P}[X - X_n \le x - \varepsilon - X_n, x - \varepsilon - X_n < -\varepsilon] \\
&\le F_n(x) + \mathbf{P}[X - X_n < -\varepsilon] \\
&\le F_n(x) + \mathbf{P}[|X_n - X| > \varepsilon].
\end{aligned}$$

Combining the two inequalities, we obtain

$$F(x - \varepsilon) - \mathbf{P}[|X_n - X| > \varepsilon] \le F_n(x) \le F(x + \varepsilon) + \mathbf{P}[|X_n - X| > \varepsilon].$$

Now let $n \to \infty$ and apply lim inf and lim sup separately to it. We use the lim inf and lim sup since we don't know whether the limit exists. From the hypothesis we get

$$F(x - \varepsilon) \le \liminf_{n\to\infty} F_n(x) \le \limsup_{n\to\infty} F_n(x) \le F(x + \varepsilon).$$

Recall that we are only interested in x the points of continuity of F. Since x is a point at which F is continuous, by taking $\varepsilon \downarrow 0$ we obtain $F(x)$ in both ends of inequality and we are done. ■

In general, convergence in distribution does not imply convergence in probability. We cannot even talk about the implication in most cases since the random variables X_n may be defined on totally different spaces. For example, X_n may be counting the total of ones in an experiment where at every n even we record 1 if a head shows up when tossing a coin, and for n odd we record 10 if we win at a game of baccarat.

However, there is one exception. The constant random variable exists in any probability space. Suppose that $X(\omega) \equiv c$ a constant. In this case the convergence in probability reads

$$\lim_{n\to\infty} \mathbf{P}[|X_n - c| > \varepsilon] = 0,$$

and this statement makes sense even if each X_n is defined in a different probability space.

Theorem 12.9 *If $X_n \xrightarrow{D} c$ where c is a constant, then $X_n \xrightarrow{P} c$.*

Proof: If $X(\omega) \equiv c$—in other words, the constant c is regarded as a random variable—its distribution is the distribution of the delta function at c, that is,

$$\mathbf{P}[X \le x] = \mathbf{1}_{(c,\infty)}(x) = \begin{cases} 0, & x \le c, \\ 1, & x > c. \end{cases}$$

We may write

$$\begin{aligned} \mathbf{P}[|X_n - c| > \varepsilon] &= \mathbf{P}[X_n > c + \varepsilon] + \mathbf{P}[X_n < c - \varepsilon] \\ &= 1 - \mathbf{P}[X_n < c + \varepsilon] + \mathbf{P}[X_n < c - \varepsilon] \\ &\le 1 - F_n(c + \varepsilon) + F_n(c - \varepsilon). \end{aligned}$$

By hypothesis the distribution functions of X_n (F_n) converges to this function at all $x \neq c$ (the only discontinuity point is c). So, taking $n \to \infty$, we obtain that the limit above is zero for all $\varepsilon > 0$. Done, since this is the definition of convergence in probability. ■

Finally, here is a fundamental result that explains why even though the weakest, paradoxically convergence in distribution (or law) is also the most remarkable convergence type.

Theorem 12.10 (Skorohod's representation theorem) *Suppose $X_n \xrightarrow{\mathcal{D}} X$. Then, there exist a probability space $(\Omega', \mathcal{F}', \mathbf{P}')$ and a sequence of random variables Y, Y_n on this new probability space, such that X_n has the same distribution as Y_n, X has the same distribution as Y, and $Y_n \to Y$ a.s. In other words, there is a representation of X_n and X on a single probability space, where the convergence occurs almost surely.*

Thus we can represent the random variables with the same distributions on a particular space, and in that space the convergence is almost sure.

In general, if $X_n \xrightarrow{D} X$ and $Y_n \xrightarrow{D} Y$, it does not necessarily mean that $(X_n, Y_n) \xrightarrow{D} (X, Y)$. Recall that the convergence in law is characterized by the distribution and so what we see above directly relates to the fact that knowing the marginals does not tell me anything about the joint distribution unless the variables are independent.

However, for the convergence in probability, this is in fact true. Specifically, if $X_n \xrightarrow{P} X$ and $Y_n \xrightarrow{P} Y$, then we have to have $(X_n, Y_n) \xrightarrow{P} (X, Y)$. This result is easy to prove since we may write

$$\begin{aligned}\mathbf{P}(|(X_n, Y_n) - (X, Y)| > \varepsilon) &= \mathbf{P}\big((X_n - X)^2 + (Y_n - Y)^2 > \varepsilon^2\big) \\ &\leq \mathbf{P}(|X_n - X| > \varepsilon) + \mathbf{P}(|Y_n - Y| > \varepsilon),\end{aligned}$$

and both terms on the right go to zero because of the individual convergence.

However, just as in the case of convergence to a constant presented above, we have an exception and this is quite a notable exception too. The following theorem is one of the most widely used and useful in all of probability theory.

Theorem 12.11 (Slutsky's Theorem) *Suppose that $\{X_n\}_n$ and $\{Y_n\}_n$ are two sequences of random variables such that $X_n \xrightarrow{D} X$ and $Y_n \xrightarrow{D} c$, where c is a constant and X is a random variable.*

Suppose that $f : \mathbb{R}^2 \to \mathbb{R}$ is a continuous functional. Then

$$f(X_n, Y_n) \xrightarrow{D} f(X, c).$$

Proof: We shall use the Characterization Theorem for convergence in distribution (Theorem 12.4) throughout this proof. We will also use the Proposition 12.12 stated below. That proposition shows that the limit in distribution is the same for any two sequences that have the same limit in probability:

We start by showing the weak convergence of the vector (X_n, c) to (X, c). Specifically, let $\phi : \mathbb{R}^2 \to \mathbb{R}$ be a bounded continuous function. It we take $g(x) = \phi(x, c)$, then g is itself continuous and bounded and using the Theorem 12.4 we have

$$\mathbf{E}[\phi(X_n, c)] = \mathbf{E}[g(X_n)] \to \mathbf{E}[g(X)] = \mathbf{E}[\phi(X, c)],$$

where we use the bounded convergence theorem for the convergence of the expectations.

By the representation theorem, we obtain that $(X_n, c) \xrightarrow{D} (X, c)$. Next we have

$$|(X_n, Y_n) - (X_n, c)| = |Y_n - c| \xrightarrow{D} 0$$

because $Y_n \xrightarrow{D} c$. However, from Theorem 12.9 we also have

$$|(X_n, Y_n) - (X_n, c)| \xrightarrow{P} 0.$$

Finally, using the Proposition 12.12 we immediately obtain that

$$(X_n, Y_n) \xrightarrow{D} (X, c).$$

Thus, according to Theorem 12.4 we have

$$\mathbf{E}[\phi(X_n, Y_n)] \to \mathbf{E}[\phi(X, c)],$$

for every continuous bounded function ϕ. Now let $\psi : \mathbb{R} \to \mathbb{R}$ be any continuous bounded function. Since f in the hypothesis of the theorem is a continuous functional, then $\psi \circ f$ is continuous. It is also bounded since ϕ is bounded. Applying the derivation above to $\psi \circ f$, we obtain

$$\mathbf{E}[\psi \circ f(X_n, Y_n)] \to \mathbf{E}[\psi \circ f(X, c)],$$

which, once again using Theorem 12.4, proves that $f(X_n, Y_n) \xrightarrow{D} f(X, c)$. ■

Proposition 12.12 *Let X_n and Y_n be two random sequences such that $X_n \xrightarrow{D} X$ and $|X_n - Y_n| \xrightarrow{P} 0$. Then, we must have $Y_n \xrightarrow{D} X$. The conclusion holds even if the two sequences are random vectors.*

Proof: Note that if $|X_n - Y_n| \xrightarrow{P} 0$, then it also converges in distribution, and therefore by the weak convergence characterization theorem (Theorem 12.4) we have for any continuously bounded function g:

$$\int_{\mathbb{R}^d} g(x)\, dF_{Y_n - X_n}(x) \to \int_{\mathbb{R}^d} g(x)\delta_{\{0\}}(dx) = g(0),$$

where $\delta_{\{0\}}$ is the delta measure at 0 and we used the definition of this particular measure to obtain $g(0)$ at the end. In fact, owing to this particular measure, the same conclusion applies if $\mathbb{R}^d$ is replaced with any interval in $\mathbb{R}^d$ containing 0.

To recall the Dirac delta measure, this is a measure denoted $\delta_{\{a\}}$ defined on $\mathbb{R}^d$ and the attached Borel σ-algebra for any constant $a \in \mathbb{R}^d$. The measure of any set A in the Borel sets ($A \in \mathcal{B}(\mathbb{R}^d)$) is defined as

$$\delta_{\{a\}}(A) = \begin{cases} 1 & \text{if } a \in A, \\ 0 & \text{if } a \notin A. \end{cases}$$

This is a proper probability measure. If a variable (the constant a) has this distribution, we have

$$\mathbf{E}[a] = \int_{\mathbb{R}^d} x\delta_{\{a\}}(dx) = a,$$

which confirms the obvious property of the expectation of a constant. However, since now it is a properly defined random variable, for any measurable function g we can define the random variable $g(a)$. This will have the expectation

$$g(a) = \mathbf{E}[g(a)] = \int_{\mathbb{R}^d} g(x)\delta_{\{a\}}(dx),$$

which is just the property stated above. Now let us return to the proof.

Let x a continuity point of $F_X(x)$. We may write

$$Y_n = X_n + (Y_n - X_n)$$

and therefore we can calculate the distribution of Y_n as a convolution:

$$\begin{aligned} F_{Y_n}(x) &= \mathbf{P}(Y_n \leq x) = \mathbf{P}(X_n + (Y_n - X_n) \leq x) \\ &= \int_{\mathbb{R}^d} F_{X_n}(x - s) dF_{Y_n - X_n}(s). \end{aligned}$$

The function $F_X(x - s)$ is bounded but not necessarily continuous at $x - s$. However, since F_X is continuous at x we may find a neighborhood of 0, say denoted by U, such that $F_X(x - s)$ is continuous for any $s \in U$. If $s \notin U$, then $F_{Y_n - X_n}(s)$ may be made arbitrarily close to a constant (either 1 or 0) and thus the integral may be made negligible. Finally, using the characterization of convergence in distribution and picking the particular function F_{X_n} as g, we may write

$$F_{Y_n}(x) = \int_U F_{X_n}(x - s)\, dF_{Y_n - X_n}(s) \longrightarrow \int_U F_X(x - s)\delta_{\{0\}}(ds) = F_X(x).$$

Since x was an arbitrary continuity point of F_X, using the definition we conclude $Y_n \xrightarrow{D} X$. ■

EXAMPLE 12.2

With the hypothesis stated in Slutsky's theorem (Theorem 12.11), let $X_n \xrightarrow{D} X$ and $Y_n \xrightarrow{D} c$ where c is a constant and X is a random variable. Now take

$$f(X_n, Y_n) = X_n + Y_n$$

and

$$g(X_n, Y_n) = X_n Y_n.$$

Applying the theorem, we immediately obtain

$$X_n + Y_n \xrightarrow{D} X + c$$

and

$$X_n Y_n \xrightarrow{D} Xc.$$

12.3.2 THE LAW OF LARGE NUMBERS

In this part we will give several variants of the law of large numbers.

Let us start with the following, so-called weak law of large numbers (WLLN). The law says that the empirical mean of a sequence of independent r.v. with common expectation converges in probability to the common expectation. It is called weak since the convergence is only in probability.

Theorem 12.13 (WLLN) *Let $X_1, \dots, X_n, \dots$ be a sequence of independent random variables with common expectation $\mathbf{E}X = \mu$. For every $n \geq 1$ denote S_n by*

$$S_n := X_1 + \cdots + X_n.$$

Then

$$P\left(\left|\frac{S_n}{n} - \mu\right| \geq \varepsilon\right) \to_{n\to\infty} 0$$

for every $\varepsilon > 0$.

Proof: The proof follows from the Markov inequality (Proposition 14.1 in Appendix B) since

$$\begin{aligned} P\left(\frac{1}{n}\left(\sum_{i=1}^{n}(X_i - \mathbf{E}X)\right) \geq \varepsilon\right) &\leq \frac{1}{n^2\varepsilon^2}\mathbf{E}\left|\sum_{i=1}^{n}(X_i - \mathbf{E}X))\right|^2 \\ &= \frac{1}{n\varepsilon^2}\mathbf{E}(X_1 - \mathbf{E}X)^2 \\ &= \frac{1}{n\varepsilon^2}\mathbf{E}(X_1 - \mu)^2, \end{aligned}$$

and this clearly converges to zero when $n \to \infty$. Note that we do not need the variables to have the same distribution, only be independent and have common mean and variance (which needs to be finite). ■

The next result uses a stronger hypothesis (identically distributed random variables), but it has a stronger conclusion. The result is called the strong law of large numbers (SLLN).

Theorem 12.14 (SLLN) *Let $X_1, X_2, \dots$ be independent identically distributed random variables. Suppose that X_i are integrable and denote by μ the common expectation. Then, with $S_n = \sum_{i=1}^{n} X_i$ we have*

$$\frac{1}{n}S_n \to \mu \qquad \textit{almost surely and in } L^1$$

We will actually write the proof of the theorem in several steps. The first step is to prove the SLLN in the stronger hypothesis when the random variables X_i have a fourth moment which is finite. We will relax this assumption in the later results. Please note that the mean μ being substituted by 0 is no restriction; in fact, replacing the original random variables X_i with $Y_i = X_i - \mu$ and proving the results for Y_i with mean 0 will automatically give all the results for X_i with a general μ.

Proposition 12.15 *Let $X_1, X_2, \ldots$ be independent identically distributed random variables such that there exists a constant $K > 0$ with*

$$\mathbf{E}X_i = 0 \quad \text{and} \quad \mathbf{E}X_i^4 \leq K \qquad \text{for every } i \geq 1.$$

Then,

$$P\left(\frac{1}{n}S_n \to 0\right) = 1$$

and

$$\lim_{n\to\infty} \frac{S_n}{n} = 0 \text{ in } L^4.$$

Proof: We note that since the fourth moment is finite, we must also have

$$\mathbf{E}X^2 < \infty \quad \text{and} \quad \mathbf{E}X^3 < \infty.$$

Then, by using the independence of X_i, we have

$$\mathbf{E}X_iX_j^3 = \mathbf{E}X_iX_jX_k^2 = \mathbf{E}X_iX_jX_kX_l = 0$$

for all distinct indices i, j, k, l.

Therefore, after expanding the sum, we obtain

$$\begin{aligned}
\mathbf{E}S_n^4 &= \mathbf{E}\left(\sum_{i=1}^n X_i \sum_{j=1}^n X_j \sum_{k=1}^n X_k \sum_{l=1}^n X_l\right) \\
&= \mathbf{E}\left(\sum_{k=1}^n X_k^4 + 6 \sum_{i<j;i,j=1}^n X_i^2 X_j^2\right) \\
&= \sum_{k=1}^n \mathbf{E}[X_k^4] + 6 \sum_{i<j;i,j=1}^n \mathbf{E}\left[X_i^2 X_j^2\right].
\end{aligned} \tag{12.1}$$

Next, we bound the expectation of the second sum above (clearly the expectation of the first sum is less than Kn). Using the independence and the Cauchy–Schwarz inequality, we can write for every $i < j$

$$\begin{aligned}
\mathbf{E}X_i^2X_j^2 &= \mathbf{E}X_i^2\mathbf{E}X_j^2 \\
&\leq \left(\mathbf{E}X_i^4\right)^{\frac{1}{2}} \left(\mathbf{E}X_j^4\right)^{\frac{1}{2}} \\
&\leq K.
\end{aligned} \tag{12.2}$$

Relations (12.1) and (12.2) imply the following bound:

$$\mathbf{E}S_n^4 \leq nK + 3n(n-1)K \leq 3Kn^2 \tag{12.3}$$

(we used the fact that there are $\frac{n(n-1)}{2}$ terms in the second sum over i, j). Relation (12.3) immediately implies the L^4 convergence of $\frac{S_n}{n}$ to zero since

$$\mathbf{E}\left(\frac{S_n}{n}\right)^4 = \frac{1}{n^4}\mathbf{E}S_n^4 = \frac{K}{n^3} + \frac{3(n-1)K}{n^3} \leq \frac{3K}{n^2} \to 0.$$

as $n \to \infty$. In order to obtain the almost sure convergence, we use again (12.3) and we form the series

$$\mathbf{E}\left(\sum_{n\geq 1}\left(\frac{S_n}{n}\right)^4\right) = \sum_{n\geq 1}\mathbf{E}\left(\frac{S_n}{n}\right)^4 \leq 3K\sum_{n\geq 1}\frac{1}{n^2} < \infty.$$

Since the expectation of the random variable is finite, we must have the random variable finite with the exception of a set of measure 0 (otherwise the expectation will be infinite), hence

$$\sum_{n\geq 1}\left(\frac{S_n}{n}\right)^4 < \infty \qquad \text{a.s.}$$

But a sum can only be convergent if the term under the sum converges to zero. and thus

$$\frac{S_n}{n}^4 \longrightarrow 0 \qquad \text{a.s.,}$$

and immediately $S_n/n \to 0$ almost surely. ■

There exists a useful variant for uncorrelated random variables—that is, random variables such that $Corr(X_i, X_j) = 0$ for all $i \neq j$, but which are not necessarily independent.

Proposition 12.16 *Let $X_1, X_2, \ldots$ be random variables with the same distribution. Assume that they are square integrable and denote by*

$$\mu = \mathbf{E}X_1$$

and

$$\sigma^2 = Var(X_1).$$

Suppose that for all indices $i \neq j$, we have

$$Cov(X_i, X_j) = 0.$$

Then

$$P\left(\frac{1}{n}S_n \to 0\right) = 1, \qquad \left(or\ \frac{S_n}{n} \xrightarrow{a.s.} 0\right).$$

Proof: Again, we can assume without loss of generality that $\mu = 0$. Recall that once we prove the result for $\mu = 0$, we can obtain the same result for a general μ by considering the random variables $Y_i = X_i - \mu$.

Set

$$Z_m = \sup_{1\le k\le 2m+1} |X_{m^2+1} + \cdots + X_{m^2+k}|.$$

We claim that it suffices to show that

$$\lim_{m\to\infty} \frac{S_{m^2}}{m^2} = 0 \qquad \text{a.s.} \tag{12.4}$$

and

$$\lim_{m\to\infty} \frac{Z_m}{m^2} = 0 \tag{12.5}$$

to prove the theorem. Indeed, this holds because of the inequality

$$\left|\frac{S_n}{n}\right| \le \left|\frac{S_{m(n)^2}}{m(n)^2}\right| + \left|\frac{Z_{m(n)}}{m(n)^2}\right|,$$

where $m(n)$ is the integer such that

$$m(n)^2 \le n \le (m(n)+1)^2.$$

We next prove that relation (12.4) is true. Using Markov's inequality, we obtain

$$\mathbf{P}\left(\frac{S_{m^2}}{m^2} \ge \varepsilon\right) \le \frac{\sigma^2}{m^2\varepsilon^2}$$

and therefore

$$\sum_m \mathbf{P}\left(\frac{S_{m^2}}{m^2}\right) < \infty.$$

Since the sum of probabilities is finite, the relation (12.4) is obtained by applying the first Borel–Cantelli lemma.

Next we show that (12.5) holds. We set

$$a_k^m = X_{m^2+1} + \cdots + X_{m^2+k}.$$

We have

$$P\left(\frac{Z_m}{m^2} \ge \varepsilon\right) \le \sum_{k=1}^{2m+1} P\left(|a_k^m| \ge m^2\varepsilon\right)$$

and

$$P\left(|a_k^m| \geq m^2\varepsilon\right) \leq \frac{(2m+1)\sigma^2}{m^4\varepsilon^2},$$

again using the Markov's inequality. Therefore

$$\sum_m P\left(\frac{Z_m}{m^2} \geq \varepsilon\right) < \infty,$$

and finally the relation (12.5) follows from the same application of the first Borel–Cantelli lemma. ■

The following inequality is sometimes referred to as the Kolmogorov inequality.

Lemma 12.17 *Let $X_1, \ldots, X_n$ be centered (mean 0) independent r.v.'s. Set*

$$S_n = X_1 + \cdots + X_n$$

and denote by s_n the variance of S_n. Then

$$P(\exists k \in \{1, \ldots, n\}, |S_k| > t) \leq s_n t^{-2}.$$

Proof: Left as an exercise. ■

The next result is the so-called Kronecker's lemma.

Lemma 12.18 (Kronecker's lemma) *Let (b_n) be an increasing sequence of strictly positive real numbers diverging to infinity, that is,*

$$\lim_{n\to\infty} b_n = \infty.$$

Let (x_n) be another sequence of real numbers, and denote

$$s_n = x_1 + \cdots + x_n,$$

the partial sums formed with the sequence (x_n). Then, if

$$\sum_n \frac{x_n}{b_n} < \infty,$$

we have that

$$\lim_n \frac{s_n}{b_n} = 0.$$

Proof: The first step is to prove the following result about real numbers, known as the Césaro's lemma. If b_n is a sequence as in the theorem with $b_0 = 0$ and v_n

is a sequence converging to a finite limit denoted v_∞, then

$$\frac{1}{b_n}\sum_{k=1}^{n}(b_k - b_{k-1})v_k \to v_\infty \tag{12.6}$$

To prove (12.6), we take $\varepsilon > 0$ and we find an N such that

$$v_k \geq v_\infty - \varepsilon$$

for all $k \geq N$. Such an N exists from the convergence of v_n to v_∞. Then, for any $n > N$, we obtain

$$\begin{aligned}
\liminf_n \frac{1}{b_n}\sum_{k=1}^{n}(b_k - b_{k-1})v_k &= \liminf_n \left(\frac{1}{b_n}\sum_{k=1}^{N}(b_k - b_{k-1})v_k \right. \\
&\quad \left. + \frac{1}{b_n}\sum_{k>N}(b_k - b_{k-1})v_k\right) \\
&\geq \liminf_n \left(\frac{1}{b_n}C_N + \frac{(v_\infty - \varepsilon)}{b_n}\sum_{k>N}(b_k - b_{k-1})\right) \\
&= 0 + v_\infty - \varepsilon
\end{aligned}$$

Similarly, by picking an M such that $v_k \leq v_\infty + \varepsilon$, we obtain

$$\limsup_n \frac{1}{b_n}\sum_{k=1}^{n}(b_k - b_{k-1})v_k \leq v_\infty + \varepsilon$$

and (12.6) follows. Next, define

$$u_n = \sum_{k=1}^{n}\frac{x_k}{b_k}.$$

By hypothesis, we know that

$$u_\infty := \lim_n u_n$$

exists and we have

$$u_n - u_{n-1} = \frac{x_n}{b_n}, \quad \text{or } x_n = b_n(u_n - u_{n-1}).$$

Since

$$\begin{aligned}
s_n = \sum_{k=1}^{n} x_k &= \sum_{k=1}^{n} b_k(u_k - u_{k-1}) \\
&= u_n b_n - u_0 b_1 - \sum_{k=1}^{n}(b_k - b_{k-1})u_k,
\end{aligned}$$

we deduce from (12.6) that

$$\frac{s_n}{b_n} \to 0.$$

■

In the next step we prove the strong law of large numbers under a certain hypothesis on the variance.

Proposition 12.19 *Let $(W_n)_{n\geq 1}$ be a sequence of centered independent random variables and denote*

$$\sigma_n^2 = Var(W_n) < \infty.$$

Assume that

$$\sum_{n\geq 1} \frac{\sigma_n^2}{n^2} < \infty.$$

Then

$$\lim_n \frac{1}{n} \sum_{k=1}^{N} W_k = 0 \qquad a.s.$$

Proof: For every $n \geq 1$, let us denote X_n by

$$X_n = \frac{W_n}{n}.$$

These X_n are also centered (mean zero); and using the property of the σ_n's, we obtain

$$\sum_n Var\, X_n < \infty.$$

Set

$$S_n = X_1 + \cdots + X_n.$$

The proof of the theorem is done in the following steps. The form of the proof is done this way on purpose. Rarely, one proves a theorem the way it is presented in books or journal articles. Typically, one reduces the problem to prove to a (hopefully) simpler problem. The structure below presents this idea in a sequence of steps.

Step 1. In order to get the result about W_i's, it suffices to show that S_n converges almost surely.

Indeed, if S_n converges a.s., then we apply Kronecker's lemma with $b_n = n$ and $X_n = x_n$ and the result about W_n will follow.

Step 2. To show that S_n converges almost surely it is enough to prove that for every $\varepsilon > 0$

$$\lim_{n_0 \to \infty} \mathbf{P}(\exists m, n \geq n_0, |S_n - S_m| > \varepsilon) = 0.$$

Indeed, suppose that we showed that the above limit is zero. We have

$$\begin{aligned} &\mathbf{P}(\exists \varepsilon > 0, \forall n_0, \exists m, n \geq n_0, |S_n - S_m| > \varepsilon) \\ &\quad \leq \sum_{\varepsilon \in Q_+} \mathbf{P}(\forall n_0, \exists m, n \geq n_0, |S_n - S_m| > \varepsilon) \end{aligned}$$

and using the Cauchy criterium of convergence for sequences, it is enough to show that

$$\mathbf{P}(\forall n_0, \exists m, n \geq n_0, |S_n - S_m| > \varepsilon) = 0.$$

However, the probability above may be written as

$$\mathbf{P}(\forall n_0, \exists m, n \geq n_0, |S_n - S_m| > \varepsilon) = \mathbf{P}\left(\cap A_{n_0}\right),$$

where the set is

$$A_{n_0} = (\exists n, m \geq n_0, |S_n - S_m| > \varepsilon)$$

Note that these sets form a decreasing family of events ($A_m \supseteq A_n$ if $m < n$). Therefore, since the intersection is empty,

$$\mathbf{P}(\forall n_0, \exists m, n \geq n_0, |S_n - S_m| > \varepsilon) = \lim_{n_0 \to \infty} \mathbf{P}(A_{n_0}) = 0.$$

Step 3. We prove that

$$\mathbf{P}\left(\exists k \in \{n_0, \ldots, m\}, |S_{n_0} - S_k| > \frac{\varepsilon}{2}\right) \geq \mathbf{P}(\exists m, n \geq n_0, |S_n - S_m| > \varepsilon),$$

and therefore if we show that the probability on the left converges to zero, so will the probability on the right.

Using the triangle inequality, we can write

$$A_{n_0} \subset \bigcup_{m \geq n_0} \left(|S_m - S_{n_0}| > \frac{\varepsilon}{2}\right).$$

But

$$\bigcup_{m \geq n_0} \left(|S_m - S_{n_0}| > \frac{\varepsilon}{2}\right) = \bigcup_{m \geq n_0} \bigcup_{m \geq k \geq n_0} \left(|S_k - S_{n_0}| > \frac{\varepsilon}{2}\right).$$

Step 3 is now obtained since the sequence of events

$$B_m = \bigcup_{m \geq k \geq n_0} \left(|S_k - S_{n_0}| > \frac{\varepsilon}{2} \right)$$

is increasing.

Step 4. We conclude the proof. By the Kolmogorov's inequality (Lemma 12.17), we have

$$\begin{aligned} \mathbf{P}\left(\exists k \in \{n_0, \ldots, m\}, |S_{n_0} - S_k| > \frac{\varepsilon}{2} \right) \\ \leq \frac{4 Var(X_{n_0+1} + \cdots + X_m)}{\varepsilon^2} \\ = \frac{4}{\varepsilon^2} \sum_{k=n_0+1}^{m} Var(X_k). \end{aligned}$$

To conclude the proof, we take the limit when $m \to \infty$ and then when $n_0 \to \infty$. ■

The proof presented above can in fact be extended to a more general context. Specifically, the sequence n can be replaced by any a_n sequence increasing and diverging to infinity.

Proposition 12.20 *Let $(W_n)_{n \geq 1}$ be a sequence of centered independent random variables and denote*

$$\sigma_n^2 = Var(W_n) < \infty.$$

assume that

$$\sum_{n \geq 1} \frac{\sigma_n^2}{a_n^2} < \infty,$$

where a_n is a sequence of positive real numbers that increases to infinity. Then

$$\lim_n \frac{1}{a_n} \sum_{k=1}^{N} W_k - \mathbf{E} W_1 = 0 \qquad a.s.$$

Proof: The proof follows the lines of the proof of Proposition (12.19). ■

In the case when the variables are i.i.d., we obtain an interesting consequence.

Corollary 12.21 *If X_n, $n \geq 1$ are centered i.i.d. square integrable r.v.'s, then*

$$\frac{S_n}{\sqrt{n}(\ln n)^{\beta}} \to 0$$

almost surely for every $\beta > \frac{1}{2}$.

Proof: It suffices to apply Proposition (12.20) to the sequence

$$a_k := k^{\frac{1}{2}}(\ln n)^{\beta}.$$

■

To prove the general version of the strong law of large numbers, we need another auxiliary results (Kolmogorov truncation lemma).

Lemma 12.22 (Truncation) *Let $X_1, X_2, \ldots$ be independent identically distributed random variables. Assume that X_1 is integrable and let $\mu = \mathbf{E}X_1$. Define the truncation of X_n up to n as*

$$Y_n = X_n 1_{\{|X_n| \leq n\}}.$$

Then

1.

$$\mathbf{E}Y_n \to \mu.$$

2.

$$P\big(X_n = Y_n \text{ except a finite number of } n\big) = 1.$$

3.

$$\sum_n \left(\frac{VarY_n}{n^2}\right) < \infty.$$

Proof: Let us consider the variables

$$Z_n := X_1 1_{\{|X_1| \leq n\}}.$$

Since the variables have the same distribution, the law of Z_n is the same as the law of Y_n and therefore $\mathbf{E}[Y_n] = \mathbf{E}[Z_n]$. However, the Z_n sequence defined above all in terms of X_1 converges almost surely to X_1 when $n \to \infty$. We also note that

$$\mathbf{E}|Z_n| \leq |X_1|, \qquad \forall n \geq 1.$$

The dominated convergence theorem imply that .

$$\mathbf{E}Z_n \to \mathbf{E}[X_1] = \mu.$$

But since Y_n and Z_n have the same distribution, part 1 follows.

Next,

$$\begin{aligned}\sum_n P(X_n \neq Y_n) &= \sum_n P(|X_n| > n)\\ &= \sum_n P(|X_1| > n)\\ &= \sum_n \mathbf{E}\big(\mathbf{1}_{\{|X_1|>n\}}\big)\\ &= \mathbf{E}\left(\sum_n \mathbf{1}_{\{|X_1|>n\}}\right)\\ &= \mathbf{E}\left(\sum_{\{|X_1|>n\}} 1\right)\\ &\leq \mathbf{E}|X_1| < \infty.\end{aligned}$$

Thus

$$\sum_n P(X_n \neq Y_n) < \infty$$

and the Borel–Cantelli lemma gives the conclusion of part 2.

For part 3, set

$$f(z) = \sum_{n \geq \sup(1,z)} \frac{1}{n^2}.$$

It is not difficult to show that

$$f(z) \leq \frac{2}{\sup(1,z)} \qquad \text{for every } z. \tag{12.7}$$

Moreover,

$$\sum_n \frac{VarY_n}{n^2} \leq \sum_n \frac{|X_1|^2 \mathbf{1}_{\{|X_1|<n\}}}{n^2} = \mathbf{E}\big(X_1^2 f(|X_1|)\big)$$

and using (12.7), we get

$$\sum_n \frac{VarY_n}{n^2} \leq 2\mathbf{E}|X_1| < \infty.$$

■

Let us finally give the general version of the strong law of large numbers.

Theorem 12.23 *Let $X_1, X_2, \ldots$ be independent identically distributed random variables. Assume that X_1 is integrable and let $\mu = \mathbf{E}X_1$. Then*

$$\frac{1}{n}S_n \xrightarrow{a.s.} \mu.$$

Proof: To prove this result we will use the above Kolmogorov truncation lemma and a version of the strong law of large numbers with a condition on the variance (Proposition 12.20). Set

$$Y_n = X_n \mathbf{1}_{\{|X_n| \leq n\}}.$$

Then, by the part 2 of the previous truncation lemma, $X_n = Y_n$ except for a finite number of n's. Therefore,

$$\lim_{n\to\infty} \frac{1}{n}\sum_{k=1}^{n} Y_k - \frac{1}{n}\sum_{k=1}^{n} X_k = 0 \text{ a.s.}\,.$$

By the above relation, it is enough to show that

$$\lim_{n\to\infty} \frac{1}{n}\sum_{k=1}^{n} Y_k = \mu.$$

We have the following decomposition of this sum

$$\lim_{n\to\infty} \frac{1}{n}\sum_{k=1}^{n} Y_k = \lim_{n\to\infty} \frac{1}{n}\sum_{k=1}^{n} \mathbf{E}Y_k + \frac{1}{n}\sum_{k=1}^{n}(Y_k - \mathbf{E}Y_k).$$

The first point in the Kolmogorov truncation lemma implies that

$$\mathbf{E}Y_k \to \mu$$

and then by Césaro's lemma we have

$$\lim_{n\to\infty} \frac{1}{n}\sum_{k=1}^{n} \mathbf{E}Y_k = \mu.$$

It is sufficient to show that the term

$$\frac{1}{n}\sum_{k=1}^{n}(Y_k - \mathbf{E}Y_k)$$

converges to zero almost surely when n goes to infinity.

To show that this term goes to zero, we apply Proposition 12.19 and once again the Kolmogorov truncation lemma. To apply these results, we note that the random variables $W_k = Y_k - \mathbf{E}Y_k$ are centered i.i.d. and

$$Var(W_k) = Var(Y_k).$$

■

Finally, we conclude this section by showing that the convergence of S_n/n also holds in L^1 sense. To prove this claim, we need the Sheffé's theorem stated next.

Theorem 12.24 (Sheffé's theorem) *If $(X_n)_{n\geq 1}$ is a sequence of random variables such that $X_n \in L^1$ for every n, $X_n \geq 0$ for every n and $X_n \to_{n\to\infty} X$ almost surely where $X \in L^1$. Also assume that $\mathbf{E}X_n \to_{n\to\infty} \mathbf{E}X$, then*

$$X_n \to_n X \qquad in\ L^1.$$

We are now in position to give a proof that the Law of Large numbers hold in L^1 as well. The proof is due to Etemadi.

Theorem 12.25 *Let $X_1, X_2, \ldots$ be independent identically distributed random variables. Assume that X_1 is integrable and let $\mu = \mathbf{E}X_1$. Then*

$$\frac{1}{n}S_n \longrightarrow \mu \qquad in\ L^1.$$

Proof: For an r.v. Z, we denote for an outcome ω

$$Z^+(\omega) = \max(Z(\omega), 0)$$

and similarly

$$Z^-(\omega) = \max(-Z(\omega), 0).$$

Then, we clearly have

$$Z = Z^+ - Z^-.$$

Apply this notation to our sequence of i.i.d. random variables and consider the sequence of i.i.d. integrable r.v.'s $(X_n^+)_n$. Then form its partial sum $S_n = \frac{1}{n}(X_1^+ + \cdots + X_n^+)$. We know that

$$S_n^+ \longrightarrow \mathbf{E}X_1^+ \qquad \text{a.s.}$$

since the variables are integrable (being positive) and using Theorem 12.23. Also, $\mathbf{E}S_n^+ = \mathbf{E}X_1^+ < \infty$ for all n. Therefore, Sheffé's theorem (Theorem 12.24) says that the convergence of S_n^+ to $\mathbf{E}X_1^+$ also holds in L^1. Repeating the argument for the sequence X_n^- and combining the two results provides the proof. ■

The next theorem provides the strongest form of the Law of Large Numbers (under the weakest hypotheses).

Theorem 12.26 *Let $\{X_n\}_n$, $n \geq 1$ be a sequence of pairwise independent, identically distributed and integrable random variables. Then*

$$\frac{1}{n} S_n \longrightarrow \mathbf{E}X_1 \qquad a.s.$$

Proof: Without loss of generality, we can assume $X_i \geq 0$ for every i (otherwise, we can prove the result for X_i^+ and X_i^- and the general case will follow). Let

$$Y_i = X_i 1_{\{|X_i| \leq i\}}$$

and let

$$S_n^* := \sum_{i=1}^{n} Y_i.$$

Fix an $\varepsilon > 0$ and an $\alpha > 1$, and let $k_n = [\alpha^n]$ denote the integer part of α^n (the biggest integer smaller that α^n). Throughout the proof, C will denote a generic positive constant that may change from line to line. Since we are interested in convergence, the constants will not influence the conclusion. Let

$$\mu = P_{X_i} \text{ the distribution of } X_i.$$

Recall that all random variables have the same distribution so the subscript i is not needed. Then

$$\begin{aligned}
\sum_{n\geq 1} P\left(\frac{\left|S_{k_n}^* - \mathbf{E}S_{k_n}^*\right|}{k_n}\right) &\leq C \sum_{n\geq 1} \frac{VarS_{k_n}^*}{k_n^2} = C \sum_{n\geq 1} \frac{1}{k_n^2} \sum_{i=1}^{k_n} VarY_i \\
&\leq C \sum_{i\geq 1} \frac{\mathbf{E}Y_i^2}{i^2} = C \sum_{i\geq 1} \frac{1}{i^2} \int_0^i x^2 \mu(dx) \\
&= C \sum_{i\geq 1} \frac{1}{i^2} \sum_{k=0}^{i-1} \int_k^{k+1} x^2 \mu(dx) \\
&\leq C \sum_{k=0}^{\infty} \frac{1}{k+1} \int_k^{k+1} x^2 \mu(dx) \\
&\leq C \sum_{k=0}^{\infty} \int_k^{k+1} x\mu(dx) = C\mathbf{E}X_1 < \infty.
\end{aligned}$$

On the other hand,

$$\lim_n \frac{\mathbf{E}S_{k_n}^*}{k_n} = \lim_n Y_n = \lim_n \int_0^n x\mu(dx) = \mathbf{E}X_1$$

and therefore Borel–Cantelli lemma allows us to conclude that

$$\lim_n \frac{S^*_{k_n}}{k_n} = \mathbf{E}X_1 \qquad \text{a.s.} \tag{12.8}$$

Now, since $P(X_n \neq Y_n) = P(X_n > n)$, we can write

$$\begin{aligned}\sum_{n\geq 1} P(X_n \neq Y_n) &= \sum_{n\geq 1} P(X_n > n) \\ &= \sum_{n\geq 1} \int_n^{\infty} \mu(dx) = \sum_{n=1}^{\infty}\sum_{i=n}^{\infty} \int_i^{i+1} \mu(dx)\end{aligned}$$

and by changing the order of summation, we obtain

$$\begin{aligned}\sum_{n\geq 1} P(X_n \neq Y_n) &= \sum_{i=1}^{\infty} i \int_i^{i+1} \mu(dx) \\ &\leq \mathbf{E}X_1 < \infty.\end{aligned}$$

Again, we apply Borel–Cantelli and we obtain

$$P(X_n = Y_n \text{ except a finite number of } n\text{'s }) = 1.$$

This relationship together with (12.8) gives

$$\lim_n \frac{S_{k_n}}{k_n} = \mathbf{E}X_1 \qquad \text{a.s.}$$

For every $n \geq 1$, consider $m(n) \geq 0$ such that

$$k_{m(n)} \leq n \leq k_{m(n)+1}.$$

Obviously such $m(n)$ exists for given n. Since the function $n \to S_n$ is increasing (because the X_i are positive), we can write, with probability one,

$$\begin{aligned}\liminf_n \frac{S_n}{n} &\geq \liminf_n \frac{S_{k_{m(n)}}}{k_{m(n)}} \frac{k_{m(n)}}{k_{m(n)+1}} \\ &\geq \frac{1}{\alpha} \lim_n \frac{S_{k_{m(n)}}}{k_{m(n)}} = \frac{1}{\alpha}\mathbf{E}X_1 < \infty.\end{aligned}$$

In the same way, we obatin

$$\limsup_n \frac{S_n}{n} \leq \alpha \mathbf{E}X_1.$$

Thus, with probability one, we have

$$\frac{1}{\alpha}\mathbf{E}X_1 \leq \liminf_n \frac{S_n}{n} \leq \limsup_n \frac{S_n}{n} \leq \alpha \mathbf{E}X_1.$$

Taking into account that $\alpha > 1$ is arbitrary, we get the conclusion by taking the limit $\alpha \to 1$. ■

The strong law of large numbers gives immediately the almost sure convergence of the characteristic functions.

Proposition 12.27 *Let Y_k, $k \geq 1$ be a sequence of random vectors in $\mathbb{R}^d$, i.i.d., with characteristic function*

$$\varphi(u) = \mathbf{E}e^{i\langle u, Y_1\rangle}, \qquad u \in \mathbb{R}^d.$$

Then the empirical characteristic function

$$\varphi_n(u) := \frac{1}{n}\sum_{k=1}^{n} \mathbf{E}e^{i\langle u, Y_k\rangle}, \qquad u \in \mathbb{R}^d$$

converges for every u almost surely to $\varphi(u)$.

Proof: To obtain the result, we apply the Law of Large Numbers (LLN) to the random variables

$$X_k' = \cos(\langle u, Y_k\rangle)$$

and

$$X_k'' = \sin(\langle u, Y_k\rangle).$$

■

Remark 12.28 *In fact, since the characteristic function is bounded, the proposition above is a consequence of the LLN for bounded variables.*

Applying the strong LLN in the case of Bernoulli random variables, we obtain the so-called strong law of large numbers for frequencies.

Corollary 12.29 *Let X_n, $n \geq 1$, be a sequence of i.i.d. Bernoulli distributed r.v.'s with parameter $p \in (0, 1)$. Then*

$$\frac{1}{n}\sum_{i=1}^{n} X_i \longrightarrow p \qquad a.s.$$

Proof: We apply the LLN and we note that $\mathbf{E}X_1 = p$. ■

EXAMPLE 12.3 Coin tossing

Let us consider the special case of tossing a fair coin n times and let S_n denote the number of heads that turn up. Then the random variable S_n/n represents the fraction of times heads show up and will have values between 0 and 1. The LLN predicts that the outcomes for this random variable will, for large n, be near $1/2$.

EXAMPLE 12.4

Consider n rolls of a die. Let X_j be the outcome of the jth roll. Then construct

$$S_n = X_1 + \cdots + X_n.$$

Since each roll is independent of any other one, we obtain

$$\mathbf{E}X_j = \mathbf{E}X_1 = \frac{7}{2} \qquad \text{for every } j.$$

By the LLN,

$$\frac{S_n}{n} \longrightarrow \frac{7}{2} = 3.5 \qquad \text{a.s.}$$

EXAMPLE 12.5

Suppose that we choose at random n numbers in the interval $[0, 1]$. Let us denote by X_i the ith choice. Then X_i, $i = 1, \ldots, n$, are independent r.v's, each uniformly distributed on $[0, 1]$. Thus

$$\mathbf{E}X_i = \frac{1}{2} \quad \text{and} \quad Var X_i = \frac{1}{12}.$$

As usual, let $S_n = X_1 + \cdots + X_n$. Then

$$\mathbf{E}\left(\frac{S_n}{n}\right) = \frac{1}{2} \quad \text{and} \quad Var\left(\frac{S_n}{n}\right) = \frac{1}{12n}.$$

The LLN says that S_n/n converges to $1/2$ almost surely.

12.4 Central Limit Theorem

The second fundamental theorem of probability is the Central Limit Theorem (CLT). This theorem says that if S_n is the sum of n independent random variables, then the distribution function of S_n is well-approximated by the Gaussian distribution.

The theorem was first stated in the context of the Bernoulli random variables. It was soon proven for general random variables. We will discuss the theorem in the case when the individual random variables are identically distributed. However, the theorem remains true, under certain conditions, even if the individual random variables have different distributions.

Theorem 12.30 (Central Limit Theorem) *Let $X_1, X_2, \dots$ be a sequence of i.i.d. random variables with finite mean μ and finite nonzero variance σ^2. Let $S_n = X_1 + \cdots + X_n$. Then*

$$\frac{S_n - n\mu}{\sqrt{n\sigma^2}} \xrightarrow{D} N(0, 1).$$

Equivalently, if we denote $\bar{X} = \frac{S_n}{n}$, we may rewrite

$$\frac{\bar{X} - \mu}{\sigma/\sqrt{n}} \xrightarrow{D} N(0, 1).$$

Proof: We first reduce the proof to the case when the random variables in the i.i.d. sequence have mean 0 and variance 1. To this end, let $Y_i = \frac{X_i - \mu}{\sigma}$. Also, let $T_n = \sum_{i=1}^{n} Y_i$. This gives in terms of the original S_n and X_i:

$$T_n = \sum_{i=1}^{n} \frac{X_i - \mu}{\sigma} = \frac{\sum X_i - n\mu}{\sigma} = \frac{S_n - n\mu}{\sigma}.$$

Note that, as promised, the new variables Y_i in the sequence are i.i.d. mean zero and variance one. Due to this reduction, if we prove that that $T_n/\sqrt{n} \xrightarrow{D} N(0, 1)$, then the CLT would be demonstrated.

With this idea in mind, let Y_i be i.i.d. random variables such that $\mathbf{E}(Y_i) = 0$ and $Var(Y_i) = 1$. We need to show that

$$\frac{T_n}{\sqrt{n}} = \frac{1}{\sqrt{n}} \sum_{i=1}^{n} Y_i \xrightarrow{D} N(0, 1).$$

The next idea is to use the Continuity Theorem (Theorem 12.7) and show that the characteristic function of $T_n/\sqrt{n}$ converges to the characteristic function of the $N(0, 1)$.

Let us calculate the characteristic function of $\frac{T_n}{\sqrt{n}}$. We denote by $\varphi(t)$ the characteristic function of any of the Y variables.

$$\varphi_{T_n/\sqrt{n}}(t) = \varphi_{Y_1+\ldots+Y_n}\left(\frac{1}{\sqrt{n}}t\right) = \left(\varphi\left(\frac{1}{\sqrt{n}}t\right)\right)^n$$
$$= \left(\varphi(0) + \frac{t}{\sqrt{n}}\varphi'(0) + \frac{1}{2!}\left(\frac{t}{\sqrt{n}}\right)^2 \varphi''(0) + o\left(\left(\frac{t}{\sqrt{n}}\right)^2\right)\right)^n,$$

where we used the Taylor expansion of the function $\varphi(t)$ around 0. We can do this since the characteristic function is well-behaved around zero, and for an n large enough the term $t/\sqrt{n}$ becomes sufficiently small.

Furthermore, the terms within parentheses are

$$\varphi(0) = 1,$$

and

$$\varphi'(t) = \frac{d}{dt}\mathbf{E}[e^{itY_1}] = i\mathbf{E}(Y_1 e^{itY_1}),$$

which gives $\varphi'(0) = i\mathbf{E}[Y_1] = 0$. For the second derivative, we have

$$\varphi''(t) = i^2\mathbf{E}[Y_1^2 e^{itY_1}].$$

Therefore,

$$\varphi''(0) = i^2\mathbf{E}[Y_1^2] = -1.$$

Thus, substituting these terms, we obtain

$$\varphi_{\frac{T_n}{\sqrt{n}}}(t) = \left(1 - \frac{t^2}{2n} + o\left(\frac{t^2}{n}\right)\right)^n$$

The above sequence converges to $e^{-\frac{t^2}{2}}$ as n goes to infinity, which is the characteristic function of a $N(0, 1)$.

As a result, we obtain $\varphi_{T_n/\sqrt{n}}(t) \to \varphi_{N(0,1)}(t)$ and therefore applying the Continuity Theorem (Theorem 12.7), we obtain

$$T_n/\sqrt{n} \xrightarrow{D} N(0, 1),$$

This concludes the proof of the CLT. ■

There is one weak point in the proof above. Note that to calculate derivatives of the characteristic function we had to switch the order of the derivative and the expectation. Is this even possible? The next theorem shows that exchanging the order is a valid operation under very general conditions. In particular, the exchange of derivative and integral is permitted in any finite measure space, provided that the result after the exchange results in finite integrals. In our case, this is perfect since both conditions apply.

Lemma 12.31 (Derivation under the Lebesgue integral) *Let $f : \mathbb{R} \times \mathbf{E} \to \mathbb{R}$, where $(\mathbf{E}, \mathscr{K}, P)$ is a finite measure space such that:*

(i) *For all $t \in \mathbb{R}$ fixed, $x \mapsto f(t, x) : E \to \mathbb{R}$ is measurable and integrable.*

(ii) *For all $x \in E$ fixed; $t \mapsto f(t, x) : \mathbb{R} \to \mathbb{R}$ is derivable with respect to t and the derivative is $f_{1,0}(t, x)$, a continuous function. Furthermore, there exists a g integrable on $(E, \mathscr{K}, P)$ such that*

$$|f_{1,0}(t, x)| \le g(x), \qquad \forall t$$

Then the function

$$\mathscr{K}(t) = \int f(t, x)\, dP(x) = \mathbf{E}[f(t, X)]$$

is derivable and its derivative is

$$\mathscr{K}'(t) = \int f_{1,0}(t, x)\, dP(x).$$

Proof: We have

$$\frac{\mathscr{K}(t+h) - \mathscr{K}(t)}{h} = \frac{1}{h} \int [f(t+h, x) - f(t, x)]\, dP(x)$$

By the Lagrange theorem (mean value theorem) we continue:

$$= \int f_{1,0}(t + \theta h, x)\, dP(x)$$

for some $\theta \in (0, 1)$.

Now if we let $h \downarrow 0$ and we use the continuity of $f_{1,0}$, we obtain that

$$f_{1,0}(t + \theta h, x) \to f_{1,0}(t, x).$$

Further use of the dominated convergence theorem provides the result we need. ∎

We state a "local" variant of the CLT.

Theorem 12.32 *Let X_n, $n \ge 1$ be centered (mean zero) i.i.d. random variables with variance equal to one. Suppose that the common density g_1 is integrable. Denote by φ the common characteristic function. We assume that this function is also integrable.*

Then, the density function g_n of

$$U_n = \frac{1}{\sqrt{n}}(X_1 + \cdots + X_n)$$

exists and satisfies

$$g_n(x) \longrightarrow \frac{1}{\sqrt{2\pi}} e^{-\frac{1}{2}x^2}$$

uniformly with respect to $x \in \mathbb{R}$.

Proof: Note that this theorem is identical with the one above except is in the specific context when the variables have densities (g_1). We will only give the main line of the proof. We do this theorem since the proof is different than the previous one. We leave the details to the reader.

Step 1. One can prove that g_n exists and satisfies

$$g_n(x) = \sqrt{n}(g_1 * \cdots * g_1)\left(\frac{x}{\sqrt{n}}\right).$$

The symbol $*$ means the convolution of the functions.

Step 2. Let φ_n be the characteristic function of U_n. The next step is to show that

$$\varphi_n(t) = \varphi\left(\frac{t}{\sqrt{n}}\right)^n$$

and then that

$$g_n(x) = \frac{1}{2\pi} \int_{\mathbb{R}} \varphi_n(t)\, dt.$$

Step 3. It follows that

$$|\varphi(t)| < 1 \qquad \text{if } t \neq 0.$$

The inequality above may be proven by reduction to absurd. Assume that there exists a t such that the above is not true, put $\varphi(t) = e^{i\alpha}$, and regard the real part of the Fourier integral that defines $\varphi(0) - e^{i\alpha} - \varphi_t$.

Step 4. Show that there exists a δ such that

$$|\varphi(t)| \leq e^{-\frac{1}{4}t^2}$$

for $|t| \leq \delta$. To show this, one can use the inequality $1 - t \leq e^{-t}$, which is true for every $t \in \mathbb{R}$.

Step 5. The next step shows that for every $a > 0$ we have

$$\varphi\left(\frac{t}{\sqrt{n}}\right)^n \to e^{-\frac{1}{2}t^2}$$

uniformly on $t \in [-a, a]$.

Step 6. Show that

$$\left| g_n(x) - \frac{1}{2\pi} e^{-\frac{1}{2}x^2} \right| \leq \frac{1}{2\pi} \int_{\mathbb{R}} \left| \varphi(\frac{t}{\sqrt{n}})^n - e^{-\frac{1}{2}t^2} \right| dt = I_n.$$

Step 7. Show that

$$\int_{a \leq |t| \leq \delta\sqrt{n}} \left| \varphi(\frac{t}{\sqrt{n}})^n - e^{-\frac{1}{2}t^2} \right| dt \leq 2 \int_a^{\infty} 2e^{-\frac{1}{4}t^2} dt \to 0$$

when $a \to \infty$.

Step 8. We study the integral I_n over the region $|t| \geq \delta\sqrt{n}$. Set

$$\eta = \sup_{|t| \geq \delta} |\varphi(t)| < 1.$$

Show that

$$\int_{|t| \geq \delta\sqrt{n}} \left| \varphi(\frac{t}{\sqrt{n}})^n - e^{-\frac{1}{2}t^2} \right| dt \leq \eta^{n-1} \int_{\mathbb{R}} \left| \varphi(\frac{t}{\sqrt{n}}) \right| dt + 2 \int_{\delta\sqrt{n}} e^{-\frac{1}{2}t^2} dt \longrightarrow 0.$$

Step 9. Conclude. ■

Remark 12.33 *By adapting the previous proof, we can show that the conclusion holds for d-dimensional random variables as well.*

EXAMPLE 12.6 Estimate the value of an integral

Suppose we want to estimate the integral

$$I = \int_{[0,1]^d} f(x)\, dx,$$

where $f : [0, 1]^d \to \mathbb{R}$.

To this end, we take $\{U_1, U_2, \ldots, U_n \ldots\}$, a sequence of i.i.d. $U[0, 1]^d$ d-dimensional random vectors, with all components uniform on $[0, 1]$. We form the sequence

$$I_n = \frac{1}{n} \sum_{i=1}^{n} f(U_i).$$

By the LLN, if f is integrable, then we have

$$I_n \to I = \mathbf{E}f(U)$$

and by the CLT, if f^2 is integrable, then we obtain

$$\sqrt{n}(I_n - I) \xrightarrow{D} N(0, \sigma^2),$$

where

$$\sigma^2 = \int_{[0,1]^d} f(x)^2\, dx - \left(\int_{[0,1]^d} f(x)\, dx\right)^2.$$

Remark 12.34 *Since the result is about convergence, an obvious question is: How large n should be to get a good approximation? In practice, spanned from medical application the n is typically chosen greater than or equal to 30, but this is too simple an answer. A fuller answer is that it depends on the shape of the population, that is, the distribution of X_i, and in particular how much it deviates from the normal density. If the distribution of X_i's is fairly symmetric even though non-normal, then $n = 10$ may be large enough. If the distribution is heavily skewed (i.e., one long tail), $n = 50$ or more may be necessary.*

EXAMPLE 12.7 Normal approximation of the binomial distribution

Let X_i be i.i.d. Bernoulli random variables distributed with parameter p. In other words, X_i is the outcome of a single Bernoulli trial, that is,

$$\mathbf{P}(X_i = 1) = p \quad \text{and} \quad \mathbf{P}(X_i = 0) = 1 - p.$$

Then, the sum

$$X = X_1 + \cdots + X_n$$

is a binomial random variable with parameters n and p, $X \sim Binom(n, p)$.

Denote

$$\bar{X} = \frac{X}{n}.$$

As an immediate consequence of the CLT, for large n we have

$$\bar{X} \sim N\left(p, \frac{p(1-p)}{n}\right)$$

or

$$X \sim N\big(np, np(1-p)\big);$$

this is called the normal approximation to the binomial distribution.

This approximation is heavily dependent on the value of p. When p is close to 0 or close to 1, this simple distribution is heavily skewed (the probability of one outcome is much larger than the probability of the other). The normal approximation does badly in this case unless n is very large. To summarize this succinctly, we look at the quantities np and $n(1-p)$.

A commonly quoted rule of thumb is that the approximation can be used only when both np and $n(1-p)$ are greater than 5.

Note that when $p = 0.5$ the Bernoulli distribution is symmetric. In this case both np and $n(1-p)$ equal 5 when $n = 10$, and so the rule of thumb suggests that $n = 10$ is large enough.

As p is far away from 0.5 toward either 0 or 1 and the Bernoulli distribution becomes severely skewed, the n needed increase tremendously. For example, when $p = 0.05$ or 0.95 the rule of thumb gives $n = 100$, and if $p = 0.001$, then $n \geq 5000$.

This poses a problem if p is very small as is the case, for example, with genetic markers. In the coin tossing scenario it only costs a bit of time to toss a coin several times more but for instance in the case of clinical trials when the patient gets better or not as a result of medication getting more observations can be very prohibitive. In these cases one should perform an approximation using the Poisson distribution. This is applicable when p is very small (or $1-p$) and n is large (but not large enough to apply the normal approximation). The distribution of the sum X of n trials (the binomial) is approximated with a Poisson with rate $\lambda_n = np$.

EXAMPLE 12.8 Normal approximation for the Poisson distribution

Let X_i be i.i.d. *Poisson*(λ) random variables. Then

$$\mu = \mathbf{E}X_i = \lambda$$

and

$$\sigma^2 = VarX_i = \lambda.$$

Therefore, the CLT implies that for n large enough we have

$$\sum_{i=1}^{n} X_i \sim N(n\lambda, n\lambda)$$

But the sum of independent Poisson r.v. is a Poisson r.v. Actually

$$\sum_{i=1}^{n} X_i \sim Poisson(n\lambda) \qquad \text{for every } n,$$

so for n large the law $Poisson(n\lambda)$ is close to the law $N(n\lambda, n\lambda)$ or equivalently the law $Poisson(\lambda)$ is close to $N(\lambda, \lambda)$ for λ large.

A rule of thumb for this one is that the approximation is good if $\lambda > 5$. Normally, those are the values we are interested anyway (lambda large).

Remark 12.35 (Continuity correction when CLT is used for discrete random variables) *When using the normal distribution to approximate the binomial and the Poisson distributions, which are both discrete, a continuity correction must be employed. To understand why, suppose we need to approximate the probability that the binomial random variable takes the value 2. If we use the normal directly, then this probability will always be zero since for any continuous random variable X and any number a we have $\mathbf{P}(X = a) = 0$. For a continuous variable, only the probability that X lies in some interval are positive.*

To allow for this issue a continuity correction is used in these situations. Essentially, the correction corresponds to treating the integer values as being rounded to the nearest integer.

To exemplify, to calculate the probability $\mathbf{P}(X = 2)$ of a binomial or Poisson (or indeed any other discrete distribution) using any continuous density Y, we would calculate instead $\mathbf{P}(1.5 < X < 2.5)$.

We also note that if we use a discrete distribution to approximate another discrete (for example using Poisson to approximate the binomial), such continuity correction is not needed.

EXAMPLE 12.9 Continuity correction in practice

When $X \sim Poisson(20)$ the above approximation using the CLT tells us that X is approximately $N(20, 20)$. Therefore,

$$\frac{X - 20}{\sqrt{20}} \sim N(0, 1).$$

To exemplify the correction, let us denote with Y an $N(20, 20)$. Then we write

$$\mathbf{P}(X \leq 15) \sim \mathbf{P}(Y < 15.5) = \mathbf{P}\left(\frac{Y - 20}{\sqrt{20}} < \frac{15.5 - 20}{\sqrt{20}}\right)$$
$$= \mathbf{P}\left(Z < \frac{15.5 - 20}{\sqrt{20}}\right) = P(Z < -1.006).$$

where we standardized to $Z \sim N(0, 1)$. This probability is equal to $1 - 0.84279 = 0.1572078$ using a normal distribution table or any software (in R the command is *pnorm*(-1.006)).

To check the performance of the approximation, we also calculate the Poisson distribution directly using software (or a c.d.f. table).

$$P(X \leq 15) = 0.1565131$$

(the R command is *ppois*$(15, 20)$).

Therefore, the error is under 0.0006946504 (about 0.45 percent of the relative error).

The examples above presenting the normal approximations for the binomial and the Poisson distributions are the most commonly used in practice. They are needed as the direct calculation of probabilities is computationally hard without them. However, CLT can be applied to any distribution.

EXAMPLE 12.10 Normal approximation of the Gamma distribution

Let X_i be i.i.d. exponentially distributed with parameter $\lambda > 0$. Let

$$Y_n = \sum_{i=1}^{n} X_i.$$

The law $Exp(\lambda)$ has mean $\frac{1}{\lambda}$ and variance $\frac{1}{\lambda^2}$. On the other hand, the sum of n exponentials Y is a gamma-distributed random variable with parameters $n, \lambda)$. Using the CLT, for large n, it can be approximated by the normal law $N(\frac{n}{\lambda}, \frac{n}{\lambda^2})$.

Since the chi square distribution $\chi^2(k)$ is exactly a $Gamma(k/2, 1/2)$ random variable, an immediate consequence is that the law $\chi^2(k)$ can be approximated using a normal distribution for large values of k.

EXERCISES

Problems with Solution

12.1 Let $(X_n)_{n\geq 1}$ be a sequence of independent random variables with common distribution $Exp(\lambda)$, $\lambda > 0$.

(a) Show that

$$\frac{1}{\ln n} \max_{1\leq k\leq n} X_k \longrightarrow \frac{1}{\lambda} \qquad \text{in probability.}$$

(b) Show that

$$Z_n := \max_{1\leq k\leq n} X_k - \frac{\ln n}{\lambda}$$

converges in distribution and determine its limit.

Solution: Let $\varepsilon > 0$. We have

$$\begin{aligned}
\mathbf{P}\left(\frac{1}{\ln n} \max_{1\leq k\leq n} X_k - \frac{1}{\lambda} > \varepsilon\right) &= \mathbf{P}\left(\max_{1\leq k\leq n} X_k > \left(\varepsilon + \frac{1}{\lambda}\right)\ln n\right) \\
&= \prod_{k=1}^{n} \mathbf{P}\left(X_k > \left(\varepsilon + \frac{1}{\lambda}\right)\ln n\right) \\
&= \left(\mathbf{P}\left(X_1 > \left(\varepsilon + \frac{1}{\lambda}\right)\ln n\right)\right)^n \\
&= e^{-n(1+\lambda\varepsilon)\ln n} \\
&\longrightarrow 0.
\end{aligned}$$

We also need to estimate the probability

$$P\left(\frac{1}{\ln n} \max_{1\leq k\leq n} X_k - \frac{1}{\lambda} < -\varepsilon\right)$$

(for the absolute value) and we need to show that it converges to zero when $n \to \infty$. Note that this probability is zero if $\varepsilon \geq \frac{1}{\lambda}$.

When $\varepsilon < \frac{1}{\lambda}$, we obtain

$$P\left(\frac{1}{\ln n} \max_{1\leq k\leq n} X_k - \frac{1}{\lambda} < -\varepsilon\right) = \left(1 - e^{(\frac{1}{\lambda}-\varepsilon)\ln n}\right)^n \longrightarrow 0.$$

This finishes the convergence in probability.

Let us study the converges in distribution of the sequence $(Z_n)_{n\geq 1}$. We compute the c.d.f. of Z_n. For every $t \leq 0$, clearly

$$\mathbf{P}(Z_n \leq t) = 0$$

while for $t > 0$ we obtain

$$P(Z_n \leq t) = P\left(X_1 \leq \frac{\ln n}{\lambda} + t\right)^n$$
$$= \left(1 - \frac{e^{-\lambda t}}{n}\right)^n$$
$$\longrightarrow e^{-\lambda t}.$$

Consequently, Z_n converges in distribution to a random variable with c.d.f.

$$F(t) = e^{e^{-\lambda t}} 1_{[0,\infty)}(t).$$

As a side note, this distribution is called the Gumbel law. The density (p.d.f.) of this law is

$$f(t) = \lambda e^{-\lambda t} e^{e^{-\lambda t}} 1_{[0,\infty)}(t).$$

■

12.2 Give an example of a sequence $(X_n)_{n \geq 1}$ that converges in distribution but not in probability.

Solution: Let $X \sim N(0, 1)$. Then, we obviously have $-X \sim N(0, 1)$. Define

$$X_n = (-1)^n X$$

for every $n \geq 1$. Then clearly X_n converges in law to X since all the random variables in the sequence have the same distribution. However,

$$X_n - X = 0 \qquad \text{if } n \text{ is odd}$$

and

$$X_n - X = -2X \qquad \text{if } n \text{ is even.}$$

Consequently,

$$\mathbf{P}(|X_n - X| > 1)$$

does not converge to zero as $n \to \infty$ (the probability is strictly positive and constant for n even). ■

12.3 Show that if the sequence X_n converges in distribution to a constant c, then X_n also convergence in probability to c.

Solution: If X_n converges to c in law, then

$$\mathbf{P}(|X_n - c| > \varepsilon) = 1 - \mathbf{P}(c - \varepsilon < X_n < c + \varepsilon) \longrightarrow 0.$$

To get the final limit, we use the convergence of the c.d.f. at the points where the limit function is continuous. ■

12.4 Let X_n be a sequence of random variables with an exponential distribution $Exp(\lambda_n)$ for every $n \geq 1$. Study the convergence in distribution of the sequence X_n, $n \geq 1$ in each of the following cases:

(a) $\lim_n \lambda_n = \lambda > 0$.

(b) $\lim_n \lambda_n = \infty$.

(c) $\lim_n \lambda_n = 0$.

Solution: **Case a.** Let g be a continuous bounded function. We have

$$\mathbf{E}[g(X_n)] = \int_0^\infty \lambda_n e^{-\lambda_n x} g(x)\, dx.$$

Since λ_n is convergent, it is bounded. Therefore, there exist λ_+, λ_- finite real numbers such that

$$0 < \lambda_- < \lambda < \lambda_+.$$

Then

$$\lambda_n e^{-\lambda_n x} g(x) \leq \|g\|_\infty \lambda_+ e^{-\lambda_- x}$$

for every $x > 0$. Applying the dominated converge theorem, we obtain

$$\mathbf{E}g(X_n) = \int_0^\infty \lambda_n e^{-\lambda_n x} g(x)\, dx \longrightarrow_n \int_0^\infty \lambda e^{-\lambda x} g(x)\, dx.$$

We therefore conclude that X_n convergence in distribution to a exponential r.v. with parameter λ.

Case b. Since $\lambda_n \to \infty$ we cannot bound λ_n from above as we did in the previous case. We will use a different argument. We first perform the change of variables $y = \lambda_n x$, which implies

$$\mathbf{E}g(X_n) = \int_0^\infty e^{-y} g\left(\frac{y}{\lambda_n}\right) du.$$

Since for any y we have

$$\left| e^{-y} g\left(\frac{y}{\lambda_n}\right) \right| \leq \|g\|_\infty,$$

we can apply the dominated convergence theorem and we get

$$\mathbf{E}g(X_n) \longrightarrow_n \int_0^\infty g(0) e^{-x} dx = g(0) = \mathbf{E}g(0).$$

Therefore, X_n converges in law to 0.

Case c. In this case, $\lambda_n \to 0$ so we cannot bound λ_n from below by a strictly positive constant. We will use a characteristic function argument. Note that

$$\varphi_{X_n}(u) = \frac{\lambda_n}{\lambda_n - iu} \longrightarrow \mathbf{1}_{(u=0)}.$$

This limit is not continuous, so it cannot be the characteristic function of a random variable. Consequently, X_n does not converge in distribution. ■

12.5 Let X_n, $n \geq 1$ be i.i.d. random variables with common law $Exp(\lambda)$ where $\lambda > 0$. Denote by

$$\bar{X}_n = \frac{1}{n}(X_1 + \cdots + X_n)$$

and

$$Z_n = \frac{1}{\bar{X}_n}.$$

(**a**) Show that the sequence Z_n converges almost surely to λ when $n \to \infty$.

(**b**) For n large enough, which Gaussian distribution could approximate the law of $\bar{X}_n$?

Solution: By the strong law of large numbers, we have

$$\bar{X}_n \to \mathbf{E}X_1 = \frac{1}{\lambda}.$$

So

$$Z_n \to \lambda$$

almost surely.

By the CLT, if $S_n = n\bar{X}_n$,

$$\frac{S_n - n\mathbf{E}X_1}{\sqrt{n}} \xrightarrow{D} N\left(\frac{1}{\lambda}, \frac{1}{\sqrt{n}\lambda}\right).$$

Thus

$$\bar{X}_n \sim N\left(\frac{1}{\lambda}, \frac{1}{\sqrt{n}\lambda}\right).$$

■

12.6 Let X_n, $n \geq 1$ be i.i.d. random variables with law $N(0, 1)$. Define

$$Y_n = \frac{1}{n}\sqrt{k}X_k.$$

Prove that $(Y_n)_n$ convergence in distribution to $N\left(0, \frac{1}{2}\right)$.

Solution: Using the characteristic function of the standard normal distribution and denoting by φ_k the characteristic function of $\sqrt{k}X_k$, we get

$$\varphi_k(t) = e^{-\frac{kt^2}{2n^2}}.$$

By the independence of the terms in the sequence, we obtain

$$\begin{aligned}\varphi_{Y_n}(t) &= \prod_{k=1}^{n} \varphi_k(t) \\ &= \exp\left(\frac{1+2+\cdots+n}{2n^2}t^2\right) \\ &= \exp\left(-\frac{n+1}{2n}t^2\right)\end{aligned}$$

and as $n \to \infty$ this converges to

$$e^{-\frac{t^2}{4}},$$

which is the characteristic function of the law $N(0, 1/2)$. The Levy's continuity theorem (Theorem 12.7) gives the conclusion. ■

12.7 Let $X_1, \ldots, X_n, \ldots$ be i.i.d. with law $Exp(1)$. Define

$$Z_n = \max(X_1, ..., X_n).$$

(a) Compute the characteristic function of Z_n.

(b) Find the law of

$$\frac{X_{n+1}}{n+1}.$$

(c) Find the law of

$$Y_n = Z_n + \frac{X_{n+1}}{n+1}.$$

(d) Compare the distributions of Y_n and Z_{n+1}.

Solution: **(a)** For every $x \geq 0$ (otherwise the probability is zero) we have

$$P(Z_n \leq x) = P(X_1 \leq x)^n$$
$$= (1 - e^{-x})^n.$$

Taking the derivative, we obtain

$$f_{Z_n}(x) = ne^{-x}(1 - e^{-x})^{n-1}\mathbf{1}_{\{x \geq 0\}},$$

which is the desired density.

(b) One can prove that

$$\frac{X_{n+1}}{n+1} \sim Exp(n+1).$$

(c) Using convolutions, we obtain the density of Y_n as

$$f_{Y_n}(x) = (n+1)e^{-x}(1 - e^{-x})^n.$$

(d) It follows from above that Y_n and Z_{n+1} have the same distributions. ■

Problems without Solution

12.8 Let $(X_n)_{n \geq 1}$ be i.i.d. random variable with law $N(1, 3)$.

(a) Show that the sequence

$$Y_n := \frac{1}{n} \sum_{i=1}^{n} X_i e^{X_i}$$

converges almost surely and in distribution and find the limiting distribution.

(b) Answer the same question for the sequence

$$Z_n := \frac{X_1 + \cdots + X_n}{X_1^2 + \cdots + X_n^2}.$$

12.9 Let $(X_n)_n$ be a sequence of random variables such that

$$P(X_n = 1) = P(X_n = -1) = \frac{1}{2}.$$

Define

$$Y_n = \sum_{k=1}^{n} \frac{X_k}{2^k}.$$

Prove that the sequence $(Y_n)_n$ converges in distribution to $Y \sim U[-1, 1]$.

12.10 Consider $(Y_n, n \geq 1)$ a sequence of independent identically distributed random variables with common law $U([a, b])$ (uniform on the interval). For every $n \geq 1$ we define

$$I_n = \inf(Y_1, \dots, Y_n)$$

and

$$M_n = \sup(Y_1, \dots, Y_n).$$

Recall the expression of the c.d.f. of M_n (denoted F_n) and I_n (denoted G_n) from problem 5.13.

(a) Show that for every t the sequence $F_n(t)$ converges, and derive that the sequence of random variables M_n converges in distribution to b.

(b) Show that the sequence of functions $G_n(t)$ converges, and derive that the sequence of random variables I_n converges in distribution to a.

12.11 Let X_n, $n \geq 1$ be a sequence of i.i.d. square-integrable random variables.

(a) Show that the sample variance converges to the theoretical variance, that is,

$$\lim_n \frac{1}{n} \sum_{i=1}^{n} (X_i - \mathbf{E}X_i)^2 = VarX_1,$$

and the limit holds almost surely.

Consider the least squares estimate of the variance. Specifically, denote for $n \geq 1$

$$V_n := \frac{1}{n-1} \sum_{i=1}^{n} (X_i - \bar{X}_n)^2,$$

where

$$\bar{X}_n = \frac{X_1 + \cdots + X_n}{n}.$$

is the sample mean.

(b) Compute the variance of V_n

(c) Find the almost sure limit of the sequence $(V_n)_{n \geq 1}$.

12.12 Let the notation in problem 10.18 prevail. Express the c.d.f. of S_n in terms of F. Derive the expression of the c.d.f. F_n of T_n in terms of F and show that for every $t \in \mathbb{R}$, the sequence $(F_n(t), n \geq 1)$ converges to $F(t\sqrt{3})$. What can we deduce from this?

12.13 Suppose X_n is independent of Y_n, and X is independent of Y . Use an argument based on characteristic functions to show that, if X_n converges to X in distribution and Y_n converges to Y in distribution, then $X_n + Y_n$ converges in distribution to $X + Y$.

12.14 Let X_n, $n \geq 1$ be i.i.d. r.v. For every $n \geq 1$ we denote

$$Z_n := \left(\prod_{i=1}^{n} e^{X_i}\right)^{\frac{1}{n}}.$$

Find the almost sure limit of the sequence $(Z_n)_{n\geq 1}$.

12.15 Let $(X_n, n \geq 1)$ be a sequence of i.i.d. random variables with common distribution $N(\mu, \sigma^2)$, where $\mu \in \mathbb{R}$ and $\sigma > 0$.

For every $n \geq 1$, denote $S_n := \sum_{k=1}^{n} X_k$.

(**a**) Give the distribution of S_n.

(**b**) Compute

$$\mathbf{E}\big((S_n - n\mu)^4\big).$$

(**c**) Show that for every $\varepsilon > 0$,

$$\mathbf{P}\left(\left|\frac{S_n}{n} - \mu\right| \geq \varepsilon\right) \leq \frac{3\sigma^4}{n^2\,\varepsilon^4}.$$

(**d**) Derive that the sequence $\left(\frac{S_n}{n}, n \geq 1\right)$ converges almost surely and identify the limit.

(**e**) Do you recognize this result?

12.16 Consider the sequence X_n defined in exercise 11.10.

(**a**) Show that the sequence

$$T_n = \sqrt{n}\left(X_n + \frac{1}{4}\right)$$

converges in distribution.

(**b**) Identify the limiting distribution.

12.17 Let $X_1, \ldots, X_n$ be independent random variables with the same distribution given by

$$\mathbf{P}(X_i = 0) = \mathbf{P}(X_i = 2) = \frac{1}{4}$$

and

$$\mathbf{P}(X_i = 1) = \frac{1}{2}.$$

Let

$$S_n = X_1 + \cdots + X_n.$$

(a) Find $\mathbf{E}(S_n)$ and $Var(S_n)$.

(b) Give a necessary and sufficient condition on $n \in \mathbb{N}$ to have

$$\mathbf{P}\left(\frac{1}{2} \leq \frac{S_n}{n} \leq \frac{3}{2}\right) \geq 0.999.$$

Hint: For the first part, see exercise 4.6.

12.18 Prove Lemma 12.17.

12.19 Let X_n, $n \geq 0$ be i.i.d. Bernoulli-distributed random variables with parameter $p \in (0, 1)$. Define

$$Y_n = X_n X_{n-1}$$

and

$$Z_n = \frac{1}{n} \sum_{k=1}^{n} Y_k$$

for every $n \geq 1$.

(a) Give the distribution of Y_k.

(b) Calculate $\mathbf{E}Y_k$ and $VarY_k$.

(c) Show that Y_k and Y_{k+1} are not independent.

(d) Prove that Y_k and Y_{m+k} are independent for any $m > 1$.

(e) Calculate $\mathbf{E}Z_n$ and $VarZ_n$.

(f) Study the convergence in distribution of the sequence $(Z_n)_{n \geq 1}$.

12.20 Consider a sequence of iid r.v.'s with Bernoulli distribution on $\{-1, 1\}$. That is

$$P(X_1 = 1) = p \quad \text{and} \quad P(X_1 = -1) = 1 - p.$$

Denote

$$S_n = X_1 + \cdots + X_n.$$

(a) Show that if $p \neq \frac{1}{2}$, then the sequence $(|S_n|)_{n \geq 1}$ converges almost surely.

For the rest of the problem we take $p = \frac{1}{2}$. Let $\alpha > 0$.

(b) Compute

$$\mathbf{E}e^{\alpha X_1}, \text{ then } \mathbf{E}e^{\alpha S_n}.$$

Deduce that for every $n \geq 1$ we have

$$\mathbf{E}\left[(\cosh \alpha)^{-n} e^{\alpha S_n}\right] = 1.$$

(c) Show that

$$\lim_{n \to \infty} \left[(\cosh \alpha)^{-n} e^{\alpha S_n}\right] = 0 \qquad \text{a.s.}$$

12.21 Let $(X_n)_{n\geq 1}$ be i.i.d. random variables. Suppose that

$$\mathbf{P}(X_1 = 0) < 1.$$

(**a**) Show that there exists $\alpha > 0$ such that

$$\mathbf{P}(|X_1| \geq \alpha) > 0.$$

(**b**) Show that

$$P(\limsup_n \{|X_n| \geq \alpha\} = 1.$$

Derive that

$$P(\limsup_n X_n = 0) = 0.$$

(**c**) Show that

$$\mathbf{E}|X_1| = \int_0^\infty P(|X_1| > t)\, dt.$$

(**d**) Deduce that X_1 is integrable if and only if

$$\sum_k P(|X_1| > k) < \infty.$$

(**e**) Show that X_1 is integrable if and only if

$$P(\limsup_n \{X_n > n\}) = 0.$$

(**f**) Deduce that if X_1 is integrable, then almost surely

$$\limsup_n \frac{1}{n} X_n \leq 1.$$

12.22 Let X_n be i.i.d. random variables with $\mathbf{E}X_1 = 0$ and assume that X_n are bounded. That is, there exists a $C > 0$ such that

$$|X_1| \leq C \qquad \text{a.s.}$$

Show that for every $\varepsilon > 0$, we obtain

$$P\left(\left|\frac{S_n}{n}\right| \geq \varepsilon\right) \leq 2\exp\left((-n\frac{\varepsilon^2}{2c^2})\right).$$

Deduce that S_n/n converges in probability to zero.

12.23 Let $X, X_1, \ldots, X_n, \ldots$ be i.i.d. with

$$P(X = 2^k) = \frac{1}{2^k}$$

for $k = 1, 2, \ldots$.

(a) Show that $\mathbf{E}X = \infty$.

(b) Prove that

$$\frac{S_n}{n} \longrightarrow \frac{1}{\log 2}$$

in probability as $n \to \infty$.

12.24 Let $X \sim N(100, 15)$. Find four numbers x_1, x_2, x_3, x_4 such that

$$P(X < x_1) = 0.125, \qquad P(X < x_2) = 0.25$$

and

$$P(X < x_3) = 0.75, \qquad P(X < x_4) = 0.875.$$

12.25 A fair coin is flipped 400 times. Determine the number x such that the probability that the number of heads is between $200 - x$ and $200 + x$ is approximately 0.80.

12.26 In an opinion poll it is assumed that an unknown proportion p of people are in favor of a proposed new law and a proportion $1 - p$ are against it. A sample of n people is taken to estimate p. The sample proportion $\hat{p}$ of people in favor of the law is taken as an estimate of p. Using the Central Limit Theorem, determine how large a sample will ensure that the estimate $\hat{p}$ will, with probability 0.95, be within 0.01 of the true p.

12.27 Write a statement explaining why Skorohod's theorem (Theorem 12.10) does not contradict our earlier statement that convergence in distribution does not imply convergence a.s.

Chapter Thirteen

Appendix A: Integration Theory. General Expectations

In this appendix we formalize the theory of calculating expectations. We learned about random variables and their distribution. This distribution completely characterizes a random variable. But in general, distributions are very complex functions. The human brain cannot comprehend such things easily. So the human brain wants to talk about one typical value. For example, one can give a distribution for the random variable representing player salaries in the NBA. Here the variability (probability space) is represented by the specific player chosen. However, suppose we simply want to know the typical salary in the NBA. We probably contemplate a career in sports and want to find out if as an athlete we should go for basketball or baseball. Thus, a single number corresponding to each of these distributions would be much easier to compare. In general, if the distribution is discrete or continuous, then calculating expectations by means we have seen already (summation or integration using probability mass function ir probability density function) will suffice. However, suppose the distribution is more complex so that its c.d.f. is not continuous. For example, suppose that the salary of said athlete depends on whether or not he/she gets injured, whether or not is a male or female, the severity of the injury, and so on. To be able to calculate expectations in this more realistic case, we need to introduce a more complex theory to deal with this situation.

In this appendix we present the theory which allows us the calculation of any number we want from a given distribution. Paradoxically, to calculate a simple number, we need to understand a very complex theory.

Handbook of Probability, First Edition. Ionuţ Florescu and Ciprian Tudor.

13.1 Integral of Measurable Functions

Not all random variables have p.d.f.'s. However, all random variables have c.d.f.'s. The integration theory is centered on learning how to integrate this function. Recall that any random variable is a measurable function from Ω into $\mathbb{R}$. The integration theory is constructed for any measurable function.

To this end, let $(\Omega, \mathscr{F}, P)$ be a probability space. We want to define, for any measurable function f on this space, a notion of integral of f with respect to the measure P.

Notation. We shall use the following notations for this integral:

$$\int_{\Omega} f(\omega)\mathbf{P}(d\omega) = \int f d\mathbf{P},$$

$$\text{for } A \in \mathscr{F} \quad \text{we have} \quad \int_{A} f(\omega)\mathbf{P}(d\omega) = \int_{A} f d\mathbf{P} = \int f \mathbf{1}_A d\mathbf{P},$$

where $\mathbf{1}_A$ denotes the indicator function of the set A in Ω.

Recall the Dirac delta we have defined previously? With its help summation is another kind of integral. To exemplify, let $\{a_n\}$ be a sequence of real numbers. Let $\Omega = \mathbb{R}$, $\mathscr{F} = \mathscr{B}(\mathbb{R})$ and let the measure on this space for any set A be defined using

$$\delta(A) = \sum_{i=1}^{\infty} \delta_i(A).$$

Then the function $i \mapsto a_i$ is integrable if and only if $\sum a_i < \infty$, and in this case we have

$$\sum_{n=1}^{\infty} a_n = \sum_{n=1}^{\infty} \int_{-\infty}^{\infty} a_x d\delta_n(x) = \int_{-\infty}^{\infty} a_x \sum_{n=1}^{\infty} d\delta_n(x) = \int_{-\infty}^{\infty} a_x d\delta(x)$$

What is the point of this?

This simple argument above shows that any "discrete" random variable may be treated as a "continuous" random variable. Thus the unifying theory presented here will apply to any random variable regardless of the form of its distribution.

13.1.1 INTEGRAL OF SIMPLE (ELEMENTARY) FUNCTIONS

If $A \in \mathscr{F}$, we know that we can define a measurable function by its indicator $\mathbf{1}_A$. We define the integral of this measurable function

$$\int \mathbf{1}_A \, d\mathbf{P} = \mathbf{P}(A).$$

We note that this variable has the same distribution as that of the Bernoulli random variable. The variable takes values 0 and 1 and we can easily calculate the probability that the variable is 1 as

$$\mathbf{P} \circ \mathbf{1}_A^{-1}(\{1\}) = \mathbf{P}\{\omega : \mathbf{1}_A(\omega) = 1\} = \mathbf{P}(A).$$

Therefore the variable is distributed as a Bernoulli random variable with parameter $p = \mathbf{P}(A)$.

Definition 13.1 (Simple function) *f is called a* simple *(elementary) function if and only if f can be written as a finite linear combination of indicators. More specifically, f is a simple function if and only if there exist sets $A_1, A_2, \ldots, A_n$ all in $\mathscr{F}$ and constants $a_1, a_2, \ldots, a_n$ in $\mathbb{R}$ such that*

$$f(\omega) = \sum_{k=1}^{n} a_k \mathbf{1}_{A_k}(\omega).$$

If the constants a_k are all positive, then f is a positive simple function.

Note that the sets A_i do not have to be disjoint, though we can show that any simple function f can be rewritten in terms of disjoint sets.

For any simple function f we define its integral using

$$\int f d\mathbf{P} = \sum_{k=1}^{n} a_k \mathbf{P}(A_k) < \infty.$$

We adopt the conventions $0 * \infty = 0$ and $\infty * 0 = 0$ in the above summation.

We will need to check that the above definition is proper. Specifically, since there exist many representations of a simple function, we need to make sure that any and all such representations produce the same integral value.

Furthermore, using the definition above we can prove that the integral is linear and monotone. We leave the proof of these results to the reader; however, we state them in a lemma.

Lemma 13.2 *The integral of a simple function has the following properties.*

1. *The integral is linear; that is, if f_1 and f_2 are two simple functions and $a_1, a_2 \in \mathbb{R}$, then*

$$\int_\Omega (a_1 f_1 + a_2 f_2)\, d\mathbf{P} = a_1 \int_\Omega f_1\, d\mathbf{P} + a_2 \int_\Omega f_2\, d\mathbf{P}.$$

2. *The integral is monotone; that is, if a sequence of simple functions $\{f_n\}_n$ converges increasingly (or decreasingly) to f for all points $\omega \in \Omega$, then the integral does as well:*

$$\int_\Omega f_n\, d\mathbf{P} \longrightarrow \int_\Omega f\, d\mathbf{P}$$

also increasingly or decreasingly.

13.1.2 INTEGRAL OF POSITIVE MEASURABLE FUNCTIONS

For every f positive measurable function $f : \Omega \longrightarrow [0, \infty)$, we define

$$\int f \, d\mathbf{P} = \sup \left\{ \int h \, d\mathbf{P} : h \text{ is a simple function, } h \leq f \right\}$$

In this definition we formalize the construction of the integral for any positive measurable function. However, we still need to show that the definition is proper. To this end, can we show that for any given positive measurable function f there exists a sequence of simple functions that converge to it? The answer is yes and is provided by the next exercise:

EXAMPLE 13.1 A simple construction

Let $f : \Omega \to [0, \infty]$ be a positive, measurable function. For all $n \geq 1$, we define

$$f_n(\omega) := \sum_{k=0}^{n2^n - 1} \frac{k}{2^n} \mathbf{1}_{\left\{\frac{k}{2^n} \leq f(\omega) < \frac{k+1}{2^n}\right\}}(\omega) + n \mathbf{1}_{\{f(\omega) \geq n\}}. \tag{13.1}$$

1. Show that f_n is a simple function on $(\Omega, \mathscr{F})$, for all $n \geq 1$.
2. Show that the sets present in the indicators in equation (13.1) form a partition of Ω, for all $n \geq 1$.
3. Show that the sequence of simple functions is increasing $g_n \leq g_{n+1} \leq f$, for all $n \geq 1$.
4. Show that $g_n \uparrow f$ as $n \to \infty$. Note that this is not an a.s. statement, it is true for all $\omega \in \Omega$.

The solution to this example is not complicated and in fact may be assigned as a problem. Using this construction, the integral of positive measurable functions is a well-defined number. Suppose that f is a positive measurable function. The example provides us with a sequence of simple functions f_n which increase to f at any point ω. Since we know that the integral of simple functions is monotone by Lemma 13.2, then the sequence of integrals will converge to a number. However, since the integral of f is defined as the supremum of such integrals and since the limit exists, there is no other way than the limit being the integral of f. This will in fact be the proof of the Monotone Convergence Theorem (below).

The next lemma is a very useful tool going forward.

Lemma 13.3 *If f is a positive measurable function and $\int f \, d\mathbf{P} = 0$, then*

$$\mathbf{P}\{f > 0\} = 0 \text{ (or equivalently } f = 0 \text{ a.s.).}$$

Proof: We can write

$$\{f > 0\} = \bigcup_{n \geq 0} \left\{f > \frac{1}{n}\right\}.$$

Since the sequence of events on the right is increasing, by the monotone convergence property of measure we must have

$$\mathbf{P}\{f > 0\} = \lim_{n \to \infty} \mathbf{P}\left\{f > \frac{1}{n}\right\}.$$

Now, assume by absurd that $\mathbf{P}\{f > 0\} > 0$. Then, there must exist an n such that $\mathbf{P}\{f > \frac{1}{n}\} > 0$. However, in this case by the definition of the integral of positive measurable functions, we obtain

$$\int f\, d\mathbf{P} \geq \int \frac{1}{n} \mathbf{1}_{\{f > \frac{1}{n}\}}\, d\mathbf{P} > 0,$$

which is a contradiction and our absurd assumption is false. ■

The next theorem is one of the most useful in probability theory. In our immediate context, it tells us that the integral for positive measurable functions is well-defined.

Theorem 13.4 (Monotone Convergence Theorem) *If f is a sequence of positive measurable functions such that $f_n(\omega) \uparrow f(\omega)$ for all $\omega \in \Omega$, then*

$$\int_\Omega f_n(\omega)\mathbf{P}(d\omega) \uparrow \int_\Omega f(\omega)\mathbf{P}(d\omega)$$

where the symbol $\uparrow$ denotes convergence from below (increasing sequences).

Note: As mentioned before, this theorem concludes the construction of the integral for positive measurable function. As we shall see, this result is the key for all the integration theory.

Proof: First let us show that f is positive measurable. First off, f is clearly positive since $f(\omega)$ is a limit of positive numbers $f_n(\omega)$, and this happens for all ω. Since f is positive, we need to look at the domain space $([0, \infty), \mathscr{B}([0, \infty)))$.

Let $B = (-\infty, b]$ an interval in $\mathbb{R}$ (recall that the Borel sets in $\mathbb{R}$ are generated by these intervals). We need to show that $f^{-1}(B) \in \mathscr{F}$. Since $f_n(\omega) \leq f(\omega)$ for all ω and n, we see that if $\omega \in f^{-1}(B)$, then $f(\omega) \leq b$, which implies $f_n(\omega) \leq b$; therefore $\omega \in f_n^{-1}(B)$. This immediately implies that

$$f^{-1}(B) \subseteq f_n^{-1}(B), \qquad \text{for all } n;$$

therefore

$$f^{-1}(B) \subseteq \bigcap_n f_n^{-1}(B).$$

Similarly, if $\omega \in f_n^{-1}(B)$ for all n, then $f_n(\omega) \leq b$ for all n; therefore the limit $f(\omega) = \lim f_n(\omega) \leq b$ thus we get the other inclusion. Therefore we have

$$f^{-1}(B) = \bigcap_n f_n^{-1}(B);$$

and since each set in the right side is measurable and $\mathscr{F}$ is a σ-algebra, we immediately have that f is measurable. Since we showed that f is positive measurable, then the integral is defined using

$$\int f\, d\mathbf{P} = \sup \left\{ \int h\, d\mathbf{P} : h \text{ is a simple function, } \ h \leq f \right\}.$$

Next we will show that $\int f\, d\mathbf{P} \geq \lim_{n\to\infty} \int f_n\, d\mathbf{P}$. Since $f_n(\omega) \leq f(\omega)$ at all ω, we have

$$\left\{ \int h\, d\mathbf{P} : h \text{ is a simple function, } \ h \leq f_n \right\} \subseteq \left\{ \int h\, d\mathbf{P} : h \text{ is a simple function, } \ h \leq f \right\}$$

since the set on the right simply has more simple functions. Since the supremum cannot go over the bound, we have

$$\sup \left\{ \int h\, d\mathbf{P} : h \text{ is a simple function, } \ h \leq f_n \right\} \leq \sup \left\{ \int h\, d\mathbf{P} : h \text{ is a simple function, } \ h \leq f \right\};$$

and since this is true for any n using the definition, we obtain

$$\lim_{n\to\infty} \int f_n\, d\mathbf{P} \leq \int f\, d\mathbf{P}.$$

To end the theorem (i.e., prove that the two quantities are equal), we need to show the reverse inequality. To this end, using Example 13.1, we know there exists an increasing sequence of *simple* functions—let us denote it g_n such that $g_n(\omega) \leq f(\omega)$ and

$$\lim_{n\to\infty} \int g_n\, d\mathbf{P} = \int f\, d\mathbf{P}.$$

It is enough to show that

$$\int g_n d\mathbf{P} \leq \lim_{k\to\infty} \int f_k d\mathbf{P}, \qquad \forall n.$$

Indeed, if we show this, then by going to the limit over n we will conclude the inequality needed.

So, let us consider a simple function g such that $g(\omega) \leq f(\omega)$. We want to show that

$$\int g\,d\mathbf{P} \leq \lim_{k\to\infty} \int f_k\,d\mathbf{P}.$$

To this end, construct the sets for all k:

$$A_k = \{\omega \in \Omega \mid g(\omega) \leq f_k(\omega)\}.$$

The plan is to use the monotone convergence property of probability measures. Note:

$$\int_{A_k} g\,d\mathbf{P} \leq \int_{A_k} f_k\,d\mathbf{P} \leq \int_{\Omega} f_k\,d\mathbf{P}$$

since the functions f_k are positive. Furthermore, since $f_k(\omega) \leq f_{k+1}(\omega)$, we have $A_k \subseteq A_{k+1}$, an increasing sequence of sets. The sequence will increase to Ω. To see this, let us take an $\omega \in \Omega$.

Suppose that the ω is such $g(\omega) = 0$. Then clearly $\omega \in A_k$ for all k since f_k is positive.

Suppose that ω is such $g(\omega) > 0$. Since

$$g(\omega) \leq f(\omega) = \lim_k f_k(\omega),$$

there must exist a k such that $g(\omega) \leq f_k(\omega)$; therefore $\omega \in A_k$ for a k large enough. Therefore, all omega points belong to an A_k for a k large enough; thus $\bigcup_k A_k = \Omega$.

Next we use the fact that g is a positive simple function. We can find a representation with positive constants a_i and disjoint sets B_j such that

$$g(\omega) = \sum_j a_j \mathbf{1}_{B_j}.$$

We have

$$\int_{\Omega} f_k\,d\mathbf{P} \geq \int_{A_k} g\,d\mathbf{P} = \int_{A_k} \sum_j a_j \mathbf{1}_{B_j} d\mathbf{P} = \sum_j \int_{A_k} \mathbf{1}_{B_j} d\mathbf{P}$$
$$= \sum_j \int_{\Omega} \mathbf{1}_{B_j \cap A_k} d\mathbf{P} = \sum_j a_j \mathbf{P}(A_k \cap B_j).$$

Since $A_k \uparrow \Omega$ we have $A_k \cap B_j \uparrow B_j$ for all j, and thus taking the limit in the inequality gives:

$$\lim_{k\to\infty} \int_{\Omega} f_k\,d\mathbf{P} \geq \sum_j a_j \lim_{k\to\infty} \mathbf{P}(A_k \cap B_j) = \sum_j a_j \mathbf{P}(B_j) = \int g\,d\mathbf{P}$$

where we used the monotone convergence property of measure.

This finally concludes the proof of the Monotone Convergence Theorem. ■

13.1.3 INTEGRAL OF MEASURABLE FUNCTIONS

Let f be any measurable function. Then, we can write $f(\omega) = f^+(\omega) - f^-(\omega)$ for any point ω, where

$$f^+(\omega) = \max\{f(\omega), 0\},$$
$$f^-(\omega) = \max\{-f(\omega), 0\}.$$

These f^+ and f^- are positive measurable functions and $|f(\omega)| = f^+(\omega) + f^-(\omega)$. Since they are positive measurable, their integrals are well-defined by the previous part.

Definition 13.5 *We define $L^1(\Omega, \mathscr{F}, P)$ as being the space of all functions f such that*

$$\int |f|\, d\mathbf{P} = \int f^+ d\mathbf{P} + \int f^- d\mathbf{P} < \infty$$

For any f in this space which we will shorten to $L^1(\Omega)$ or even simpler to L^1, we define

$$\int f\, d\mathbf{P} = \int f^+ d\mathbf{P} - \int f^- d\mathbf{P}.$$

13.1.3.1 Note. With the above, it is trivial to show that $|\int f d\mathbf{P}| \leq \int |f| d\mathbf{P}$.

13.1.3.2 Linearity. If $f, g \in L^1(\Omega)$ with $a, b \in \mathbb{R}$, then

$$af + bg \in L^1(\Omega)$$
$$\int (af + bg)\, d\mathbf{P} = a\int f\, d\mathbf{P} + b\int g\, d\mathbf{P}.$$

Next we present two results for general measurable functions. These results are extremely important for probability theory; but since a random variable is just a measurable function, these results will apply immediately.

Lemma 13.6 (Fatou's lemma for measurable functions) *If one of the following is true:*

(a) *$\{f_n\}_n$ is a sequence of positive measurable functions or*
(b) *$\{f_n\} \subset L^1(\Omega)$*

then

$$\int \liminf_n f_n\, d\mathbf{P} \leq \liminf_n \int f_n\, d\mathbf{P}$$

Proof: Note that $\liminf_n f_n = \lim_{m\to\infty} \inf_{n\geq m} f_n$, where $\lim_{m\to\infty} \inf_{n\geq m} f_n$ is an increasing sequence.

Let $g_m = \inf_{n \geq m} f_n$ and $n \geq m$:

$$f_n \geq \inf_{n \geq m} f_m = g_m \Rightarrow \int f_n \, d\mathbf{P} \geq \int g \, d\mathbf{P} \Rightarrow \int g_m \, d\mathbf{P} \leq \inf_{n \geq m} \int f_n \, d\mathbf{P}.$$

Now g_m increases, so we may use the Monotone Convergence Theorem, and we get

$$\int \lim_{m \to \infty} g_m \, d\mathbf{P} = \lim_{m \to \infty} \int g_m \, d\mathbf{P} \leq \lim_{m \to \infty} \inf_{n \geq m} \int f_n \, d\mathbf{P} = \liminf_{n} \int f_n \, d\mathbf{P}.$$

■

Theorem 13.7 (Dominated Convergence Theorem) *If f_n, f are measurable, $f_n(\omega) \to f(\omega)$ for all $\omega \in \Omega$ and the sequence f_n is dominated by $g \in L^1(\Omega)$:*

$$|f_n(\omega)| \leq g(\omega), \qquad \forall \omega \in \Omega, \forall n \in \mathbb{N},$$

then

$$f_n \to f \text{ in } L^1(\Omega) \qquad \left(\text{i.e., } \int |f_n - f| \, d\mathbf{P} \to 0 \right)$$

Thus $\int f_n d\mathbf{P} \to \int f d\mathbf{P}$ and $f \in L^1(\Omega)$.

13.1.3.3 The Standard Argument. This argument is the most important argument in the probability theory. Suppose that we want to prove that some property holds for all functions h in some space such as $L^1(\Omega)$ or the space of measurable functions.

1. Show that the result is true for all indicator functions.
2. Use linearity to show the result holds true for all f simple functions.
3. Use the Monotone Convergence Theorem to obtain the result for measurable positive functions.
4. Finally from the previous step and writing $f = f^+ - f^-$, we show that the result is true for all measurable functions.

13.2 General Expectations and Moments of a Random Variable

Since a random variable is just a measurable function, we just need to use the results of the previous section in the specific context of a space with probability one. Any integral with respect to a probability measure is called an expectation. Let $(\Omega, \mathcal{F}, P)$ be a probability space.

Definition 13.8 *For X an r.v. in $L^1(\Omega)$ define*

$$\mathbf{E}(X) = \int_\Omega X\, d\mathbf{P} = \int_\Omega X(\omega)\, d\mathbf{P}(\omega) = \int_\Omega X(\omega)\mathbf{P}(d\omega).$$

This expectation has the same properties of the integral defined before and some extra ones since the space has finite measure.

For any other measurable function $f : \mathbb{R} \to \mathbb{R}$, we may construct the variable $Y = f(X)$ and therefore define the expectation:

$$\mathbf{E}[Y] = \mathbf{E}[f(X)] = \int_\Omega f(X)\, d\mathbf{P} = \int_\Omega f(X)(\omega)\, d\mathbf{P}(\omega).$$

In particular when $f(x) = x^p$ we obtain the p-moment of the random variable. Specifically:

$$\mathbf{E}[X^p] = \int_\Omega X^p\, d\mathbf{P} = \int_\Omega X^p(\omega)\, d\mathbf{P}(\omega).$$

However, the expectation needs to exist. We define the space of the variables for which these types of expectation exist in the next section.

13.2.1 MOMENTS AND CENTRAL MOMENTS. L^p SPACE

We generalize the L^1 notion presented earlier in the following way.

Definition 13.9 *We define $L^p(\Omega, \mathcal{F}, P)$ For $1 \le p \le \infty$ as being the space of all random variables X such that:*

$$\mathbf{E}[|X|^p] = \int |X|^p d\mathbf{P} = \int (X^p)^+ d\mathbf{P} + \int (X^p)^- d\mathbf{P} < \infty,$$

where $(X^p)^+$ and $(X^p)^-$ denote the positive respectively negative parts of the variable X^p. For any X in this space which we will shorten to $L^p(\Omega)$ or even simpler to L^p, we define

$$\mathbf{E}[X^p] = \int X^p d\mathbf{P} = \int (X^p)^+ d\mathbf{P} - \int (X^p)^- d\mathbf{P}$$

Mathematically,

$$L^p(\Omega, \mathcal{F}, P) = L^p(\Omega) = \left\{X : \Omega \longrightarrow \mathbb{R} : \mathbf{E}\left[|X|^p\right] = \int |X|^p d\mathbf{P} < \infty\right\}.$$

Definition 13.10 (Moments) *For a random variable $X \in L^p$ we define the p-moment of the random variable as*

$$\mathbf{E}[X^p].$$

If we denote with $\mu = \mathbf{E}[X]$, *then the central p-moment of the random variable is*

$$\mathbf{E}[(X - \mu)^p].$$

On the L^p space we define a norm called the p-norm as

$$||X||_p = \mathbf{E}\left[|X|^p\right]^{1/p}$$

Lemma 13.11 (Properties of L^p spaces) *We have the following:*

(i) *L^p is a vector space. (i.e., if $X, Y \in L^p$ and $a, b \in \mathbb{R}$, then $aX + bY \in L^p$).*
(ii) *L^p is complete (every Cauchy sequence in L^p is convergent).*
(iii) *If $p \leq q$, then $L^q \subset L^p$ and for every $X \in L^q$ we have*

$$\|X\|_p \leq \|X\|_q.$$

The last property means that if a moment p exists, then all moments for all $q < p$ exist as well. This will imply that the central p-moment exists for any variables in L^p.

13.2.2 VARIANCE AND THE CORRELATION COEFFICIENT

In the special case when $p = 2$, we obtain several particular cases of special interest.

Definition 13.12 *The variance or the Dispersion of a random variable $X \in L^2(\Omega)$ is*

$$V(X) = \mathbf{E}[(X - \mu)^2] = \mathbf{E}(X^2) - \mu^2,$$

where $\mu = \mathbf{E}(X)$.

Definition 13.13 *Given two random variables X, Y we obtain the covariance between X and Y the quantity:*

$$Cov(X, Y) = \mathbf{E}[(X - \mu_X)(Y - \mu_Y)],$$

where $\mu_X = \mathbf{E}(X)$ *and* $\mu_Y = \mathbf{E}(Y)$.

Definition 13.14 *Given random variables X, Y, we obtain the correlation coefficient:*

$$\rho = Corr(X, Y) = \frac{Cov(X, Y)}{\sqrt{V(X)V(Y)}} = \frac{\mathbf{E}[(X - \mu_X)(Y - \mu_Y)]}{\sqrt{\mathbf{E}[(X - \mu_X)^2]\mathbf{E}[(Y - \mu_Y)^2]}}.$$

If we apply the Cauchy–Schwartz inequality (Lemma 14.7) to $X - \mu_X$ and $Y - \mu_Y$, we get

$$|\rho| < 1 \quad or \quad \rho \in [-1, 1].$$

The variable X and Y are called **uncorrelated** *if the covariance (or equivalently the correlation) between them is zero.*

Proposition 13.15 (Properties of expectation) *The following are true:*

(i) *If X and Y are integrable r.v.'s, then for any constants α and β the r.v. $\alpha X + \beta Y$ is integrable and $\mathbf{E}[\alpha X + \beta Y] = \alpha \mathbf{E}X + \beta \mathbf{E}Y$.*

(ii) *$V(aX + bY) = a^2 V(X) + b^2 V(Y) + 2abCov(X, Y)$.*

(iii) *If X, Y are independent, then $\mathbf{E}(XY) = \mathbf{E}(X)\mathbf{E}(Y)$ and $Cov(X, Y) = 0$.*

(iv) *If $X(\omega) = c$ with probability 1 and $c \in \mathbb{R}$ a constant, then $\mathbf{E}X = c$.*

(v) *If $X \geq Y$ a.s., then $\mathbf{E}X \geq \mathbf{E}Y$. Furthermore, if $X \geq Y$ a.s. and $\mathbf{E}X = \mathbf{E}Y$, then $X = Y$ a.s.*

Proof (Exercise): Please note that the reverse of part (iii) above is not true; if the two variables are uncorrelated, this does not mean that they are independent. ■

EXAMPLE 13.2 Due to Erdós

Suppose there are 17 fence posts around the perimeter of a field and exactly 5 of them are rotten. Show that irrespective of which of these 5 are rotten, there should exist a row of 7 consecutive posts of which at least 3 are rotten.

Proof (Solution): First we label the posts $1, 2, \ldots, 17$. Now define

$$I_k = \begin{cases} 1 & \text{if post } k \text{ is rotten,} \\ 0 & \text{otherwise .} \end{cases}$$

For any fixed k, let R_k denote the number of rotten posts among $k+1, \ldots, k+7$ (starting with the next one). Note that when any of $k+1, \ldots, k+7$ are larger than 17, we start again from 1 (i.e., modulo 17 +1).

Now pick a post at random; this obviously can be done in 17 ways with equal probability. Then after we pick this post, we calculate the number of rotten boards. We have

$$\begin{aligned} \mathbf{E}(R_k) &= \sum_{k=1}^{17} (I_{k+1} + \cdots + I_{k+7}) \frac{1}{17} \\ &= \frac{1}{17} \sum_{k=1}^{17} \sum_{j=1}^{7} I_{k+j} = \frac{1}{17} \sum_{j=1}^{7} \sum_{k=1}^{17} I_{j+k} \end{aligned}$$

$$= \frac{1}{17}\sum_{j=1}^{7} 5 \qquad \text{(the sum is 5 since we count all the rotten posts in the fence)}$$

$$= \frac{35}{17}.$$

Now, $35/17 > 2$, which implies $\mathbf{E}(R_k) > 2$. Therefore, $\mathbf{P}(R_k > 2) > 0$ (otherwise the expectation is necessarily bounded by 2) and since R_k is integer-valued, $\mathbf{P}(R_k \geq 3) > 0$. So there exists some k such that $R_k \geq 3$.

Of course, now that we see the proof, we can play around with numbers and see that there exists a row of 4 consecutive posts in which at least two are rotten, or that there must exist a row of 11 consecutive posts in which at least 4 are rotten, and so on (row of 14 containing all 5 rotten ones). ■

13.2.3 CONVERGENCE THEOREMS

Rewriting the convergence theorems in terms of expectations, we have

(i) *Monotone Convergence Theorem:* If $X_n \geq 0$, $X_n \in L^1$ and $X_n \uparrow X$, then $\mathbf{E}(X_n) \uparrow \mathbf{E}(X) \leq \infty$.

(ii) *Fatou:* $\mathbf{E}(\liminf_{n\to\infty} X_n) \leq \liminf_{n\to\infty} \mathbf{E}(X_n)$.

(iii) *Dominated Convergence Theorem:* If $|X_n(\omega)| \leq Y(\omega)$ on Ω with $Y \in L^1(\Omega)$ and $X_n(\omega) \to X(\omega)$ for all $\omega \in \Omega$, then $\mathbf{E}(|X_n - X|) \to 0$.

CHAPTER FOURTEEN

Appendix B: Inequalities Involving Random Variables and Their Expectations

In this appendix we present specific properties of the expectation (additional to just the integral of measurable functions on possibly infinite measure spaces). It is to be expected that on probability spaces we may obtain more specific properties since the probability space has measure 1.

Proposition 14.1 (Markov inequality) *Let Z be a r.v. and let $g : \mathbb{R} \longrightarrow [0, \infty]$ be an **increasing**, positive measurable function. Then*

$$\mathbf{E}\left[g(Z)\right] \geq \mathbf{E}\left[g(Z)\mathbf{1}_{\{Z \geq c\}}\right] \geq g(c)\mathbf{P}(Z \geq c).$$

Thus

$$\mathbf{P}(Z \geq c) \leq \frac{\mathbf{E}[g(Z)]}{g(c)}$$

for all g increasing functions and $c > 0$.

Proof: Take $\lambda > 0$ arbitrary and define the random variable

$$Y = 1_{\{|X| \geq \lambda\}}.$$

Handbook of Probability, First Edition. Ionuţ Florescu and Ciprian Tudor.

Then clearly

$$\lambda Y \leq X$$

and taking the expectation, we get

$$\lambda EY = \lambda P(|X| \geq \lambda) \leq E|X|.$$

■

EXAMPLE 14.1 Special cases of the Markov inequality

If we take $g(x) = x$ an increasing function and X a positive random variable, then we obtain

$$\mathbf{P}(Z \geq c) \leq \frac{\mathbf{E}(Z)}{c}.$$

To get rid of the condition $X \geq 0$, we take the random variable $Z = |X|$. Then we obtain the classical form of the Markov inequality:

$$\mathbf{P}(|X| \geq c) \leq \frac{\mathbf{E}(|X|)}{c}.$$

If we take $g(x) = x^2$, $Z = |X - \mathbf{E}(X)|$ and we use the definition of variance, we obtain the Chebyshev inequality:

$$\mathbf{P}(|X - \mathbf{E}(X)| \geq c) \leq \frac{Var(X)}{c^2}.$$

If we denote $\mathbf{E}(X) = \mu$ and $Var(X) = \sigma$ and we take $c = k\sigma$ in the previous inequality, we will obtain the classical Chebyshev inequality presented in undergraduate courses:

Proposition 14.2 *For every $\lambda \geq 0$ and for any random variable X such that $EX^2 < \infty$ we have*

$$\mathbf{P}(|X - \mu| \geq k\sigma) \leq \frac{1}{k^2}.$$

If we take $g(x) = e^{\theta x}$, with $\theta > 0$, then

$$\mathbf{P}(Z \geq c) \leq e^{-\theta c}\mathbf{E}(e^{\theta z}).$$

This last inequality states that the tail of the distribution decays exponentially in c if Z has finite exponential moments. With simple manipulations, one can obtain Chernoff's inequality using it.

Remark 14.3 *In fact the Chebyshev inequality is far from being sharp. Consider, for example, a random variable X with standard normal distribution $N(0, 1)$.*

If we calculate the probability of the normal using a table of the normal law or using the computer, we obtain

$$P(X \geq 2) = 1 - 0.9772 = 0.0228.$$

However, if we bound the probability using Chebyshev inequality, we obtain

$$P(X \geq 2) = \frac{1}{2}P(|X| \geq 2) \leq \frac{1}{2}\frac{1}{4} = \frac{1}{8} = 0.125,$$

which is very far from the actual probability.

The following definition is just a reminder.

Definition 14.4 *A function $g : I \longrightarrow \mathbb{R}$ is called a convex function on I (where I is any open interval in $\mathbb{R}$, if its graph lies below any of its chords). Mathematically: For any $x, y \in I$ and for any $\alpha \in (0, 1)$, we have*

$$g(\alpha x + (1 - \alpha)y) \leq \alpha g(x) + (1 - \alpha)g(y).$$

A function g is called concave if the opposite is happening:

$$g(\alpha x + (1 - \alpha)y) \geq \alpha g(x) + (1 - \alpha)g(y).$$

Some examples of convex functions on the whole $\mathbb{R}$: $|x|$, x^2, and $e^{\theta x}$, with $\theta > 0$.

Lemma 14.5 (Jensen's inequality) *Let f be a convex function and let X be an r.v. in $L^1(\Omega)$. Assume that $\mathbf{E}(f(X)) \leq \infty$, then*

$$f(\mathbf{E}(X)) \leq \mathbf{E}(f(X)).$$

Proof: Skipped. The classic approach indicators → simple functions → positive measurable → measurable is a standard way to prove Jensen. ■

Remark 14.6 *The discrete form of Jensen's inequality is as follows: Let $\varphi : \mathbb{R} \to \mathbb{R}$ be a convex function and let $x_1, \ldots, x_n \in \mathbb{R}$ and $a_i > 0$ for $i = 1, \ldots, n$. Then*

$$\varphi\left(\frac{\sum_{i=1}^n a_i x_i}{\sum_{i=1}^n a_i}\right) \leq \frac{\sum_{i=1}^n a_i \varphi(x_i)}{\sum_{i=1}^n a_i}.$$

If the function φ is concave, we have

$$\varphi\left(\frac{\sum_{i=1}^n a_i x_i}{\sum_{i=1}^n a_i}\right) \geq \frac{\sum_{i=1}^n a_i \varphi(x_i)}{\sum_{i=1}^n a_i}.$$

The remark is a particular case of the Jensen inequality. Indeed, consider a discrete random variable X with outcomes x_i and corresponding probabilities

$a_i / \sum a_i$. Apply the classic Jensen approach above to the convex function φ using the expression of expectation of discrete random variables.

A Historical Remark. The next inequality, one of the most famous and useful in any area of analysis (not only probability), is usually credited to Cauchy for sums and Schwartz for integrals and is usually known as the Cauchy–Schwartz inequality. However,the Russian mathematician Victor Yakovlevich Bunyakovsky (1804–1889) discovered and first published the inequality for integrals in 1859 (when Schwartz was 16). Unfortunately, he was born in eastern Europe. However, all who are born in eastern Europe (including myself) learn the inequality by its proper name.

Lemma 14.7 (Cauchy–Bunyakovsky–Schwarz inequality) *If $X, Y \in L^2(\Omega)$, then $XY \in L^1(\Omega)$ and*

$$|\mathbf{E}[XY]| \leq \mathbf{E}[|XY|] \leq ||X||_2 ||Y||_2,$$

where we used the notation of the norm in L^p:

$$\|X\|_p = \left(\mathbf{E}[|X|^p]\right)^{\frac{1}{p}}.$$

Proof: The first inequality is clear applying Jensen inequality to the function $|x|$. We need to show

$$\mathbf{E}[|XY|] \leq (\mathbf{E}[X^2])^{1/2} (\mathbf{E}[Y^2])^{1/2}.$$

Let

$$W = |X| \quad \text{and} \quad Z = |Y|.$$

Clearly, $W, Z \geq 0$.

Truncation. Let $W_n = W \bigwedge n$ and $Z_n = Z \bigwedge n$ that is

$$W_n(\omega) = \begin{cases} W(\omega), & \text{if } W(\omega) < n, \\ n, & \text{if } W(\omega) \geq n. \end{cases}$$

Clearly, defined in this way, W_n, Z_n are bounded. Let $a, b \in \mathbb{R}$ two constants. Then

$$0 \leq \mathbf{E}[(aW_n + bZ_n)^2] = a^2 \mathbf{E}(W_n^2) + 2ab\mathbf{E}(W_n Z_n) + b^2 \mathbf{E}(Z_n^2)$$

If we let $a/b = c$, we get

$$c^2 \mathbf{E}(W_n^2) + 2c\mathbf{E}(W_n Z_n) + \mathbf{E}(Z_n^2) \geq 0, \qquad \forall c \in \mathbb{R}.$$

This means that the quadratic function in c has to be positive. But this is only possible if the determinant of the equation is negative and the leading coefficient

$\mathbf{E}(W_n^2)$ is strictly positive; the later condition is obviously true. Thus we must have

$$4(\mathbf{E}(W_nZ_n))^2 - 4\mathbf{E}(W_n^2)\mathbf{E}(Z_n^2) \leq 0$$
$$\Rightarrow (\mathbf{E}(W_nZ_n))^2 \leq \mathbf{E}(W_n^2)\mathbf{E}(Z_n^2) \leq \mathbf{E}(W^2)\mathbf{E}(Z^2) \qquad \forall n,$$

which is in fact the inequality for the truncated variables.

If we let $n \uparrow \infty$ and we use the monotone convergence theorem, we get

$$(\mathbf{E}(WZ))^2 \leq \mathbf{E}(W^2)\mathbf{E}(Z^2).$$

■

A generalization of the Cauchy–Buniakovski–Schwartz is:

Lemma 14.8 (Hölder inequality) *If* $1/p + 1/q = 1$, $X \in L^p(\Omega)$, *and* $Y \in L^q(\Omega)$, *then* $XY \in L^1(\Omega)$ *and*

$$\mathbf{E}|XY| \leq \|X\|_p \|Y\|_q = \left(\mathbf{E}|X|^p\right)^{\frac{1}{p}} \left(\mathbf{E}|Y|^q\right)^{\frac{1}{q}}.$$

Proof: The proof is simple and uses the following inequality (Young inequality): If a and b are positive real numbers and p, q are as in the theorem, then

$$ab \leq \frac{a^p}{p} + \frac{b^q}{q},$$

with equality if and only if $a^p = b^q$.

Taking this inequality as given (not hard to prove), define

$$f = \frac{|X|}{\|X\|_p}, \qquad g = \frac{|Y|}{\|Y\|_p}.$$

Note that the Holder inequality is equivalent to $\mathbf{E}[f\,g] \leq 1$. (Note that $\|X\|_p$ and $\|Y\|_q$ are just numbers which can be taken in and out of integral using the linearity property of the integral.) To finish the proof, apply the Young inequality to $f \geq 0$ and $g \geq 0$ and then integrate to obtain

$$\mathbf{E}[f\,g] \leq \frac{1}{p}\mathbf{E}[f^p] + \frac{1}{q}\mathbf{E}[g^q] = \frac{1}{p} + \frac{1}{q} = 1,$$

since $\mathbf{E}[f^p] = 1$ and similarly for g.

Finally, the extreme cases ($p = 1, q = \infty$, etc.) may be treated separately, but they will yield the same inequality. ■

This inequality and Riesz representation theorem creates the notion of conjugate space. This notion is only provided to create links with real analysis. For further details we recommend Royden (1988).

Definition 14.9 (Conjugate space of L^p) *For $p > 0$ let $L^p(\Omega)$ define the space on $(\Omega, \mathscr{F}, \mathbf{P})$. The number $q > 0$ with the property $1/p + 1/q = 1$ is called the conjugate index of p. The corresponding space $L^q(\Omega)$ is called the conjugate space of $L^p(\Omega)$.*

Any of these spaces are metric spaces with the distance induced by the norm, that is,

$$d(X, Y) = \|X - Y\|_p = \left(\mathbf{E}\left[|X - Y|^p\right]\right)^{\frac{1}{p}}.$$

The fact that this is a properly defined linear space is implied by the triangle inequality in L^p—the next theorem.

Lemma 14.10 (Minkowski inequality) *If $X, Y \in L^p$ then $X + Y \in L^p$ and*

$$\|X + Y\|_p \leq \|X\|_p + \|Y\|_p.$$

Proof: We clearly have

$$|X + Y|^p \leq 2^{p-1}(|X|^p + |Y|^p).$$

For example, to show this inequality in terms of real numbers, just use the definition of convexity for the function x^p with $x = |X|$ and $y = |Y|$ and $\alpha = 1/2$.

Integrating the inequality will impliy that $X + Y \in L^p$.

Now we can write

$$\begin{aligned}
&\|X + Y\|_p^p = \mathbf{E}[|X + Y|^p] \leq \mathbf{E}\left[(|X| + |Y|)|X + Y|^{p-1}\right] \\
&= \mathbf{E}\left[|X||X + Y|^{p-1}\right] + \mathbf{E}\left[|Y||X + Y|^{p-1}\right] \\
&\overset{\text{Holder}}{\leq} \left(\mathbf{E}\left[|X|^p\right]\right)^{1/p}\left(\mathbf{E}\left[|X + Y|^{(p-1)q}\right]\right)^{1/q} \\
&\quad + \left(\mathbf{E}\left[|Y|^p\right]\right)^{1/p}\left(\mathbf{E}\left[|X + Y|^{(p-1)q}\right]\right)^{1/q} \\
&\overset{\left(q=\frac{p}{p-1}\right)}{=} \left(\|X\|_p + \|Y\|_p\right)\left(\mathbf{E}\left[|X + Y|^p\right]\right)^{1-\frac{1}{p}} \\
&= \left(\|X\|_p + \|Y\|_p\right)\frac{\mathbf{E}\left[|X + Y|^p\right]}{\|X + Y\|_p}.
\end{aligned}$$

Finally, identifying the left and right hand after simplifications, we obtain the result. ∎

The Case of L^2. The case when $p = 2$ is quite special. This is because 2 is its own conjugate index $(1/2 + 1/2 = 1)$. Because of this, the space is quite similar to the Euclidian space. If $X, Y \in L^2$, we may define the inner product:

$$< X, Y >= \mathbf{E}[XY] = \int XY \, d\mathbf{P},$$

which is a well-defined quantity using the Cauchy–Bunyakovsky–Schwartz inequality.

The existence of the inner product and the completeness of the norm makes L^2 a Hilbert space with all the benefits that follow. In particular, the notion of orthogonality is well-defined. Two variables X and Y in L^2 are orthogonal if and only if

$$< X, Y >= 0.$$

In turn the orthogonality definition allows a Fourier representation and, in general, representations in terms of an orthonormal basis of functions in L^2. Again, we do not wish to enter into more details than necessary; please consult, (Billingsley, 1995, Section 19) for further reference.

A consequence of the Markov inequality is the Berstein inequality.

Proposition 14.11 (Berstein inequality) *Let $X_1, X_2, \ldots, X_n$ be independent random variable square integrable with zero expectation. Assume that there exists a constant $M > 0$ such that for every $i = 1, \ldots, n$ we have*

$$|X_i| \le M \qquad \textit{almost surely},$$

that is, the variables are bounded by M almost surely. Then, for every $t \ge 0$, we have

$$P\left(\sum_{i=1}^{n} X_i > t\right) \le e^{-\frac{t^2}{2\sum_{i=1}^{n} EX_i^2 + \frac{2Mt}{3}}}.$$

EXAMPLE 14.2

A random variable X has finite variance σ^2. Show that for any number c,

$$P(X \ge t) \le \frac{E[(X+c)^2]}{(t+c)^2} \qquad \text{if } t > -c.$$

Show that if $E(X) = 0$, then

$$P(X \ge t) \le \frac{\sigma^2}{\sigma^2 + t^2}, \qquad \forall t > 0.$$

Solution: Let us use a technique similar to the Markov inequality to prove the first inequality. Let $F(x)$ be the distribution function of X. For any $c \in \mathbb{R}$ we may write

$$\mathbf{E}\left[(X+c)^2\right] = \int_{-\infty}^{t} (x+c)^2 \, dF(x) + \int_{t}^{\infty} (x+c)^2 \, dF(x).$$

The first integral is always positive and if $t > -c$, then $t + c > 0$ and on the interval $x \in (t, \infty)$ the function $(x + c)^2$ is increasing. Therefore we may continue:

$$\mathbf{E}\left[(X + c)^2\right] \geq \int_t^\infty (t + c)^2 \, dF(x) = (t + c)^2 \mathbf{P}(X > t).$$

Rewriting the final expression gives the first assertion. To show the second assertion, note that if $\mathbf{E}[x] = 0$, then $V(X) = \mathbf{E}[X^2]$ and thus $\mathbf{E}\left[(X + c)^2\right] = \sigma^2 + c^2$. Thus the inequality we just proved reads in this case:

$$P(X \geq t) \leq \frac{\sigma^2 + c^2}{(t + c)^2}, \qquad \text{if } t > -c.$$

Now take $c = \frac{\sigma^2}{t}$. This is a negative value for any positive t, so the condition is satisfied for any t positive. Substituting after simplifications, we obtain exactly what we need. You may wonder (and should wonder) how we came up with the value $\frac{\sigma^2}{t}$. The explanation is simple—that is, the value of c which minimizes the expression $\frac{\sigma^2+c^2}{(t+c)^2}$; in other words, the value of c which produces the best bound. ■

14.1 Functions of Random Variables. The Transport Formula

In the previous chapters dedicated to discrete and continuous random variables, we learned how to calculate distributions—in particular, p.d.f.'s—for continuous random variables. In this appendix we present a more general result. This general result allows us to construct random variables and, in particular, distributions on any abstract space. This is the result that allows us to claim that studying random variables on $([0, 1], \mathscr{B}([0, 1]), \lambda)$ is enough. We had to postpone presenting the result until this point since we had to learn first how to integrate.

Theorem 14.12 (General Transport Formula) *Let $(\Omega, \mathbb{R}, P)$ be a probability space. Let f be a measurable function such that*

$$(\Omega, \mathscr{F}) \xrightarrow{f} (S, \mathscr{G}) \xrightarrow{\varphi} (\mathbb{R}, \mathscr{B}(\mathbb{R})),$$

where $(S, \mathscr{G})$ is a measurable space. Assuming that at least one of the integrals exists, we then have

$$\int_\Omega \varphi \circ f \, d\mathbf{P} = \int_S \varphi \, d\mathbf{P} \circ f^{-1},$$

for all φ measurable functions.

Proof: We will use the standard argument technique discussed above.

1. Let φ be the indicator function $\varphi = \mathbf{1}_A$ for $A \in \mathscr{G}$:

$$\mathbf{1}_A(\omega) = \begin{cases} 1 & \text{if } \omega \in A, \\ 0 & \text{otherwise}. \end{cases}$$

Then we get

$$\begin{aligned}\int_\Omega \mathbf{1_A} \circ f \, d\mathbf{P} &= \int_\Omega \mathbf{1}_A(f(\omega)) \, d\mathbf{P}(\omega) = \int_\Omega \mathbf{1}_{f^{-1}(A)}(\omega) \, d\mathbf{P}(\omega) \\ &= \mathbf{P}(f^{-1}(A)) = \mathbf{P} \circ f^{-1}(A) = \int_S \mathbf{1}_A \, d(\mathbf{P} \circ f^{-1}),\end{aligned}$$

recalling the definition of the integral of an indicator.

2. Let φ be a simple function $\varphi = \sum_{i=1}^n a_i \mathbf{1}_{A_i}$, where a_i's are constant and $A_i \in \mathscr{G}$.

$$\begin{aligned}\int_\Omega \varphi \circ f \, d\mathbf{P} &= \int_\Omega \left(\sum_{i=1}^n a_i \mathbf{1}_{A_i} \right) \circ f \, d\mathbf{P} \\ &= \int_\Omega \sum_{i=1}^n a_i (\mathbf{1}_{A_i} \circ f) \, d\mathbf{P} = \sum_{i=1}^n a_i \int_\Omega \mathbf{1}_{A_i} \circ f \, d\mathbf{P} \\ &\overset{\text{(part 1)}}{=} \sum_{i=1}^n a_i \int_S \mathbf{1}_{A_i} \, d\mathbf{P} \circ f^{-1} \\ &= \int_S \sum_{i=1}^n a_i \mathbf{1}_{A_i} \, d\mathbf{P} \circ f^{-1} = \int_S \varphi d\mathbf{P} \circ f^{-1}.\end{aligned}$$

3. Let φ be a positive measurable function and let φ_n be a sequence of simple functions such that $\varphi_n \nearrow \varphi$, then

$$\begin{aligned}\int_\Omega \varphi \circ f \, d\mathbf{P} &= \int_\Omega (\lim_{n\to\infty} \varphi_n) \circ f \, d\mathbf{P} \\ &= \int_\Omega \lim_{n\to\infty} (\varphi_n \circ f) \, d\mathbf{P} \overset{\text{monotone convergence}}{=} \lim_{n\to\infty} \int \varphi_n \circ f \, d\mathbf{P} \\ &\overset{\text{(part 2)}}{=} \lim_{n\to\infty} \int \varphi_n \, d\mathbf{P} \circ f^{-1} \overset{\text{monotone convergence}}{=} \int \lim_{n\to\infty} \varphi_n \, d\mathbf{P} \circ f^{-1} \\ &= \int_S \varphi \, d(\mathbf{P} \circ f^{-1}).\end{aligned}$$

4. Let φ be a measurable function then $\varphi^+ = \max(\varphi, 0)$, $\varphi^- = \max(-\varphi, 0)$. This then gives us $\varphi = \varphi^+ - \varphi^-$. Since at least one integral is assumed to exist, we get that $\int \varphi^+$ and $\int \varphi^-$ exist. Also note that

$$\begin{aligned}\varphi^+ \circ f(\omega) &= \varphi^+(f^{-1}(\omega)) = \max(\varphi(f(\omega)), 0), \\ \max(\varphi \circ f(\omega), 0) &= (\varphi \circ f)^+(\omega).\end{aligned}$$

Then

$$\int \varphi^+ d\mathbf{P} \circ f^{-1} = \int \varphi^+ \circ f d\mathbf{P} = \int (\varphi \circ f)^+ d\mathbf{P},$$
$$\int \varphi^- d\mathbf{P} \circ f^{-1} = \int \varphi^- \circ f d\mathbf{P} = \int (\varphi \circ f)^- d\mathbf{P}.$$

These equalities follow from part 3 of the proof. After subtracting both, we obtain

$$\int \varphi \, d\mathbf{P} \circ f^{-1} = \int \varphi \circ f \, d\mathbf{P}.$$

■

EXAMPLE 14.3

If X and Y are independent random variables defined on $(\Omega, \mathbb{R}, P)$ with $X, Y \in L^1(\Omega)$, then $XY \in L^1(\Omega)$:

$$\int_\Omega XY \, d\mathbf{P} = \int_\Omega X \, d\mathbf{P} \int_\Omega Y \, d\mathbf{P} \qquad (\mathbf{E}(XY) = \mathbf{E}(X)\mathbf{E}(Y)).$$

Solution: Let us solve this example using the transport formula. Let us take $f : \Omega \to \mathbb{R}^2, f(\omega) = (X(\omega), Y(\omega))$; and $\varphi : \mathbb{R}^2 \to \mathbb{R}$, $\varphi(x, y) = xy$. Then we have from the transport formula the following:

$$\int_\Omega X(\omega) Y(\omega) \, dP(\omega) \overset{(T)}{=} \int_{\mathbb{R}^2} xy \, dP \circ (X, Y)^{-1}.$$

The integral on the left is $\mathbf{E}(XY)$, while the integral on the right can be calculated as

$$\int_{\mathbb{R}^2} xy \, d(P \circ X^{-1}, P \circ Y^{-1}) = \int_{\mathbb{R}} x \, dP \circ X^{-1} \int_{\mathbb{R}} y \, dP \circ Y^{-1}$$
$$\overset{(T)}{=} \int_\Omega X(\omega) \, dP(\omega) \int_\Omega Y(\omega) \, dP(\omega) = \mathbf{E}(X)\mathbf{E}(Y).$$

■

EXAMPLE 14.4

Finally we conclude with an application of the transport formula which will produce one of the most useful formulas. Let X be an r.v. defined on the probability space $(\Omega, \mathscr{F}, \mathbf{P})$ with distribution function $F(x)$. Show that

$$\mathbf{E}(X) = \int_{\mathbb{R}} x\, dF(x),$$

where the integral is understood in the Riemann–Stieltjes sense.

Proving the formula is immediate. Take $f : \Omega \to \mathbb{R}$, $f(\omega) = X(\omega)$ and $\varphi : \mathbb{R} \to \mathbb{R}$, $\varphi(x) = x$. Then from the transport formula, we have

$$\begin{aligned}\mathbf{E}(X) &= \int_{\Omega} X(\omega)\, d\mathbf{P}(\omega) = \int_{\Omega} x \circ X(\omega)\, d\mathbf{P}(\omega) \overset{(T)}{=} \int_{\mathbb{R}} x\, d\mathbf{P} \circ X^{-1}(x) \\ &= \int_{\mathbb{R}} x\, dF(x).\end{aligned}$$

Clearly if the distribution function $F(x)$ is derivable with $\frac{dF}{dx}(x) = f(x)$ or $dF(x) = f(x)\, dx$, we obtain the lower-level classes formula for calculating expectation of a "continuous" random variable:

$$\mathbf{E}(X) = \int_{\mathbb{R}} x f(x)\, dx.$$

Bibliography

Anderson, T. W. (1958). *An Introduction to Multivariate Statistical Analysis*. New York: Wiley.

Bertrand, J. L. F. (1889). *Calcul des Probabilités*. Paris: Gauthier-Villars et fils.

Billingsley, P. (1995). *Probability and Measure*, 3rd ed. New York: Wiley.

Black, F., and M. Scholes (1973). The valuation of options and corporate liability. *Journal of Political Economy* **81**, 637–654.

Bucher, C. G. (1988). Adaptive sampling—An iterative fast Monte Carlo procedure. *Structural Safety* **5**, 119–128.

Chung, K. L. (2000). *A Course in Probability Theory Revised*, 2nd ed. New York: Academic Press.

Corporation, R. (2001). *A Million Random Digits with 100,000 Normal Deviates*. American Book Publishers. Reprint.

de Lagrange, J. L. (1770). Réflexions sur la résolution algébrique des équations. *Nouveaux mémoires de l'Académie royale des sciences et belles-lettres de Berlin, années 1770 et 1771* **3**, 205–421.

Fisher, R. A. (1925). Applications of "Student's" distribution. *Metron* **5**, 90–104.

Grimmett, G., and D. Stirzaker (2001). *Probability and Random Processes*, 3rd ed. New York: Oxford University Press.

Grinstead, C. M., and J. L. Snell (1997). *Introduction to Probability, second revised edition*. Providence, RI: AMS. p. 510.

Kenney, J. F. (1942). Characteristic functions in statistics. *Nat. Math. Mag.* **17**(2), pp. 51–67.

Khintchine, A. Y. and P. Lévy (1936). Sur les lois stables. *C. R. Acad. Sci. Paris* **202**, 374–376.

Kolmogoroff, A. N. (1973). *Grundbegriffe der Wahrscheinlichkeitsrechnung*. Berlin: Springer.

Koponen, I. (1995). Analytic approach to the problem of convergence of truncated lévy flights towards the gaussian stochastic process. *Phys. Rev. E* **52**(1), 1197–1199.

Laplace, P.-S. (1812). *Théorie analytique des probabilités*, Vol. 7. Oeuvres complétes.

Laplace, P.-S. (1886). Théorie analytique des probabilités. In *Œuvres complétes de Laplace*, 3rd ed. Paris: Gauthier-Villars.

Lévy, P. (1925). *Calcul des probabilités*. Paris: Gauthier-Villars.

Handbook of Probability, First Edition. Ionuţ Florescu and Ciprian Tudor.

Mantegna, R. N., and H. E. Stanley (1994). Stochastic process with ultraslow convergence to a gaussian: The truncated lévy flight. *Phys. Rev. Lett.* **73**, 2946–2949.

Marsaglia, G., and W. W. Tsang (2000). The ziggurat method for generating random variables. *J. Stat. Software* **5**, 1–7.

Press, W., S. Teukolsky, W. Vetterling, and B. Flannery (2007). *Numerical recipes*. In *The Art of Scientific Computing*, 3rd ed. New York: Cambridge University Press.

Roy, R. (2011). *Sources in the Development of Mathematics: Infinite Series and Products from the Fifteenth to the Twenty-first Century*. New York: Cambridge University Press.

Royden, H. (1988). *Real Analysis*, 3rd ed. Englewood Cliffs, NJ: Prentice Hall.

Rubin, D. B. Using the SIR algorithm to simulate posterior distributions. *Bayesian Statistics*, **3**, 395–402.

Rubin, H., and B. C. Johnson (2006). Efficient generation of exponential and normal deviates. *J. Stat. Comput. Simulation* **76**(6), 509–518.

Samorodnitsky, G., and M. S. Taqqu (1994). *Stable non-Gaussian Random Processes: Stochastic Models with Infinite Variance*. New York: Chapman and Hall.

Student (1908). The probable error of a mean. *Biometrika* **6**(1), 1–25.

Todhunter, I. (1865). *A History of the Mathematical Theory of Probability*. Cambridge-London.

Wheeden, R. L., and A. Zygmund (1977). *Measure and Integral: An Introduction to Real Analysis*. New York: Marcel Dekker.

Whiteley, N., and A. Johansen (2011). Recent developments inauxiliary particle filtering. In D. Barber, A. Cemgil, and S. Chiappa (eds.), Chapter 3, *Inference and Learning in Dynamic Models*. New York: Cambridge University Press, pp. 52–58.

Wichura, M. J. (1988). Algorithm as 241: The percentage points of the normal distribution. *J. Roy. Stat. Soc. Ser. C (Appl. Stat.)* **37**(3), 477–484.

Wild, C. J. and G. A. F. Seber (1999). *Chance Encounters: A First Course in Data Analysis and Inference*. New York: Wiley.

Index

Handbook of Probability, First Edition. Ionuţ Florescu and Ciprian Tudor.